THE PHYSICS OF MODERN ELECTRONICS

THE PHYSICS OF MODERN ELECTRONICS

by

W. A. GÜNTHER

Translated from the German by

DAVID ANTIN

DOVER PUBLICATIONS, INC.
NEW YORK

Published in Canada by General Publishing Company, Ltd., 30 Lesmill Road, Don Mills, Toronto, Ontario.

Published in the United Kingdom by Constable and Company, Ltd., 10 Orange Street, London WC 2.

The Physics of Modern Electronics is a new English translation of *Einführung in die Elektronik,* copyright © 1960 by Juris-Verlag, Zurich. This English edition has been extensively revised and enlarged by the author.

Standard Book Number: 486-61749-1
Library of Congress Catalog Card Number: 66-28269

Manufactured in the United States of America
Dover Publications, Inc.
180 Varick Street
New York, N.Y. 10014

PREFACE TO THE DOVER EDITION

Because of the encouraging reception obtained by the first edition both at home and abroad, the author has kept to the same arrangement of chapters in this new edition. As a result of valuable suggestions and the expansion of my own courses in electronics, the scope of the book has been enlarged. The section on tunnel diodes in Chapter VI on semiconductor elements has undergone fundamental expansion. Of special importance is the addition of Chapter VII devoted to molecular amplifiers. In connection with this it was initially necessary to introduce refinements into the concepts of atomic structure described in Chapter I. Chapter VII makes the heaviest demands upon the reader; but it represents the present state of electronics.

Here I would like to express my thanks to all those whose suggestions have contributed to the improvement of the present volume. I will always be grateful for any suggestions that may be incorporated into the future edition.

I hope that this new edition will obtain the same friendly reception received by the first.

Zurich, 1966 W. A. GÜNTHER

PREFACE TO THE FIRST EDITION

The need for a comprehensive introductory text containing the most significant fundamentals of electronics was brought home to the author in the course of teaching a class in electronics this winter semester at the Kant. Volkshochschule of Zurich. The present book is based largely upon the instruction notes used in the course and takes the same approach. It was, however, kept in mind that electronics has entered such a new and fascinating stage of development that many are interested in becoming familiar with it. The subject is treated in such a manner that it is also suited for self-instruction. It is the aim of the present volume to provide something new in this respect. It undertakes the pleasant task of providing the reader with the basic principles of electronics as well as an overall view of its present state. While giving the course, but even more so after taking up the task of writing such a book, the author was forced to acknowledge that a short introduction to atomic theory, throwing some light on the behavior of electrons, was absolutely necessary. This is taken up only insofar as it is of fundamental importance in electronics. The area of present-day electronics is so large that the author had necessarily to limit himself to the treatment of the fundamental aspects and was only able to touch upon the areas of application. The circuits for electronic installations also had to be dispensed with, since these vary in the course of development, whereas an accurate knowledge of the theoretical fundamentals always remains the condition for understanding. In writing the book the attempt was made to combine theoretical explanations with physical conceptions. The present volume, which is primarily concerned with physical and electrical phenomena, is divided into six chapters.

1. An Introduction to the Physical Bases of the Electron
2. Electronics in High Vacuum
3. Gas-Filled Tubes (the thyratron, etc.)
4. The Cathode-Ray Tube and Special High Vacuum Tubes

5. Modern Velocity-Modulated Tubes
6. Semiconductor Electronics, with Special Regard to Transistors

The book is primarily designed for those taking a course, but also for students, technicians, and others who are interested in entering the field of electronics. In view of the different educational background of the prospective readers, the volume dispenses to a great extent with the precise mathematical tools and derivations of basic equations. Thus, readers with less mathematical training will also be able to understand it easily.

In the introduction the physical properties of conductors, semiconductors, and insulators are also described. Subsequent sections discuss conduction by ions and electrons and holes. In addition to the construction and development of electron tubes, the modern transit-time tubes used in microwave engineering are also considered. The germanium and silicon diodes and transistors are discussed in considerable detail. The conclusion of Chapter VI contains a discussion of the modes of operation and physical behavior of other integrating circuit elements of electronics, such as hot conductors, cold conductors, varistors, etc.

I would like to express my heartfelt gratitude to Dr. Hardy Cristen, the publisher (Juris-Verlag) for his help in editing and reading, and no less to my assistant, J. Wyss, Dip. Eng., and Mr. Hüllsman who prepared the drawings from my sketches. It is the author's hope that the book will be used by students as a supplement to their studies and by practitioners as an introduction to the manifold aspects of the field.

Suggested improvements for a future edition will be welcomed.

Zurich, January 10, 1960 W. A. GÜNTHER

CONTENTS

THE PHYSICS OF MODERN ELECTRONICS

I. INTRODUCTION TO THE PHYSICAL BASES OF ELECTRONICS

1. Introduction

Electronics is a child of electrical science. Classical electrical science treats of apparatus and circuit elements in which electrons travel through metal or liquid conductors. Properly considered, the domain of electronics is restricted to components in which the manner of operation is essentially influenced by free electrons. The most important of these components are the tubes and a number of semiconductors. Electronic instruments make use of circuit elements of this type. Precisely speaking, the birth of the new, and yet quite old, science of electronics can be traced back to the year 1884, when Edison was experimenting with his evacuated incandescent lamp. An extra lead wire that had been fused into the evacuated glass bulb, but which was not used as a filament, led him to several observations. He found

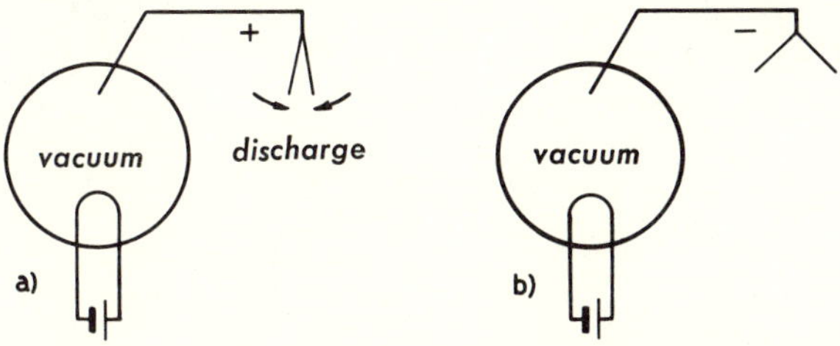

FIG. 1. Edison's experiment with an incandescent lamp containing a free lead wire. When the wire is positively charged a flow of current is observed.

that a positively charged electroscope connected to the wire (as in Fig. 1*a*) discharged as soon as the filament was caused to glow, whereas when the electroscope was negatively charged (as in Fig. 1*b*) nothing happened. Not until later was it possible to explain this phenomenon as a result of the heating of the filament, which caused negative charge carriers, the so-called electrons, to leave the solid material of the heated filament (the cathode) and cancel out the positive charge, i.e., the electron deficit of the electroscope. Conductivity of this sort, through a vacuum, was still new at that time. A practical tube with two separate electrodes, the so-called diode, was not discovered until 1904 by Fleming. However, before we concern ourselves with tubes, it seems fitting to begin by penetrating somewhat more deeply into the world of electrons, since it is to these same electrons that electronics ultimately owes its name.

2. A glance into the world of electrons

Material bodies are commonly composed of individual components called molecules. The manner in which these are assembled determines the state of the aggregate, whether solid or liquid. Molecules cannot be broken down any further by mechanical means. If, however, other methods are employed, molecules can be broken down further into atoms, and the atoms themselves are not indivisible units. Atoms consist fundamentally of two parts: a positively charged

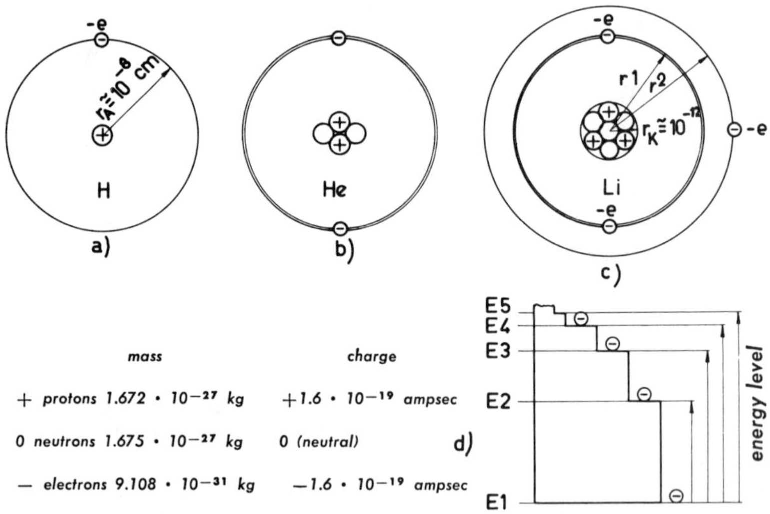

FIG. 2. Within the atoms the electrons move in quite definite spherical shells about the atomic nucleus. Atomic models: (*a*) hydrogen (H); (*b*) helium (He); (*c*) lithium (Li); (*d*) the distances between the electron shells or orbits and the nucleus are supposed to correspond to different potential energy levels (energy values).

nucleus and—separated by an intermediate zone—the negative, electron cloud. Considered as an independent unit, the atom is electrically neutral. Rutherford, first, and then Niels Bohr used spectroscopic observations to develop conceptual models of the atom, which, however, must be considered as merely analogies. In their analogy the atom is conceived as a solar system reduced to a microcosmic level (Fig. 2*a*). The element hydrogen is composed of the lightest atoms. These contain a nucleus with a single positive particle, the proton, whose charge was fixed as 1. The next largest atom, helium, possesses a positive nucleus with twice the charge of hydrogen and two electrons. The charge of the nucleus determines the atomic number. The electrons travel around the nucleus at precisely fixed distances that prevent them from being drawn into the oppositely charged nucleus in a manner similar to the laws that permit the rotation of the moon about the earth and the earth about the sun. It is assumed that the electrostatic attraction (the coulombic force) between the atomic nucleus and the electron is in equilibrium with the centrifugal force. Uranium, whose nucleus contains 92 protons, possesses 92 electrons, which rotate about it. If all the elements are arranged in the order of their atomic number, the natural sequence of elements is obtained; hydrogen occupies the first position, with an atomic number of 1, and Curium occupies the last place with the number of 96. In the U.S.A. an element known as Lawrencium with an atomic number of 103 and atomic weight of 257 has been prepared synthetically. Figures 2*b* and 2*c* show that the atomic nucleus contains so-called neutrons in addition to the electrically positive protons; these are electrically neutral particles with approximately the same weight as protons.

When the atom is in the neutral state the negative cloud-charge formed by the electrons is equal to the positive nuclear charge (the number of protons equals the number of electrons). Since the atom is composed entirely of protons, neutrons, and electrons, the atomic weights must be integral multiples of the weight of these elementary particles. However, since the electron is 1/1840 the weight of the hydrogen atom, which we have come to recognize as a proton, the electron can, therefore, be disregarded in considerations of weight. Inasmuch as protons and neutrons are of nearly the same weight, the total of protons and neutrons gives the practical atomic weight of an element, while the number of protons determines exclusively the type of element it is. Thus, it is evident that the same element may occur

period \ group	I	II	III	IV	V	VI	VII	VIII	0
1	1 ***H*** 1.008								2 ***He*** 4.00
2	3 ***Li*** 6.94	4 ***Be*** 9.02	5 ***B*** 10.82	6 **C** 12.00	7 ***N*** 14.008	8 **O** 16.000	9 ***F*** 19.00		10 ***Ne*** 20.2
3	11 ***Na*** 23.00	12 ***Mg*** 24.32	13 ***Al*** 26.97	14 ***Si*** 28.06	15 ***P*** 31.02	16 ***S*** 32.07	17 ***Cl*** 35.46		18 ***Ar*** 39.94 ←
→ 4	19 ***K*** 39.10 29 ***Cu*** 63.57	20 ***Ca*** 40.08 30 ***Zn*** 65.38	21 ***Sc*** 45.10 31 ***Ga*** 69.72	22 ***Ti*** 47.90 32 ***Ge*** 72.60	23 ***V*** 51.0 33 ***As*** 74.91	24 ***Cr*** 52.01 34 ***Se*** 78.96	25 ***Mn*** 54.93 35 ***Br*** 79.92	26 ***Fe*** 55.84 27 **Co** 58.97 ↔ 28 ***Ni*** 58.68	36 ***Kr*** 83.7
5	37 ***Rb*** 85.48 47 ***Ag*** 107.88	38 ***Sr*** 87.63 48 ***Cd*** 112.4	39 ***Y*** 88.92 49 ***In*** 114.8	40 ***Zr*** 91.2 50 ***Sn*** 118.7	41 ***Nb*** 92.91 51 ***Sb*** 121.76	42 ***Mo*** 96.0 52 ***Te*** 127.61	43 ***Ma*** ↔ 53 ***J*** 126.92	44 ***Ru*** 101.7 45 ***Rh*** 102.9 46 ***Pd*** 106.7	54 ***X*** 131.3
6	55 **Cs** 132.91 79 ***Au*** 197.2	56 ***Ba*** 137.36 80 ***Hg*** 200.6	57 ***La*** 138.9 † 81 ***Tl*** 204.4	72 ***Hf*** 178.6 82 ***Pb*** 207.2	73 ***Ta*** 180.88 83 ***Bi*** 209.0	74 ***W*** 184.0 84 ***Po*** 210	75 ***Re*** 186.3 85*	76 **Os** 191.5 77 ***Ir*** 193.1 78 ***Pt*** 195.2	86 ***Rn*** 222
7	87*	88 ***Ra*** 226.0	89 ***Ac*** (227)	90 ***Th*** 232.1 ↔	91 ***Pa*** (231)	92 ***U*** 238.2	† and rare earths		

FIG. 3. *Periodic System of the Elements.* The number to the left of the symbols for the chemical elements is the atomic number (number of protons), and the number underneath the symbol is the atomic weight. The dark lines indicate the approximate transition from electrically conductive to electrically nonconductive elements, where conduction is possible only within the atomic ensemble.

with different atomic weights (e.g., as in heavy water); in all such cases the number of protons (and the nuclear charge) remains the same, but there is a greater number of neutrons. Elements such as these, which are known as isotopes, occupy the same position in the so-called periodic system, despite their different atomic weight. The position is determined by the atomic number. The periodic system of elements (Fig. 3, p. 4) is a special table in which the elements are grouped in accordance with a law of periodicity. For over a hundred years now it has been known that the essential traits of chemical properties repeat themselves in a fairly consistent periodic fashion.

It was discovered by Pauli that the same electron orbit cannot be occupied by several electrons at one time. On the basis of considerations of energy it was found that only one electron can occupy a single orbit. When an atom possesses more than one electron, the additional electrons are arranged progressively in new and energetically more distant orbits. These orbits, which do not lie in a single plane are nevertheless relatively close to each other. After a certain number of electrons a final point is reached, after which all further electrons must be incorporated in a similar manner at a considerably greater distance from the nucleus. The final atom can be thought of roughly as being built up of a succession of individual, rotation-symmetrical shells spaced at progressively greater distances from the nucleus, each containing the orbits of several circling electrons. Even when the atom is composed of several such shells, which is the case for all of the heavier elements, the nucleus, as a result of its great distance from the periphery, is extremely small in comparison to the over-all diameter of the atom. The diameter of the nucleus is approximately 10^{-12} cm, while the diameter of the atom measures about 10^{-8} cm, or more than 10,000 times as large. An electron has an approximate magnitude of 10^{-13} cm and bears a charge equal to and opposite that of the proton, $1.59 \cdot 10^{-19}$ ampsec or coulombs. Thus, when approximately 10^{19} electrons per second pass through a cross section, a current of one ampere is said to flow. A current is understood to mean the total number of electrons passing per second through a given surface. Later we will find out how this is measured. In respect to the construction of the shells, it should be added that those elements in which all the shells are completely filled with the correct, i.e., theoretically possible, number of electrons occupy a special position. They are known as rare gases because they cannot enter into any chemical compounds. The first element of this type in the periodic system is helium. It possesses two electrons that circle the nucleus in approximately the same rotational orbit, thus both in the first shell. The first shell is, therefore, already filled when it contains two electrons. The next element, lithium, contains one more electron. This third electron travels in a new and much more distant orbit and is the single satellite in a new shell that could contain up to 8 electrons. It is easy to see that this element can easily surrender this electron. The electrons of the outermost shell are known as valence electrons. These electrons are the most significant in determining the chemical properties of an atom. The chemical activity between different types of atoms is explained by the tendency of atoms to build up or complete their shells by exchange of electrons and thus obtain

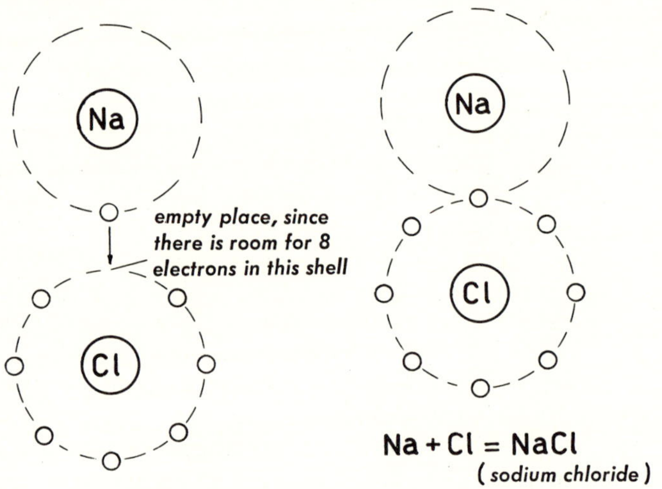

FIG. 4. Chemical bond between a sodium atom and a chlorine atom.

an energetically more stable structure for the orbits. As everywhere else in nature there is a tendency here to seek the lowest potential energy. Figure 4 shows how a chemical bond arises between one sodium and one chlorine atom to form table salt (sodium chloride). The chemical reaction occurs as the lightly bound sodium electron goes over to fill the gap in the valence shell of the chlorine atom. The shells are thereby stabilized and attain the structure of a rare gas; they are, therefore, no longer capable of further chemical reaction. Since the two atoms, which were previously neutral, now possess one positive (the sodium) and one negative (the chlorine) unit charge, the two atoms attract each other and form a stable molecule. Other chemical compounds are formed in a similar manner, e.g. $H_2 + O = H_2O$, etc. The forces holding the atoms together in the molecule are always of an electrical nature. As previously stated, the whole electron cloud is the seat of the electrical, magnetic, optical, and chemical properties of an atom. The maximum number of electrons that may occupy the successive shells is 2, 8, 18, 32 electrons. The series may be expressed in the form: $2 \cdot 1^2, 2 \cdot 2^2, 2 \cdot 3^2, 2 \cdot 4^2$. Electrons in the lowest energy orbits (H and He) are in the shell $n = 1$. The next highest shell (Li, Be, . . .) is designated as $n = 2$, etc., and the number of possible electron positions can be computed as equal to $2n^2$. It is impossible to introduce an additional electron into a completely filled shell. Thus, we have obtained a certain over-all view of the structure of an atom. The periodic system is based upon an arrangement in accordance with which the atoms are divided into seven periods according to the structure of their shells. This arrangement has taken into account the fact that as the nuclear charge is steadily increased and the number of electrons, as well, certain properties are periodically repeated, e.g., conductivity and nonconductivity. As mentioned before, the orbiting electrons are assigned to very definite, fixed, quantic orbits, each of which corresponds to a definite energy level. If the electron travels in this orbit no energy loss occurs (there is no emission of radiation). If, on the other hand, the electron passes over from its fixed orbit into another orbit (of higher or lower potential energy), radiant energy is emitted or absorbed (Fig. 5). If the electron in the outer orbit possesses the energy E_2 and the electron in the inner orbit possesses the smaller energy E_1, then the difference between these two energies, i.e., when an electron is shifted from one to another, results in a light quantum of the magnitude $E_2 - E_1 = h \cdot f$ of radiant energy. In this relation h represents Planck's action quantum, which is a natural constant equal to $6.62 \cdot 10^{-27}$ ergsec, or $6.625 \cdot 10^{-34}$ wattsec^2 in the "Giorgi" measuring system, and f is the frequency of the emitted light or electromagnetic wave. The relationship between the frequency f and the wavelength λ of the emitted radiation is given by the formula $c = f \cdot \lambda$, where c is the velocity of propagation of the electromagnetic wave (i.e., of the light) in a vacuum: $c = 300{,}000$ km/sec $= 3 \cdot 10^{-8}\ \text{m} \cdot \text{s}^{-1}$ and λ may be expressed in km, m, cm, or mm. In practice the wavelength is frequently expressed in Ångströms (one Å $= 10^{-10}$ m or 10^{-8} cm). Red light, which possesses a frequency of

$f = 5 \cdot 10^{14}$ cps, has a wavelength of $\lambda = 6000$ Å, which corresponds to a wavelength of 6.10^{-5} cm or 6/10,000 mm or 600 nanometers (nano $= 10^{-9}$).

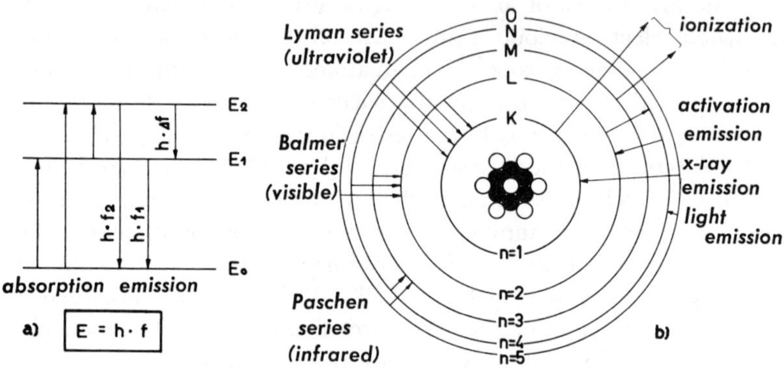

FIG. 5. Schematic diagram of the quantum transitions of electrons on absorption or emission. The horizontal lines E_0, E_1, E_2 represent energy levels. The arrow lengths correspond to the energy differences and the frequencies emitted or absorbed on transition from one level to another.

Thus, an electron requires excitation, usually from an external source, i.e., work that can be employed, to pass to a higher quantic orbit. Only after an electron drops back to the prior or a lower orbit, into a so-called electron hole, is energy released and it is possible for a light quantum, a so-called photon, to be emitted to reestablish the energy equilibrium. It is precisely in this manner that the incandescence of a heated body occurs, the excitation originating from the collisions of the atoms, since the increasing heat progressively intensifies the oscillations of the atoms and molecules. The shell electrons, as a consequence of the absorption of kinetic energy, pass over to orbits of temporarily higher energy and emit light radiation as they drop back to the lower quantum levels. If the kinetic energy of an atom of the hot body exceeds a certain minimal level, electrons can be completely ejected from the atomic field. This is already an anticipation of the nature of electron emission from incandescent metals.

As we have already mentioned, an energy E must be utilized to produce a light quantum $h \cdot f$. The activation energy required to generate visible light is less than that required for the production of roentgen radiation, but is greater than that required for production of radio waves. It is assumed as understood that all these phenomena are in essence electromagnetic waves. The production of visible light involves the quantum leaps of the loosely held, peripheral electrons. As the activation energy is increased, the frequency of the light becomes higher. If the activation energy is sufficiently strong to produce a movement between an outer shell and the innermost shell, the frequency is so high that roentgen rays are produced. If an

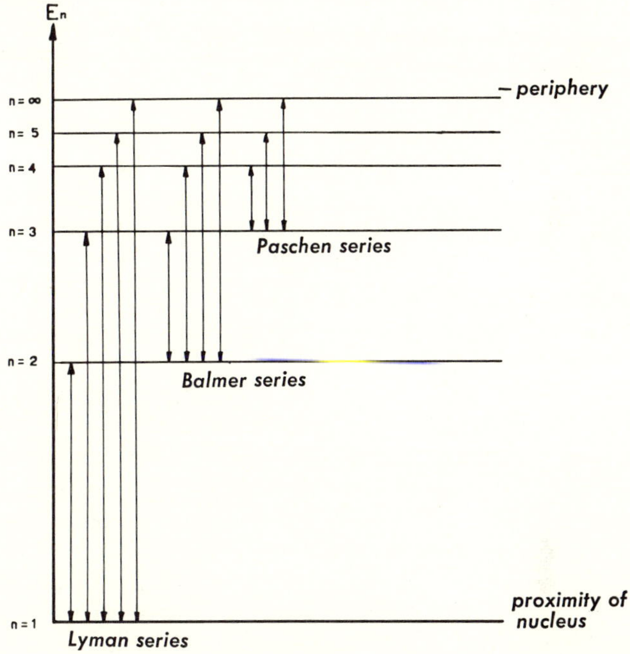

FIG. 6. Energy-level diagram of hydrogen or the hydrogen spectral series.

electron gets free of the atomic field, it is also possible for a state to occur in which an electron from one of the neighboring shells passes over into the unoccupied position. This is illustrated by the diagram of Fig. 6. This is, as it were, a representation of a section through the Bohr model of the atom. The relationships are seen at their simplest in the hydrogen atom, which possesses only one electron. When the nuclear charge is increased the radii of the orbits become smaller and the emission spectra are shifted in the direction of shorter wavelengths and higher frequencies as a result of the increased forces. This is very important, for example, in the choice of anode material for a roentgen tube; this is why heavy elements are selected for this purpose. In an atomic clock and generally in microwave spectroscopy only the small energy-level transitions $E_n - E_{n-1} = h \cdot f$ are utilized in order to produce radio waves of approximately the centimeter wave range. In the atomic clock this frequency, or spectral line, if one will, is used as the time standard. Electronics has a special interest in free electrons. Thus, measures must be employed to free the loosely held, peripheral electrons from the material. This can be accomplished, as previously indicated, by such means as heating. This method is used in all radio tubes. Inasmuch as thermal radiation also belongs to the spectrum of electromagnetic waves, it is easily understandable that light as well as heat should be capable of causing emission of electrons. As we will see, however, the physical explanation for this is somewhat differently conceived. In the photoelectric cell, an important electronic element that will be studied more thoroughly later, this electrical effect of light is employed. In this case activation is effected by light rays. Since these behave like small particles or a kind of small projectile, they are called photons, or, as they are often referred to, corpuscular light quanta. Like these, electrons themselves are also capable of effecting activation, or producing additional free electrons. This effect, which is known as secondary emission, is employed in electron multipliers. If the surface of a suitable metal is bombarded by electrons (or by ions, atoms or molecules) at high velocity, their energy is transmitted by the collision to the electrons

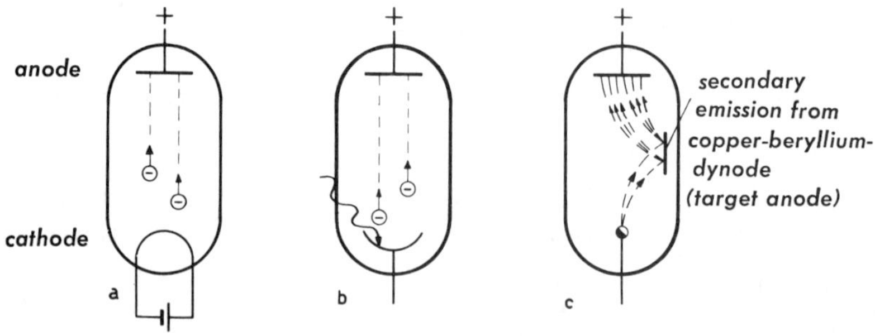

FIG. 7. (*a*) Hot cathode tube. Electron emission produced by heat energy (thermal emission). (*b*) Photoelectric cell. Electron emission produced by light energy (photoemission). (*c*) Electron multiplier. Electron emission produced by kinetic energy of bombarding particles (secondary emission).

within the metal, which results in their emission. Since the number of secondary electrons freed will be greater than the number of bombarding electrons if the proper material is selected—as illustrated in Fig. 7*c*—this multiplication effect is of great value in electronics.

Figure 7 shows the three essential types of emission: thermal emission, photoemission, and secondary emission. Electrons can be freed from every material in all aggregate states, specifically, by incandescent emission, the photoelectric effect, electron and ion impact, field emission, thermal ionization, etc.

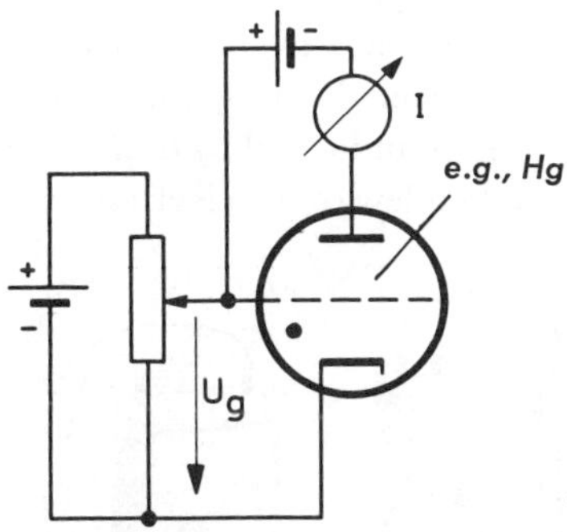

FIG. 8. Experimental installation used to demonstrate the quantization of energy values (after J. Franck and G. Hertz, 1914).

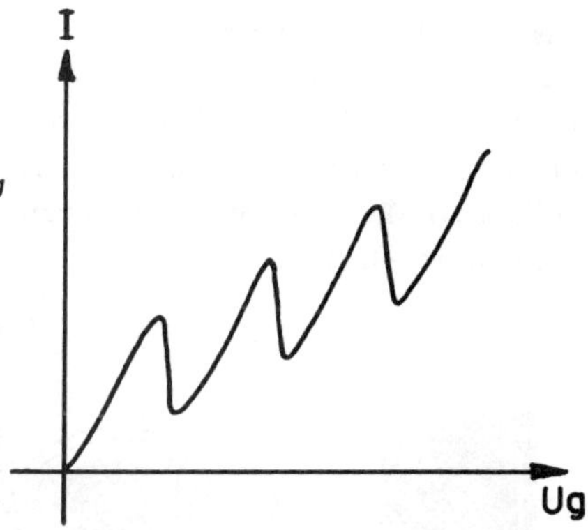

FIG. 9. The curve shows the inconstant increase in anode current resulting from increase of the positive grid bias.

The experiment of Franck and Hertz proved that an electron can in reality only assume certain discrete energy values, represented in the atomic model as orbital radii. This experiment makes use of a hot-cathode triode, which is filled at low pressure with the type of gas the atoms of which one wishes to investigate. As Fig. 8 shows, in contrast to normal use the anode is given a negative potential and the controllable grid is given a positive bias with respect to the cathode. The electrons emitted by the cathode pass through the positively biased grid mesh and, if they are traveling at sufficient speed, strike the negative target anode. For the electrons passing the grid there are two possible cases: (1) in the course of their flight they will be elastically deflected by the gas atoms, which will produce no other result than a change in their direction; (2) they will collide with the gas atoms with such great force that electrons of the gas atoms are raised to higher energy states. The original electrons of the gas atoms will then be reduced in velocity to an extent corresponding to the energy expenditure required for activation.

Increase of the positive grid bias voltage from an initial zero value potential results in a progressive increase of the electron velocity, which is at first reflected in a constant increase in the anode current, as seen in Fig. 9 (case 1). If the potential is raised further and the so-called critical electron velocity is reached, there is a sudden transition to case 2. As this energy level is reached the gas atoms become activated. This is also indicated by the appearance of certain spectral lines in the gas glow. The electrons are slowed down by the amount of energy consumed by the collisions. This is indicated by the sudden collapse of the anode current, because only a few of the electrons continue to reach the anode.

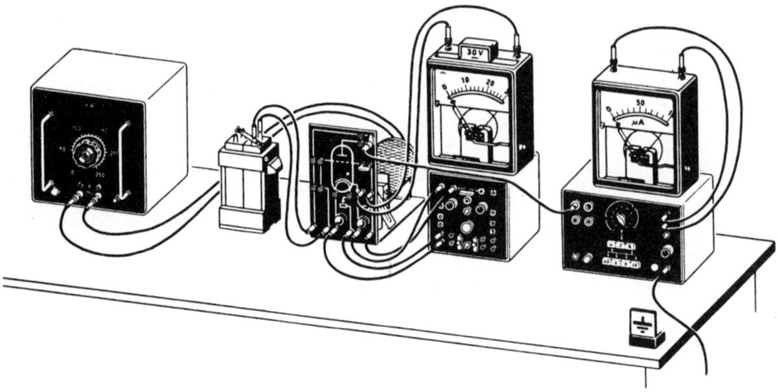

Fig. 10. The Franck-Hertz mercury experiment (with d.c. amplifier).

If the grid bias voltage is raised still further, the velocity of the electrons is increased once more. The current again increases, until, on further increase of the control voltage, a critical velocity is again achieved and the current collapses once more. The extremely sharp current drops indicate that the atom-bound electrons can absorb or emit only very definite quantified energy values. Each new critical activation potential is marked by the simultaneous appearance of new spectral lines in the ionized gas glow. Mercury vapor triodes are particularly well suited for these experiments. Figure 10 shows the experimental installation, as manufactured for schools by the firm Phywe. A mercury tube is installed in the plug panel. If a stream of electrons is passed through mercury vapor, they lose no energy as a consequence of elastic collision with the Hg atoms. If a potential of 4.9 v is exceeded, they lose their energy, which is transmitted to the mercury atom, and this is emitted as a light quantum (=2537 Å). It can thereby be demonstrated that energy = e.v. = $h \cdot f$.

3. Refined atomic models

Before we take up photoelectric cells, radio tubes, electron multipliers, etc., as circuit elements in greater detail, we shall take a closer look at electrons and point out several atomic models containing certain refinements. The French scientist de Broglie discovered in 1924 that when electrons pass through thin layers of material, e.g., silver foils, they exhibit exactly the same type of beat (i.e., interference) phenomena as are observed under corresponding conditions when diffracted light waves or x-rays of various wavelengths are superimposed on each other; this is the typical characteristic of wave radiation. The conception of an atom becomes very difficult to visualize if one keeps in mind the remarkable double character this gives them. On the one hand, an electron is represented as a material corpuscle, a so-called particle, with a definite weight, specifically $m = 0.909 \cdot 10^{-30}$ kg, which means that 10^{30} electrons are required to achieve a weight of 1 kg or 10^{27} for 1 gram; and, on the other hand, the electron also appears to be a wave. On the basis of his observations de Broglie assigned to each material particle an oscillation state of the magnitude $h \cdot f$ corresponding to its kinetic characteristic energy $m \cdot v^2$. Since $\lambda = \frac{v}{f}$ (in which λ = wavelength, v = propagation velocity, f = frequency), the wavelength of a moving particle of mass m and velocity v becomes

$$\lambda = \frac{h}{m \cdot v}$$

The model of the atom with its electron orbits then undergoes modification to the form shown in Fig. 11. Here the only possible discrete orbits are those whose periphery can be contained within a single wavelength or a whole number of wavelengths. The rule for angular momenta is, accordingly, that only electron orbits with angular momenta of $1h, 2h, 3h, 4h, \ldots$, are possible. Thus the electron orbits are, from the standpoint of wave mechanics, largely fixed. The visual conception of the atom is considerably impaired by this wave picture, but this model agrees more closely with experience. Coordination of this new condition with the motion or equilibrium equations of the orbiting electrons gives the radii of the different fixed orbits and the various possible energy states. Still greater similarity to a wave

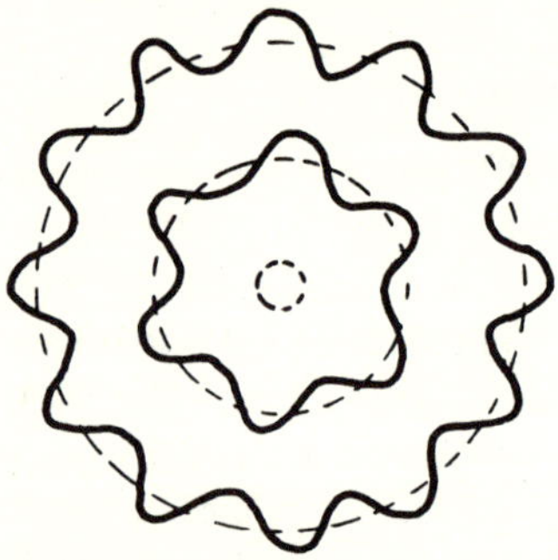

FIG. 11. Wave model of two electron orbits. The existence of discrete, waveless Bohr orbits can be reconciled with the wave conception if the periphery of an orbit is considered as equal to an integral multiple of the de Broglie wavelength. A standing wave does not lead to a loss by radiation but results only in a periodically varying, translational wave.

phenomenon, and to a standing wave phenomenon, in particular, is offered by the assumption of a density distribution of electricity in the vicinity of the nucleus. If we imagine standing waves of negatively charged electrical material that oscillate about the nucleus, the waves, too, would correspond to a definite density distribution of this material. The de Broglie matter waves can therefore also be interpreted in such a manner that the square of the wave amplitudes at a given position would correspond to a definite density of this matter. Schrödinger postulated standing waves as so-called charge clouds in a medium of variable density in the vicinity of the atomic nucleus. This assumption formed the basis of his wave mechanics. The standing waves explain the stationary distribution of electricity in the vicinity of the nucleus and make understandable the existence of the nonradiant, fixed states of the atom. This would eliminate one of the problems of the Bohr model of the atom, inasmuch as the electrons would otherwise, in accordance with classical laws, be expected to emit electromagnetic radiation and thus undergo a corresponding energy loss. Born replaced the charge cloud with the degree of probability of finding a given electron at a given place at a given time; this brings us back to the punctiform charge, as Bohr called it. Werner Heisenberg abandoned this type of visual atomic model and collected all of the magnitudes factually obtained and observed experimentally in the form of a calculating system, in a so-called matrix analysis equation. This equation contains the experimentally obtained wavelengths of the spectral lines, that we will speak of later, the intensity values, the polarity states, the electrically determinable energy levels, etc. The result was a mathematical scheme rather than a model based upon a visual conception, but one that was in agreement with all recorded observations.

If an element is excited in an electric light arc, the electrons undergo a change of place. As they fall back to a lower energy quantum orbit, the well-known emission of light takes place. Spectral analysis with prismatic equipment makes it possible to measure the individual frequencies (spectral lines). The values $h \cdot f$ then correspond exactly to the differences between the different energy levels of the electron leaps that have occurred. If such an element is placed in an electric or magnetic field, it becomes evident that the light frequencies can be split into a so-called fine structure. The splitting of the energy levels in an electrical or magnetic field was most precisely investigated by Stark and Zeeman. This demonstrated splitting of spectral line frequencies could not occur with an orbit of the type represented by our previous model; consequently we must again modify our model, which is, after all, only an aid to our understanding. The electrons, which revolved about the nucleus in the Bohr model of the atom, must now also be understood to rotate on their own axis, if these new phenomena are to be explained. The electron can be imagined as a rotating, charged sphere something like the earth. This rotation about an axis is called spin. In addition to this mechanical rotational moment, the electron also possesses a magnetic moment, since the motion of an electrical charge gives rise without fail to a surrounding magnetic field. The spin therefore also shares in the quantum characteristics of the electron. In order to interpret these harmonic oscillations, which were also demonstrated by spectral analysis, the original orbit, which was circular, must be changed to an elliptical model. In the Bohr atom model all electrons in orbits with the same n are said to be in the same shell. The shell $n = 1$ is called the K-shell, the $n = 2$ shell is called the L-shell, the $n = 3$ shell is called the M-shell, and the $n = 4$ shell is called the N-shell. In order to determine the ellipse, the so-called azimuthal quantum number l is used. This number characterizes the form of the ellipse, i.e., it shows the degree of flattening (Fig. 12). A discussion of further refinements of the atom model would take us too far beyond the scope of our theme; we must, however, keep in mind the following, already indicated, fact. As long as the sum of the negative elementary charges of the electrons orbiting in the electron cloud is equal to the positive charge of the nucleus, the atom is externally neutral. If, however, measures are taken to cause a loss of electrons from the atom, the atom remains with an excess of positive electrical charge—and the atom becomes a positive ion. If an atom absorbs one or

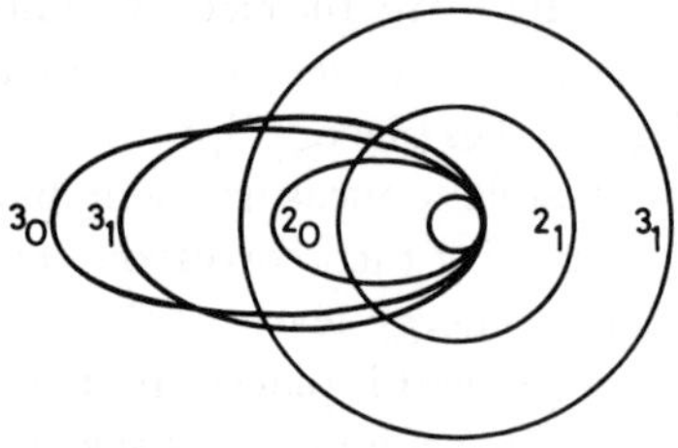

FIG. 12. Elliptical model.

more electrons, it is called a negative ion. The positive charges of the nucleus are firmly bound to the matter of the atom, so that, generally, electrical currents, e.g., in metals, are conducted by the mobile electrons alone. It is, however, also possible to dissolve the matter into ions, which will take over the transport of electrical charges, i.e., they will effect the flow of current, as, for example, in electrolytic baths, gas discharge tubes, etc. Ionization of gases can be accomplished by the most various methods of activation, e.g., by the use of ultraviolet or x-ray irradiation to produce a photoelectric effect, or visible light for certain elements, and also by the use of flame or heating of any type, in short, by the use of energy from any source whatever. As we have already seen, each of these methods frees electrons from the bonds of the atom and molecule. Particularly in ionized gases free electrons tend to collect at other previously neutral atoms, giving rise to a condition in which there are electrically positive, negative, and neutral groups of atoms, in addition to the free electrons. Filling gases always contain individual ions or chance electrons, possibly as a result of cosmic rays. On the average about 6 ion pairs are formed per second per cubic centimeter of air. If an electric potential is applied to the electrodes of a glow discharge tube that has been constructed in accordance with this principle (Fig. 13) and filled with gas at low pressure, the individual chance ions, being charge carriers, are attracted to the electrodes of the opposite charge in accordance with the tendency toward neutralization. The free negative electrons are accelerated in the direction of the positive anode, and, as they attain a certain velocity, they no longer undergo mere elastic collision with the atoms in their path, but can also dislodge electrons from the atomic shells. These initially neutral atoms now give rise to two new charge carriers, the newly freed electron and the remainder of the atom, which now, having an excess of positive charge, is a positive ion. Collisions of this type may also dislodge several

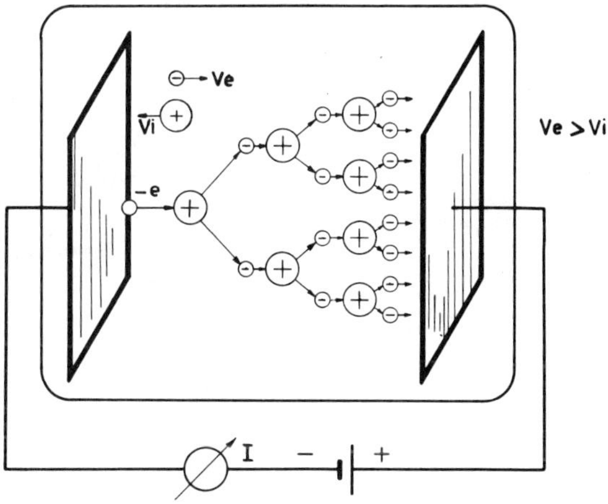

FIG. 13. The ionization process between two electrodes in a vessel filled with gas the pressure of which is only a fraction of atmospheric pressure.

electrons from an atom at one time. Both the primary and secondary electron—or electrons—that have been dislodged from the gas atom, then move toward the positive electrode (anode). As a result of the acceleration they receive from the electric field, they are enabled to free new electrons on collision with a gas atom. The positive ion resulting from the collision moves in the direction of the negative electrode (cathode), where it is capable of dislodging from the cathode new electrons as a result of its kinetic energy. The possibility of this type of impact ionization depends on both the voltage and the "correct" pressure of the gas. If the pressure is too great, as it is, for example, at atmospheric pressure, as soon as the electrons leave the cathode they immediately strike a gas atom. The length of the free path between two collisions is not great enough to provide a sufficient velocity or kinetic energy to bring about ionization. If the gas pressure is too low there are too few gas atoms present to provide

It should be mentioned, if only in passing, that there is an enormous difference between the stability of the atomic nucleus and that of its electron cloud. The cloud can be influenced by charge carriers that have been accelerated by only a few volts, while to alter the structure of the nucleus particles possessing a kinetic energy of several million volts are required.

We will not go any further into the question of the atomic nucleus since we are concerned with electronics rather than atomic physics, though it must be admitted that both areas have underlying principles in common.

4. Metal conductors, insulators or semiconductors

Figure 6 shows the energy levels of the free hydrogen atom as obtained from spectroscopic studies. The diagram shows that in order to dislodge one electron, and thus ionize the atom, an energy of 13.5 e.v. is required. The work is usually expressed in electron volts (e.v.). This can be considered in the light of an analogy of potential energy expressed in weight times height. Figuratively, the charge weight of an electron (the attraction of the positive nucleus) can be raised by a potential difference of 13.5 volts. Scientifically expressed, this is the energy of an electron which has passed through a voltage of 13.5 volts. Since the energy or the requisite work is related to a unit charge, it is also permissible to express the potential simply in volts. The work to be expended is, however, always the product of the electron charge and, for example, the output potential (1 e.v. $= 1.6 \cdot 10^{-19}$ wattsec). The energy—as a result of the relation $E = h \cdot f$, in which h is a constant—can also be expressed in frequency or in number of vibrations over a 1 cm path.

Up to now we have been primarily concerned with free atoms, i.e., atoms in an ideal gas. In solid materials the atoms are close together. Metal atoms are arranged at regular intervals as a result of their mutual forces. Arrangements of this sort are called lattices. In a lattice the atoms are so "close" to each other that the orbits of the outer electrons overlap each other. In this situation an outer electron can easily "forget" which atom it belongs to. It can change places with an electron from a neighboring atom. After a short time it migrates to a third atom. Consequently, the outer electrons are considered as relatively free to wander about. These atoms therefore lack one or more firmly bound electrons, and are, consequently, positive ions or atom fragments. A lattice is often referred to as an ionic lattice. Lattice arrangements exhibit a low specific resistance, i.e., a high conductivity, since the electrons are highly mobile current carriers.

Metals that are good conductors are made up of atoms with only one electron in the outermost shell, having, therefore, an almost empty valence shell. Moreover, the structure of the solid material must be closely packed. This condition is fulfilled by a crystal structure. Now it is easy to see how the so-called extra electron can pass quite easily from one atom to the next and yet still belong energetically to the overall atomic aggregate. The quasi-free electrons can move with a certain freedom among the metal ions. If an electric field is set up in the metal, these quasi-free electrons in the middle move toward the positive side. Thus, an electrical current arises in the metal. The sum of many mobile individual electrons results in good electrical conduction. In the previous sections we considered primarily the physical properties of free atoms, because they formed particularly good models for illustration purposes. Consequently, the energy levels we discussed were valid for atoms in the gaseous state. In the gaseous state the neighboring atoms are so far apart that they no longer exert any effect upon the electrons of any other atom; whereas in the solid state, where the atoms are relatively close together, the electrons are affected by the forces of the neighboring atoms. Since the individual atoms are coupled together, the separate parallel lines (Fig. 6) corresponding to the energy levels of the individual shells become blurred, and an individual electron can with equal probability be situated at any point of the now continuous and spread energy band (Fig. 15). If, for example, a million of the same atoms are situated next to each other in a lattice arrangement, the electrons of the same energy level will no longer exhibit the same energy value, but will all exhibit somewhat different energy values. Naturally these energy values will lie very close to each other. We must, therefore, conclude that our earlier, narrower energy bands will widen and that the bands themselves will exhibit a kind of continuum of closely situated energy levels in the energy level diagram. It should be emphasized that the orbital distances or radii are to be considered merely as figurative. Basically, it is their potential and kinetic energy content that is important. In atomic aggregates, i.e., in solid materials, the bands of the atomic model are split up, and this leads to the formation of molecular energy bands. For the electrons

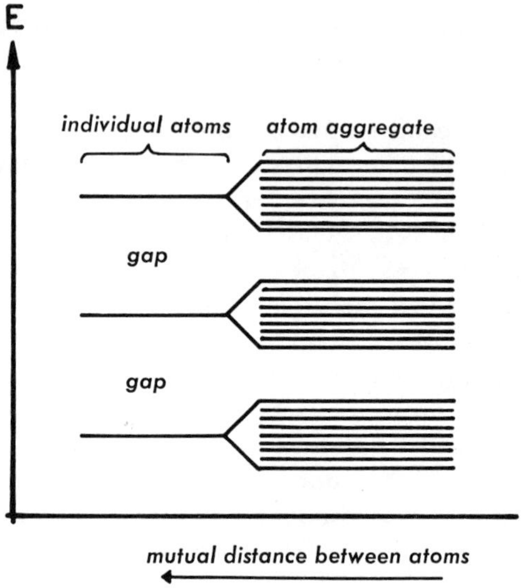

FIG. 15. Splitting up of the discrete energy walls on reduction of the interatomic distances. On the left, free atoms; on the right, atoms in an aggregate (solid bodies). The relationships between atoms situated close to one another are similar to electric oscillatory circuits of the same type that are moved close to each other. With an increase in coupling the resonance lines are spread out.

that possess a high energy level in a free atom, and are therefore somewhat further from the nucleus, the energy band is wider because the electrons are subjected to an increasing degree to the influence of the forces of neighboring atoms. The inner bands are narrow and grow progressively wider as their distance from the nucleus increases. These bands are said to be filled when all their levels are occupied by electrons. This generally holds true of all the bands with the exception of the outermost. An electron current cannot, therefore, flow through the inner bands. It can only be created by the quasi-free electrons of the outermost band under suitable circumstances (e.g., in metals). The gaps between the bands—assuming that the atoms are all the same—can be occupied not at all, or else only temporarily in the course of passage. The valence shell, like the other inner shells, also becomes somewhat blurred. These energy levels are widened to bands, as shown in Figs. 16 and 17. The interaction of the atoms gives rise to other bands outside the valence band (which was previously considered the last band), e.g., the conduction band. The conduction band is common to all the atoms of the aggregate. The lower edge of this band marks the beginning of the energy level that is no longer sufficient to bind electrons to individual atoms. In this band an electron can move from one atom to another. Here the electron can, therefore, be described as relatively mobile and serves to transport current. A substance in which such a phenomenon

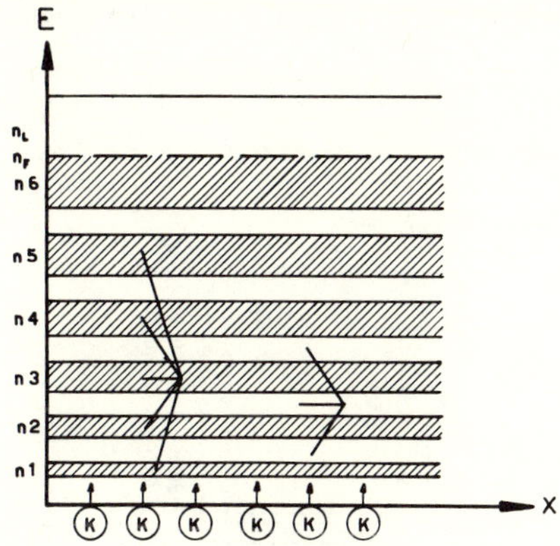

FIG. 16.

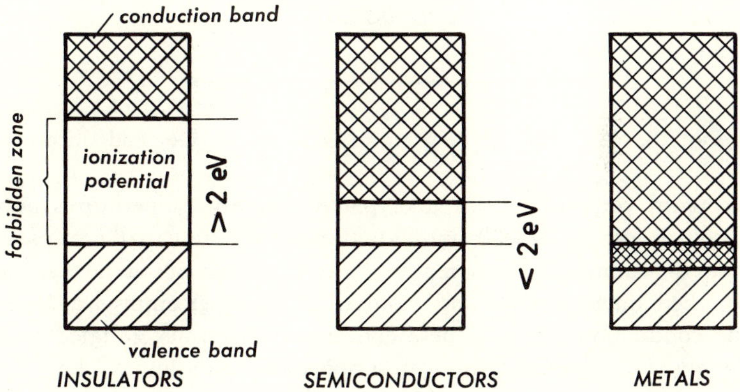

FIG. 17.

Solid body energy-level diagram. In isolated atoms the electrons can occupy only certain definite, sharply defined energy levels. One speaks of electron orbits, electron shells. The interaction of the atoms in a crystal lattice leads to wide energy bands, which replace the individual lines. In the energy-level diagram the energy is represented by the vertical lines and the locus coordinate by the horizontal.

occurs is known as a conductor, and the metals are especially good conductors. An insulator differs in that the outer electrons of the valence shell do not overlap each other, i.e., do not enter the conduction band, and consequently remain in their own atoms. There are insulators that are composed of molecules with only one type of atom, but there are also some made up of molecules composed of different types of atoms. Figure 17 shows a simplified solid body energy-level diagram that illustrates the differences between conductors, semiconductors, and insulators. Here also the ordinate represents the energy level. If there are only completely filled valence and conduction bands there are no "wanderers" and no

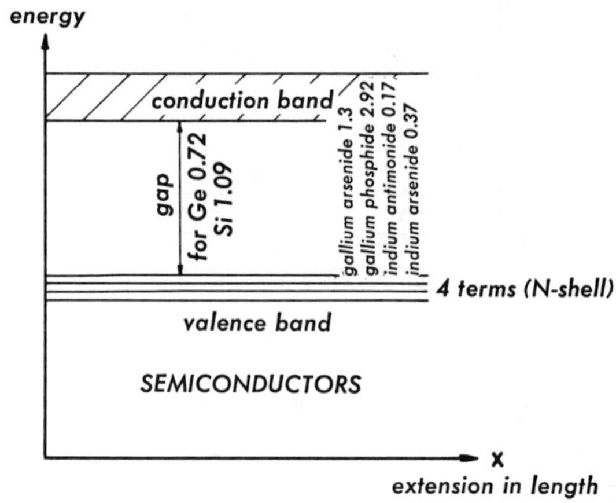

FIG. 18. Energy-level diagram of germanium (Ge) and silicon (Si).

electrical conduction can take place. At all temperatures above absolute zero (−273°C), individual electrons are capable of passing from the completely occupied stages, as a consequence of the thermally supplied kinetic energy, and wandering into other empty stages, e.g., the conduction band. These conduction electrons, situated in the energy or conduction band, can support a current flow. The same is true of the so-called holes, which are dealt with more thoroughly in Chapter VI. In the latter case, however, the movement is in the opposite direction as a result of their reversed sign. Thus, we have seen especially that in a metal the outermost and, at the same time, the widest energy bands overlap each other, and that the outermost, so-called conduction band is, moreover, not completely filled. We have also seen that in insulators the bands are either completely filled or empty. In insulators there are relatively large empty or, as one says, forbidden intermediate zones (forbidden zones or gaps) separating the valence band from the conduction band. Materials are known as insulators when an energy of more than 2 electron volts is necessary to transfer the electrons across the dead zone into the conduction band. This is an ionization energy that cannot be attained under ordinary heating or at normal room temperatures. Those materials with forbidden zones requiring less than 2 e.v. are known as semiconductors. For germanium (Fig. 18), which is very important in electronics, this required energy is about 0.72 e.v. and for silicon about 1.09 e.v. In these cases individual conduction electrons can be converted into conduction electrons at room temperature. A semiconductor is an insulator only at low temperatures. Logically the usual energy-level diagram for electrical conductors and, consequently, metals, exhibits no gap. In the latter case, the energy level of the conduction and valence band flow into one another.

Semiconductors are of special importance in electronics, and, consequently, all of Chapter VI will be devoted to this vast area. However, inasmuch as Chapter I is intended to cover all the fundamentals, the physical behavior of homogeneous semiconductors belongs here. As we have seen, in the semiconductor the gap between the valence and conduction band is so small that even at room temperature (absolute zero lies at $-273°C$) electrons pass from the valence band to the conduction band. In germanium the amount of energy required for this is less than in silicon. The higher the temperature, the more electrons that reach the conduction band. This temperature-dependent conductivity is called intrinsic conductivity. When an electron from the valence band enters the conduction band, an electron is missing from the valence band. This can be imagined as a hole remaining in the valence band. The atom from whose valence band an electron is missing and in which there is such a hole seeks to capture an electron from a neighboring atom—and is successful. As a result, the neighboring atom then lacks an electron, and the neighboring atom also captures an electron; thus, the so-called hole conduction takes place. If voltage is applied to such a semiconductor, the electrons migrate toward the positive pole and the holes toward the negative pole of the battery. Thus, the current that flows through such a semiconductor is formed by two different types of charge carriers flowing in two different directions. In an ideal semiconductor the number of electrons is exactly equal to the number of

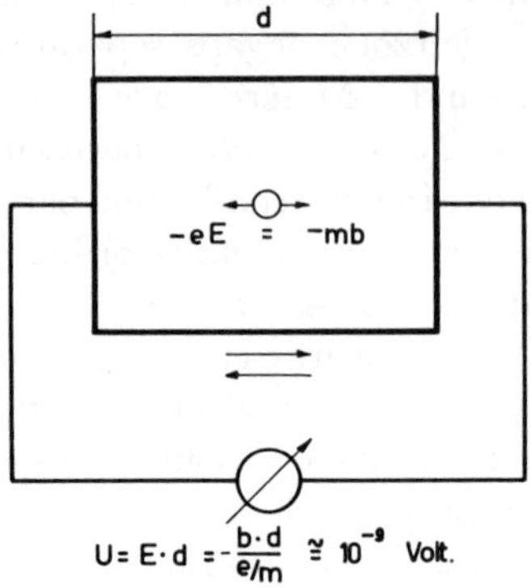

FIG. 19. Tolman's experiment with a piece of metal accelerated alternately to the left and right. The slipping of the electrons results in an alternating field.

holes. Later we will see when we consider a doped semiconductor that there need not be an equal number of the two different types of charge carriers.

The experiment conducted by C. R. Tolman (1916) shows very clearly that metallic conductors contain free electrons with a definite mass m_e. A piece of metal as in Fig. 19 is mechanically rocked, i.e., it is accelerated alternately from left to right and conversely.

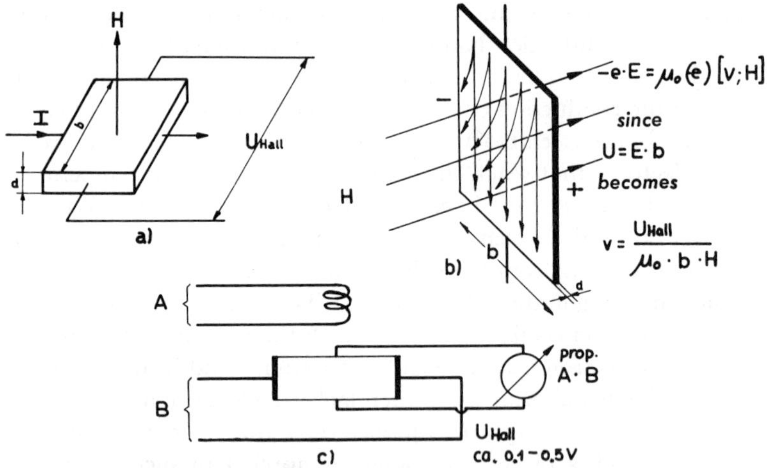

FIG. 20. (*a*) and (*b*): The Hall effect. An extended metal plate of thickness *d* is subjected to a longitudinal current *I* (control current) and to a perpendicular magnetic field *H* (control field). The simultaneous effect of these two control magnitudes gives rise to a potential difference (no-load voltage U_{Hall}) across the plate. (*c*): Electronic multiplication by means of a Hall element.

As a consequence of the mechanical acceleration b a force opposed to the direction of acceleration is applied to the electrons. On acceleration toward the right, electrons logically concentrate on the left side, while the positive charges (ionized atoms) bound to the material remain on the right. Equilibrium is attained as soon as the inertial force $m \cdot b$ equals the attractive force of the field $e \cdot E$. At the terminal faces of the metal piece, separated by a distance d equal to the length of the accelerated piece, a voltage $U = E \cdot d$, a potential difference of the order of magnitude of 10^{-9} v, is measured. This method makes it possible to determine the specific charge e/m. Another very instructive illustration of the presence of mobile electrons within metals is provided by the Hall method. Here the metal piece remains stationary, but is subjected to a magnetic field H at right angles to it (Fig. 20). As a result of the Lorentz force, which we will discuss further later on, $K = \mu_0 e[v, H]$, in which μ_0 = induction constant, v = electron velocity, H = strength of the magnetic field and $[v, H]$ is the vector product of v and H (see p. 33); the electrons are deflected laterally from their straight paths and again form a concentration. Equilibrium is attained when the Lorentz force is once more equal to the attractive force of the electric field. From the formula $\mu_0 \cdot e \cdot v \cdot H = e \cdot E$ the transverse voltage is obtained:

$$U_{\text{Hall}} = E \cdot b = b \cdot \mu_0 [v, H]$$

Initially the electron velocity v is unknown, but the potential U_{Hall} can be obtained from a voltmeter. The result of this experiment enables us to determine the electron velocity v, which is equal to

$$v = \frac{U_{\text{Hall}}}{\mu_0 \cdot b \cdot H}$$

This in turn is a function of the current, which we know from the experiment. The current I flowing through a conductor can generally be derived from the formula $I = n \cdot e \cdot v\ (b \cdot d)$, in which $b \cdot d$ refers to the surface through which the current flows. If the velocity v is replaced by I we obtain the convenient equation for the Hall effect:

$$U_{\text{Hall}} = \frac{\mu_0}{e \cdot n} \cdot \frac{I \cdot H}{d} = C_{\text{Hall}} \cdot \frac{I \cdot H}{d}$$

From U_{Hall} it is possible to determine the unknown n, i.e., the number of charge carriers per m^3 of the metal in question. By comparing this with the number of atoms per m^3—which can be calculated from the known values of the atomic weight and the density—we find that in copper, for example, there are approximately as many free electrons as there are atoms. The solution for the electron velocity $v \approx F/(n \cdot l \cdot q)$ in a copper wire, for example, shows the velocity to be astonishingly low; it equals only 0.7 cm/sec at a current density of 100 amp/mm^2, when the wire is in a bright red-hot state. Under normal conditions such a wire is placed under a maximum load of 10 amp, so that the electron gas hardly travels 1 mm through the metal. In passing it should be mentioned that this electron velocity has nothing whatever to do with the propagation velocity of the electrical signal. The behavior is like that of a rod in which a slow vibration is produced, which is nevertheless quickly noticed at the other end. In addition, it should be mentioned that the previously described Hall effect is put to use in Hall elements. For these elements materials possessing a high Hall constant, such as bismuth, indium arsenide, etc., are used.

An appropriate material for technical utilization of the Hall effect must fulfill the following requirements:

1. The no-load voltage must be high. This condition is satisfied when the Hall constant is high. Semiconductors are very well suited for this as a result of their low carrier concentration.

2. The specific resistance (great carrier mobility) must be low. This allows a high current or low losses and a high output energy. In accordance with Chapter VI the elements of groups III and V of the periodic system are particularly well suited for this purpose.

3. The above values 1 and 2 should not be temperature-dependent. This condition is well satisfied by the semiconductors indium arsenide, indium antimonide, and indium arsenide-phosphide. The material is most effectively prepared in the form of small plates about 0.1 mm thick and surrounded by a casing to protect it against mechanical disturbances. New materials have made it possible to produce Hall generators with Hall potentials of up to 1 v and Hall outputs of up to 200 mW.

Quite generally it is true that a magnetic field generates in a conductor through which a current is flowing a transversal voltage (Hall) and thereby increases the specific resistance. From the last equation on page 27 it follows that by means of the Hall effect it is possible to obtain a product of two electrical magnitudes (current I and magnetic field strength H), the result of which also appears in the form of an electrical magnitude (voltage U). By the combination of such so-called Hall multipliers it is possible to obtain products from several factors, reciprocal values, and quotients. If, meanwhile, the current is maintained at a constant control value, it is possible to determine the strength of magnetic fields, e.g., of a magnet of a dynamic loudspeaker. Hall elements of this type are valuable as field probes. The operation of a Hall element is due to the freely mobile electrons inside the metal of a semiconductor. In order to dislodge electrons from the metal, however, more work must be expended. The necessary work factor varies according to the element, or its atomic structure. Let us assume that the electric field outside the metal is of such a nature that every electron emitted escapes completely and does not return to the surface of the metal. Only then can we obtain a closer look at the free electrons.

5. How can electrons be influenced?

Let us begin with an elementary experiment illustrating the Biot-Savart force, as shown in Fig. 21. A wire through which a

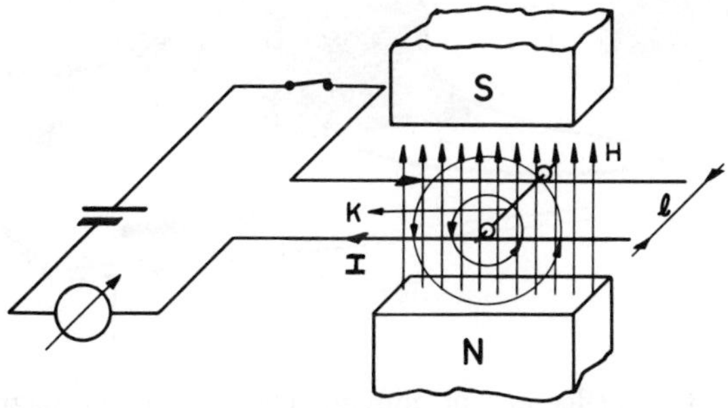

FIG. 21. Demonstration of the Biot-Savart force of a movable conductor through which a current is passed in a magnetic field.

current is passed is placed in a transverse magnetic field. The electrons flowing through the conductor also create a magnetic field (rotation field line). To make it mobile the conductor is provided with wheels and tracks, which act as leads. The lines of magnetic force are perpendicular to the conductor. It turns out that the conductor moves to the right or left in accordance with the direction of the current,[1] and consequently the conductor element must be worked upon by a force that is perpendicular to both the magnetic field and the direction of the current flow; this force is known after

[1] The direction ("the corkscrew rule") is as shown in Fig. 21. A magnetic field would exercise a trivially small force on a stationary charge.

its discoverers as the Biot-Savart force. The Biot-Savart force law is easily derived. The force exerted upon the conductor is:

$$K = \mu_0 \cdot I \cdot [l \cdot H]$$

in which K = force, μ_0 = induction constant, H = magnetic field strength, I = strength of the current traversing the conductor and l = length of the conductor. This is a typical example of classical electrical science, since the electrons flow through a conductor. How does this state of affairs appear from the standpoint of electronics? We have generated free electrons; these fly, for example, in a straight beam across a vacuum tube. An electron beam is a stream of electrons. Consider a Braun tube (Fig. 22). It then appears that each

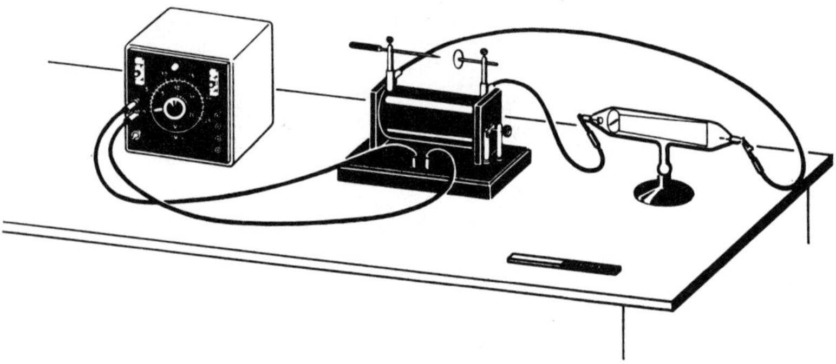

FIG. 22. The deflection of moving electrons in a magnetic field, as used, for example, in the thread-ray tube, can be clearly seen with the cathode-ray tube. The electron path in the tube is made visible on a fluorescent screen. For deflection a bar magnet or a horseshoe magnet can be used. (Experimental installation by the firm Phywe, Göttingen.)

individual electron, or the cathode-ray composed of such electrons, is also deflected in a magnetic field in precisely the same manner as the electron-traversed conductor. The current flowing through the conductor can be obtained from the already familiar formula:

$$I = n \cdot e \cdot v \cdot F$$

in which n = the number of charge carriers per m^3, e = the unit charge $1.6 \cdot 10^{-19}$ ampsec, v = electron velocity and F = the surface to be traversed. By substituting this into the Biot-Savart formula we obtain the so-called Lorentz force for an individual electron moved in a magnetic field:

$$K = \mu_0 \cdot e[v, H]$$

This derivation shows that an electron beam that passes through a magnetic field is subjected to the same force as a conductor traversed by a current. An electron moving in a direction perpendicular to the direction of the magnetic field changes its direction without changing the absolute value of its velocity. The magnetic field can do no work upon the electron because the force is perpendicular to the path. If, as in Fig. 23, an electron flies across a magnetic field,

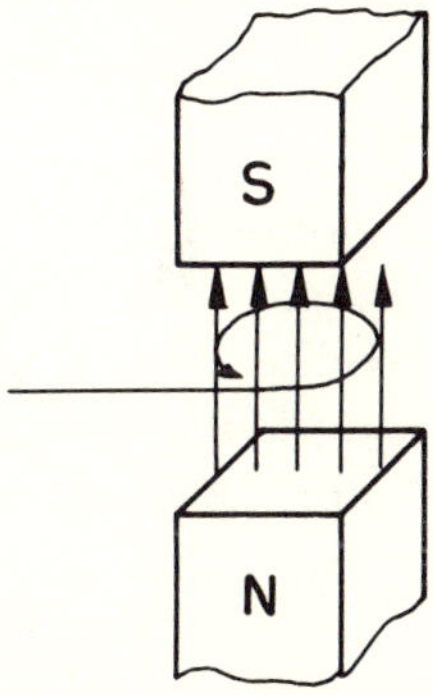

FIG. 23. The path of an electron flight across a homogeneous magnetic field. Since the force is perpendicular to the direction of movement of the electrons, the electron path is bent.

with a velocity perpendicular to the lines of the magnetic field, a circular path of radius r is produced. For the circular motion the Lorentz force must be in equilibrium with the centripetal force. Accordingly,

$$\mu_0 \cdot e \cdot v \cdot H = \frac{m \cdot v^2}{r}$$

From this formula we can obtain the rotation velocity v or the rotation period T of a charged particle. We will hear more about this when we discuss the magnetron of a radar tube.

Up to now the electron was considered as located in a space not subjected to an electrical field. If a homogeneous field is applied, the electron acquires a definite velocity v in the direction of the field. This is equally true for an electron which leaves a cathode and flies toward an anode of potential U. To this condition the following relation applies:

$$v = \sqrt{2 \frac{e}{m} \cdot U} = 594\sqrt{U} \text{ km/sec} < c$$

in which c = the velocity of light. (In the literature the potential is represented by the symbol V as well as U.) The present formula is obtained from the consideration that the kinetic energy

$$\frac{m \cdot v^2}{2}$$

must be equal to the potential energy $e \cdot U$. The equation $e \cdot U = mv^2/2$ is obviously also true for ion paths, but with corrections made for the mass m and the charge e. In the acceleration of a particle (electron or proton) in a vacuum the final energy is once again a function of potential traversed and can therefore be expressed by this potential. The energy unit is, according to page 20, that energy which an electron or any particle charged with 1 elementary quantum acquires by traversing a potential of 1 volt. This energy unit is designated as 1 electron volt (1 e.v.) = $1.6 \cdot 10^{-19}$ wattsec. It is possible for the sign of the velocity v to be reversed, in particular when the electrons are directed toward a negatively charged plate, retarded in front of it and possibly even made to return. Certain television camera tubes employ this technique.

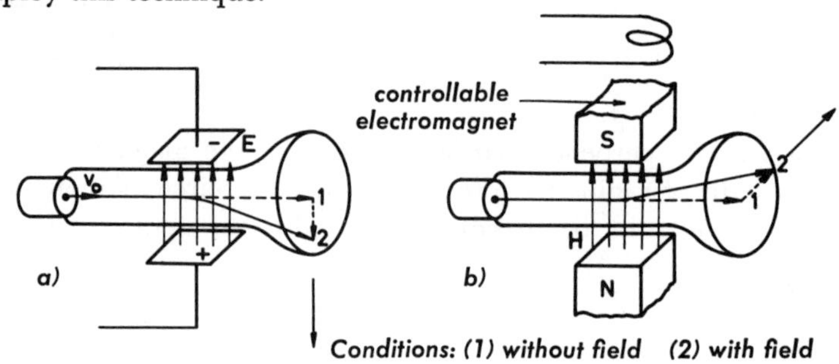

FIG. 24. The behavior of an electron, or a whole beam of electrons, in a magnetic field (right) and electrical field (left).

Figure 24 shows very clearly how electrons can be deflected in an electric field (Fig. 24*a*), as well as in a magnetic field (Fig. 24*b*), if we remember the simple fact that similar charges of electricity repel and opposite attract. This fact can be expressed by the elementary law: $K = -e \cdot E$, where the electric field E is regarded as a vector (a directed magnitude), i.e., is capable of being plus or minus. From this it follows clearly that the electron beam, or the electrons in a cathode-ray tube, must move away from the negative deflecting electrode and approach the positive electrode. Normally the electrons in a cathode-ray tube are accelerated in an "electron gun" (EK; cathode of the positive field plate or anode), then leave the electrical field by means of a small hole and pass with a constant velocity v through a field perpendicular to the direction of their movement.

The electron path, which in the field-free area is a straight line, in the homogeneous transverse electric field (between the deflection plates) undergoes a constant acceleration $b = (-1/m) \cdot e \cdot E$. This results in a parabola-shaped deflection.[1] Within a homogeneous magnetic field (see Fig. 23) it undergoes a circular deflection. The force K acts perpendicularly to the magnetic field H and to the direction of movement and the velocity; this is expressed mathematically by the "vectorial multiplication," i.e., by the bracket $[v, H]$. In a homogeneous magnetic field H the path is a circle (or, Fig. 24*b*, a circular arc) and becomes helical when the direction of an electric field E coincides with H (e.g., when there are both a magnetic and an electric longitudinal field in a tube.)

6. Magnetism

If a conductor is traversed by an electric current a magnetic field is created around the conductor. Naturally, the same is also true for the motion of free electrons. A typical example is offered by the ferromagnetism of a permanent magnet, as illustrated in Fig. 25. For these reasons the electrons rotating within an atom also form small elementary magnets. The researcher P. Weiss showed that in unmagnetized iron the circular currents of the electrons are partially parallel within very small zones, but that in the overall picture their movements are scattered in different directions. Externally such a piece of iron appears nonmagnetic. Subjection of the piece of iron to an external magnetic field causes the numerous Weiss domains to align themselves, possibly by individual leaps, in a single direction. If an exciter coil is placed around a piece of steel and an additional inductive monitor coil is provided, and these are connected across an amplifier with a loudspeaker, the turnabout of the magnetic domains becomes audible. With each abrupt change of magnetization a current impulse is induced in the coil. These abrupt changes are known

[1] The electrons fly like a horizontally thrown stone, which describes a trajectory parabola as a result of the earth's attraction.

as Barkhausen leaps. In soft iron these numerous leaps are heard as a rustling sound. In hard iron the individual leaps are heard as clicks. When all the circular currents are aligned the iron is completely magnetized. Thus we finally come to saturation, as it is known from the magnetization curves of iron or other such ferromagnetic materials.

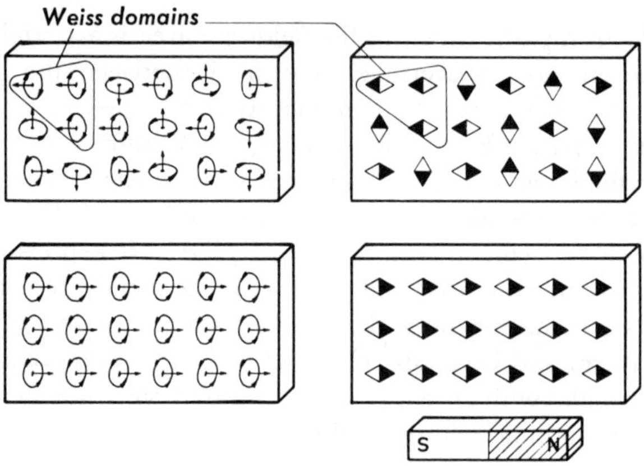

FIG. 25. Illustration of an unmagnetized (above) and a magnetized piece of iron (below).

Since we have sufficient information about the electrons rotating within an atom, we can calculate the magnetic saturation value for a given material from the number of atoms and other physical values, such as the number of rotations per second of the electrons, the atomic diameter, etc.

The so-called Curie point fixes the temperature at which the alignment of the Weiss domains is no longer possible because of the disorder resulting from thermally supplied kinetic energy.

7. What is the relation of electrons to static electricity?

If two different bodies are rubbed against each other one acquires a positive charge, i.e., several of its electrons are rubbed off by the close contact and pass over to the other body. Fundamentally the result of the friction is not generation but separation of the charge. A ball of amber (amber is *electron* in Greek), when it is submerged in water, becomes negative and the water positive. This shows that the substance with the higher dielectric constant ε (see page 38) becomes positively charged, while that with the lower dielectric constant becomes negatively charged. In an electric insulator the charges remain at the contact site after separation. However, on separation the two substances attract each other in accordance with Coulomb's Law with the force $K = (q_1 \cdot q_2)/r^2$. Thus, as a result of the displacement of the charge in the electric field and along the lines of force, a certain amount of work (force K times path d) must be accomplished against the acting force. Since $U = A/q$, the potential U rises in a manner somewhat similar to the way one imagines the origin of a lightning discharge, where the lines of force lengthen between the large, rapidly falling positive raindrops and the small negative particles, with consequent increase in potential. The requisite work A is here accomplished by the gravitational force of the earth, or by the sun's heat, which raised the water to a certain height. When the potential collapses, possibly as a result of the discharged lightning burst, the opposite attraction disappears also, and there is a sudden increase in the rate of fall of the drops.

The influence machine works in a similar manner: the mechanical rotary motion continuously lengthens the lines of force, thus increasing the potential, which charges a condenser (a Leyden jar).

8. Induction phenomena

Up to now we have concerned ourselves with obtaining a preliminary view of the conduction of electricity in metals, semiconductors and insulators, as well as in gases and in the vacuums of tubes. We will now consider the basis of the phenomenon of conduction of one or more charge carriers in the light of Fig. 26. Figure 26 shows an electrically negative sphere (which may also be an electron) displaced from right to left when placed between two condenser plates. In this experiment the current flow is registered by a measuring instrument. If the charge is moved in the opposite direction, the current flow is reversed. If the sphere is subjected to the opposite charge, i.e., a positive one (or a positive ion is utilized) and is displaced as initially, the current direction is again reversed on account of the reversed sign. The charged sphere produces so-called induction charges on the plates. In our example, as the density of the induced charges on the right plate drops, that of the left rises. The charges that disappear from the one plate are transported by the external circuit through the measuring instrument to the other plate. The magnitude of the current is a function of the charge of the moving sphere, of the velocity with which the charge carrier is moved, and of the distance between the condenser plates. Thus it can be seen that the density of the current flowing in a circuit is

$$I = \frac{q \cdot v}{d}$$

where q is the charge, v the velocity of the charge carrier and d the distance between the plates. As a rule it is not one particle alone that is involved as current carrier but the whole space of surface F

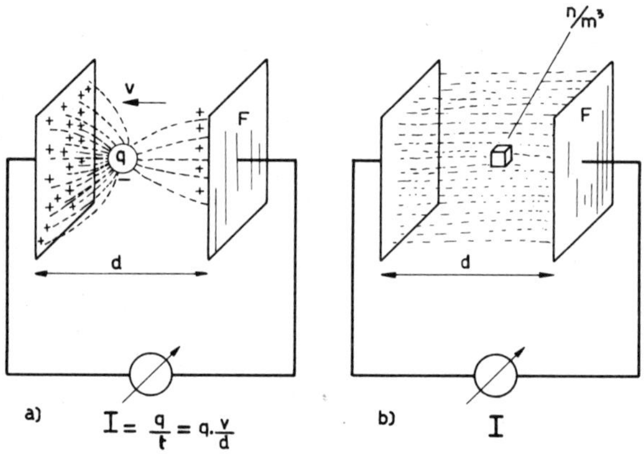

FIG. 26. The displacement of an electrical charge gives rise to induction phenomena and a current flow I.

times the distance d. There are, therefore, $n \cdot F \cdot d$ particles present, with n being equal to the number of particles per m^3. The total current can then be computed in accordance with previous considerations by the formula $I = q \cdot v \cdot n \cdot F$. In the event of ionization we will have positive charge carriers as well as negative ones. In considering the current conduction we will have to add the current carriers traveling from left to right to the current carriers traveling from right to left. In gas discharge tubes (see Fig. 13) there are charged molecules or atoms, which account for the current transport in a similar manner. The charge carriers in liquids also consist of positive and negative ions. For current conduction it is thus essential that the charge carriers travel. Normally the traveling of the carriers requires an electric field. But even a partial movement of charge carriers results in a current flow. This is very clearly illustrated by the displacement current through the dielectric[1] of a condenser. Normally the points of concentration of the charge of an atom, i.e., the negative cloud and the positive nucleus, are superimposed one upon the other (Fig. 27*a*). Then, if a potential is

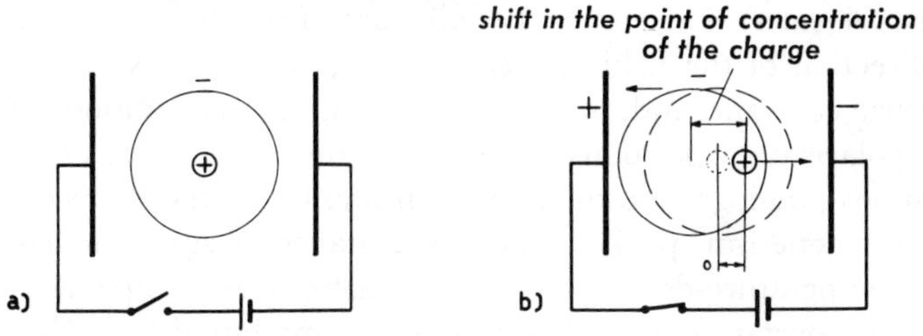

FIG. 27. Elementary event in one of the atoms of the dielectric of a condenser: (*a*) "uncharged"; (*b*) "charged." If ions are present in the dielectric the dipole moment is increased.

applied to the condenser, so that the left plate is made positive and the right negative, the atomic shells are displaced to the left and the nuclei to the right (Fig. 27*b*). The concentration points of both charges undergo displacement, though they have a tendency to return to their normal position as if attached to a spring. The atoms become electrical dipoles (Fig. 28) and polarization phenomena are manifested.

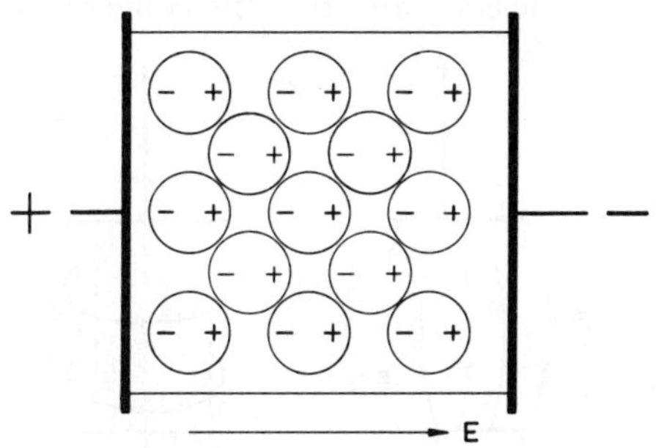

FIG. 28. Diagram representing polarization in a charged condenser. Each atom or molecule becomes a sort of electrical dumbbell consisting of two charges of equal magnitude but opposite sign.

If an a.c. potential is applied to the condenser plates, the points of charge concentration are displaced in the same rhythm, and here the "bound" electrons are of special importance. A so-called displacement current begins to flow. There are two different types of dielectric materials:

1. Symmetrical, temperature-dependent, non-polar molecules, where the charge concentration points of the electron cloud and the

[1] A dielectric is a substance containing a great number of dipoles.

atomic nucleus coincide (Fig. 27*a*). Under the influence of the electric field the charge concentration points are deformed in opposite directions.

2. Asymmetrically constructed, so-called dipole molecules (or polar molecules) that, under the influence of the field, are rotated in the direction of the field and consequently offer an especially high dielectric constant, and, on the other hand, are easily dislodged from their polarization positions by slight heating as a result of thermal vibrations (analogous to the thermal instability of the Weiss domains in ferromagnetism, p. 34). The capacitance of such condensers is very temperature-dependent. With increasing temperature the dielectric constant ε and the capacitance C are diminished. This can be expressed by the relationship

$$C = \varepsilon_0 \cdot \varepsilon \cdot \frac{F}{d},$$

where ε_0 = the induction constant, F = the surface of the condenser plate and d = the distance between the two plates. A dielectric reduces to a certain extent the effective length of the lines of force and thereby increases the capacitance.

It should be mentioned supplementarily that as long ago as 1860 Maxwell proved that every electric field altering in time generates a magnetic eddy and that the converse is also true (Fig. 29). Thus, a

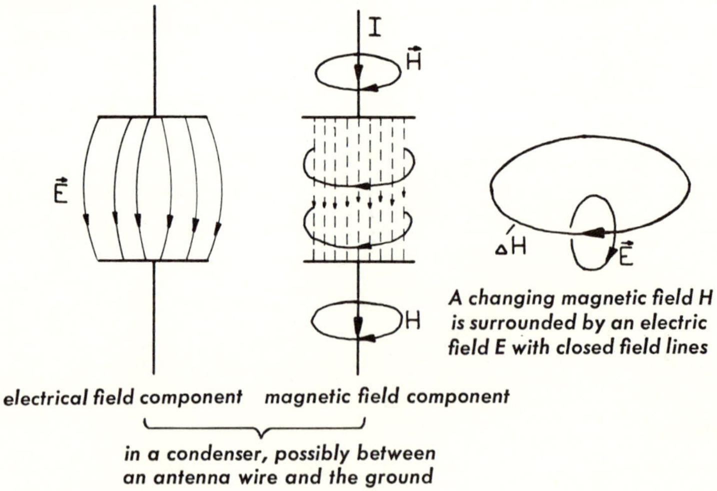

FIG. 29. An electric field altering in time generates a magnetic eddy, and vice versa.

displacement current also generates a magnetic field. The propagation of electromagnetic waves is based upon this interaction.

Let us now turn to the question of how the electric charge and mass of an electron can be determined. We learned from the Edison experiment with the incandescent lamp that the electron is negative. Moreover, we saw that a stream of electrons (cathode-rays) in a cathode-ray tube can be deflected by a magnetic as well as an electrostatic field. To determine the magnitude e/m, the so-called specific charge, we make use of an arrangement with a cathode-ray tube, as shown in Fig. 30. We will now put to use the knowledge we have

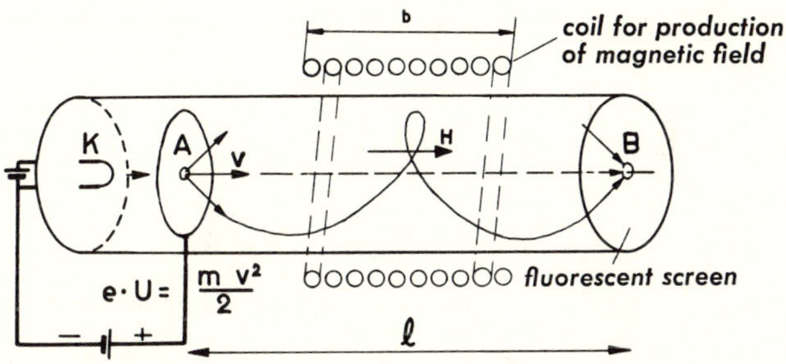

FIG. 30. Arrangement to determine the specific charge of an electron. Left, next to the perforated anode plate A, is the heating filament of the cathode.

obtained. Electrons are emitted at the cathode and fly toward the positively charged electrode A. At the exit slit (hole in the anode) the electrons diverge in all directions. A current-carrying coil that is wound about the cathode-ray tube and generates the axial magnetic field H, compels all the diverging electrons, as a result of the Lorentz force, to follow a helical path. The only electrons that are not deflected are those whose paths follow exactly the axis of the magnetic field and fly directly to the opposite point B. By an appropriate choice of electric and magnetic fields all the divergent electrons can be made to follow spiral paths and converge again at point B. An arrangement of this type results in a magnetic lens of the type used, for example, in an electron microscope.[1] Let us now consider the following equations, with which we are already familiar. The electrical work done on the electron (potential energy) $e \cdot U$ must be energetically equal to the kinetic energy: $mv^2/2$, in which m is the electron mass, as always, and v the electron velocity. Let us now

[1] An exact image of the form of the exit slit is projected upon the face of the tube at B.

keep in mind the theorem concerning the deflection in a magnetic field of an electron that follows a circular path. The formula reads

$$\mu_0 \cdot e \cdot v \cdot H = \frac{mv^2}{r} \text{ (see p. 31)}$$

Solving for v and transforming gives the rotation period $T = 2r\pi/v$ of an electrically charged particle:

$$T = \frac{2\pi}{\mu_0(e/m)H}$$

An optical-axis correct image is obtained only if exactly one whole rotation (or an integral number of rotations) along the helix is completed in the same period of time during which the particle flies from point A to point B. Since l/v also must be equal to T, one obtains from both formulas for the length l

$$l = \frac{2\pi\sqrt{2(e/m)U}}{\mu_0(e/m)H}$$

From this we obtain the specific charge we were seeking:

$$\frac{e}{m} = \frac{8\pi^2 \cdot U}{\mu_0^2 \cdot l^2 \cdot H^2}$$

A demonstration experiment of Professor Scherrer gives the following practical values. A cathode-ray tube with a fixed focal length $l = 0.28$ m is used and the anode potential is 3000 v. The magnetic field strength to be varied H proved to be correct when set at 3300 amp/m.

Only at this magnetic field strength[1] is a clear electron-optical image of the exit slit obtained on plate B. The precise value of the specific charge of the electron is

$$\frac{e}{m} = 1.76 \cdot 10^{11} \frac{\text{ampsec}}{\text{kg}}$$

This value gives the charge of 1 kg of electrons. In relation to its mass the electron's charge is inconceivably great. One might try to visualize the relationship by imagining a situation in which 1 kg of electrons were placed at the North Pole of the earth and 1 kg of

[1] The strength of the magnetic field within a coil is directly proportional to the current intensity I and the number of windings w and indirectly proportional to the coil length b. In the Giorgi system this is therefore expressed as $H = (I \cdot w)/b$ (ampere windings per meter or amp/m).

electrons at the South Pole, they would repel each other with a force of 300,000,000,000,000,000 kg. To obtain the values for the charge or mass per electron alone, one of the two values must still be found in isolation. A very elegant method is provided by the floating technique of R. A. Millikan (1916). In accordance with Fig. 31,

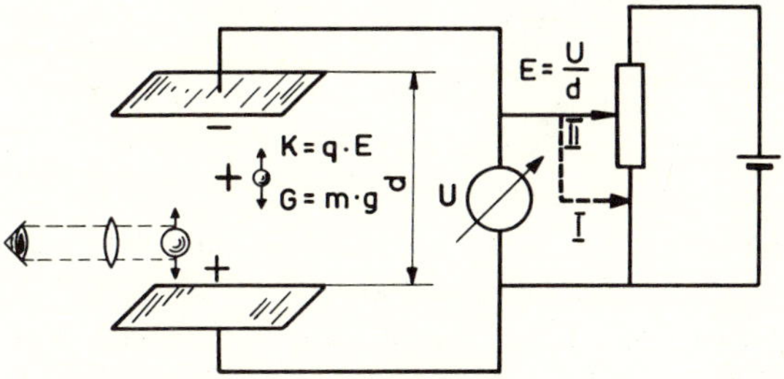

FIG. 31. Millikan's floating technique for determination of the electrical elementary charge of an electron.

microscopically small, or atomized, oil droplets[1] are introduced into an air condenser. The droplets under the influence of gravity, i.e., their weight, fall slowly downward. By means of a dark field microscope the individual oil droplets are seen as small points of light against a dark background. The first problem is to determine the mass m from the observed rate of fall. The greater the rate of fall of the oil droplet, the greater the mass and, consequently, the radius r. The mathematical relationship is yielded by the Stokes friction formula, which sets the viscosity of the air, i.e., the friction, in equilibrium with the gravitational force:

$$m \cdot g = \tfrac{4}{3}\pi r^3 \cdot \rho \cdot g = 6\pi r \cdot \eta \cdot v$$

where r equals the radius of the oil droplets, ρ the density of the particles, η the viscosity of the air and g the gravitation constant. The velocity is measured by a micrometer eyepiece or a raster optical system. If a potential is then applied to the condenser plates after the velocity measurement has been completed, some of the droplets begin rising or falling in accordance with their charge. Frequently the droplets are ionized beforehand by irradiation or by radioactive preparations. The next task is to change the condenser potential until one of the particles selected for measurement, bearing a charge q,

[1] The experiment can also be conducted with other light bodies, e.g., soap bubbles, down, etc.

is observed to float like a small aerial balloon. The electric field in the condenser $E = U/d$ works in opposition to the gravitational force pulling the oil droplets downward. This gives us the force equilibrium: $K = -G = -mg = qE$. From the floating field strength E, which we have now obtained, and from the mass m, or the weight of the particle G, which consists of numerous atoms or electrons, we obtain the charge

$$q = -\frac{mg}{E}$$

The obtained charge q, which is borne by one oil droplet, must be an integral multiple of one unit charge e and can never fall below it. As a rule these particles are missing individual electrons and consequently bear a corresponding excess of positive charge. For the most part the droplets are found to be excessive or deficient by up to forty unit charges. A number of such quantized measurements $q = n \cdot e$ enables us to obtain graphically or by computation the value of the elementary charge, which is

$$e = 1.598 \cdot 10^{-19} \text{ ampsec}$$

The electrical charge, i.e., the electrical elementary quantum of a proton is the same in magnitude but possesses a positive sign. The mass of the electron can be determined by simple division.

$$m = \frac{e}{e/m} = 0.909 \cdot 10^{-30} \text{ kg}$$

The mass of a proton can be determined by comparing the deflection of ionized hydrogen in a magnetic field with the path of an electron. The mass of a proton is $1.661 \cdot 10^{-27}$ kg.

We have devoted this first section to an exposition of the basic principles of electronics, since without a certain intellectual mastery of the behavior and nature of an electron it is impossible to obtain any accurate understanding of complicated electronic instruments such as the magnetron of the radar unit, the maser amplifier of an atomic clock, microwave spectroscopy, or the electronics of a modern television camera tube like the superorthicon.

Since in electronics we frequently have to deal with larger and smaller units of ohms, volts, amps, watts, units of capacitance and inductance, we often have recourse to prefixes of Greek and Latin origin, to represent multiples and fractions of these quantities; these are given in the following table:

10^{12}	Trillion times	Tera	T
10^{9}	Billion times	Giga	G
10^{6}	Million times	Mega	M
10^{3}	Thousand times	Kilo	K
10^{2}	Hundred times	Hekto	h
10	Ten times	Deka	dk
1	Basic unit		
10^{-1}	One tenth	Deci	d
10^{-2}	Hundredth	Centi	C
10^{-3}	Thousandth	Milli	m
10^{-6}	Millionth	Micro	μ
10^{-9}	Billionth	Nano	n
10^{-12}	Trillionth	Pico	p

II. CLASSICAL INTENSITY-MODULATED TUBES

1. Diodes, tubes, photoelectric cells, etc.

The especially good electrical and thermal conductivity of metals can be attributed to the presence within the metals, i.e., in the empty spaces between the ionic lattices, of free, mobile electrons. These are electrons occupying energy levels such that they are no longer bound to individual atoms but to the entire atomic aggregate. For the purpose of discussion we may assume that these quasi-free electrons are comparable to an ideal gas in a sealed chamber. Given this assumption, the free, mobile electrons will exhibit the most varied velocities. There will be only a few electrons with a zero velocity and equally few with an extremely high velocity. Most of the electrons will possess an average velocity that is determined by the absolute temperature of the metal. The quasi-free electrons will be able to emerge from the surface of the metal only when their kinetic energy exceeds a certain definite value, which is characteristic for any given metal; this value is known as the work function. This is analogous to an ordinary vaporization process, in which a definite energy is also required to free the particles from their original formation (solid or liquid). But the electrons that emerge from the metal cannot attain an unlimited distance from the surface, because each electron that emerges necessarily produces within the metal a deficit of one negative unit charge. The potential of the metal becomes positive, and it exerts an attractive force upon the electron.

If a conductor is placed in an evacuated glass bulb and brought to incandescence in any manner whatsoever, an electron cloud forms around it. A stable equilibrium condition is established, in which the same number of electrons leave and return to the metal. The electrons of very high velocity attain greater distances, and the fastest may even escape from the surface of the cathode. It follows from this that it is necessary to increase the kinetic energy by raising the temperature of the metal, if enough electrons are to emerge from the metal surface to give rise to a space concentration or cloud.

This would represent the phenomenon of emission in a vacuum tube with one hot cathode. Normally, however, additional electrodes are incorporated into the glass tube. The simplest type of tube contains within an evacuated glass envelope a heated filament and an opposed metal plate. A tube of this type, consisting of an anode and cathode is called a dipole tube or diode (the suffix -ode means path). If the anode is given a positive charge in respect to the cathode (as the name would indicate), an electric field is created, that no longer allows the emitted electrons that are in the space charge cloud to

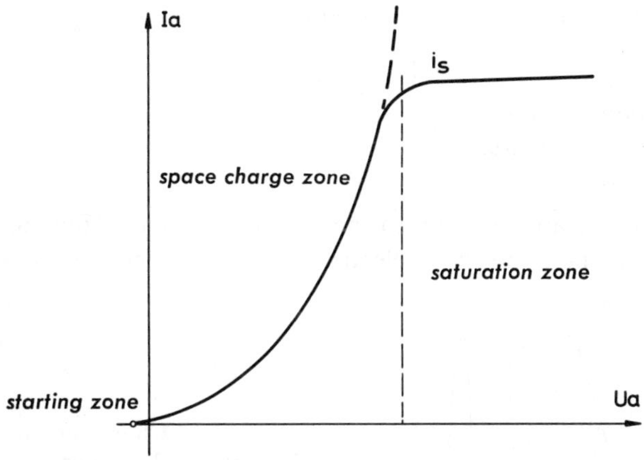

FIG. 32. Anode current as a function of anode potential. Three zones are distinguished.

return to the cathode, but attracts them to the positive anode. Thus, an electron stream flows through the vacuum. After a definite field intensity is attained, all the free electrons, the number of which has been partially determined by the temperature of the filament, are absorbed by the anode. As shown in Fig. 32 the current passes over from the space charge zone into the saturation zone by an abrupt inflection. The saturation current is reached when further increase of the anode potential ceases to produce any significant increase in the current. The saturation current density i_s, i.e., the current intensity per surface unit of the cathode, varies from metal to metal even under conditions of equal temperature. This variation is taken into account by the material exponent B in the Richardson formula.

$$i_s = A \cdot T^2 \cdot e^{-B/kt}$$

$A = 6 \cdot 10^5$ amp/m² deg²
$k =$ Boltzmann constant $= 1.38 \cdot 10^{-23}$ wattsec/deg
$B =$ vaporization heat per electron
$T =$ absolute temperature (0 at -273°C)

The emission current is lower than i_s when the anode potential is insufficiently high in relation to the saturation potential. Thus, we return to the so-called space charge zone, the range over which the tubes are normally used. The increase in the anode current in the space charge zone has the form of an exponential curve. It can, however, be computed from the formula

$$I_a = C \cdot U_a \cdot \sqrt{U_a} = C \cdot U_a^{3/2}$$

I_a = anode current
U_a = anode potential
C = constant

In this zone the anode current is nearly independent of the temperature, because there are excess electrons in the space charge cloud. If

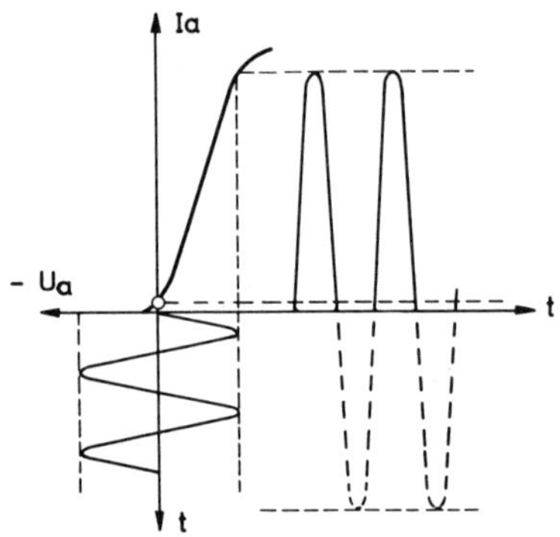

FIG. 33. Characteristic curve like Fig. 32. Anode current curve with an a.c. anode potential.

the anode is situated close to the cathode in the space charge cloud zone, it will be bombarded by the swift electrons even without the necessity of an anode potential. In order to bring about disappearance of the anode current it would even be necessary to apply a small negative potential to the anode. This current, which bucks against a negative field, is called the starting current, and this whole range is known as the starting zone. Figure 33 shows a characteristic curve for a diode. A diode can be used as a current valve. When the anode is positive the tube conducts, but if the anode potential becomes negative, the current drops to zero at low negative potentials and remains at zero even if this value is increased. If an a.c. potential is applied to the anode, the current flows only during the positive half of the wave and is suppressed during the negative half. The result is a pulsating unidirectional current. The rectifier is considerably more efficient when the negative half wave is also put to use, i.e., brought into the rectified phase position. To accomplish this two diodes must be used, and this arrangement is known as a full-wave rectifier. In this case it makes no difference whether the two diode systems are enclosed in a single evacuated glass envelope or whether they are situated in two separate tubes. Figure 34 shows the wiring diagram

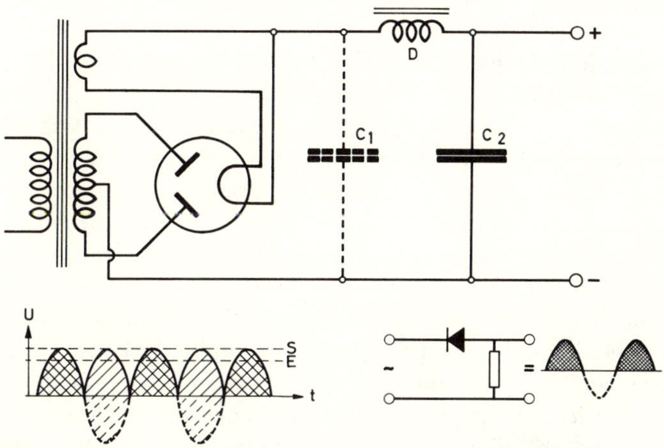

FIG. 34. ABOVE: Full-wave rectifier circuit. BELOW LEFT: Rectified current, in which the crosshatched curves represent the half-wave rectifier, in accordance with the symbolic representation at the right. In the double or full-wave rectifier, on the other hand, both half-waves are used.

for such a full-wave rectifier. Though the pulsation factor of the unidirectional current thus produced is considerably smaller than that of a half-wave rectifier (the crosshatched curve), it is usually still necessary to further smooth out the unidirectional current with a choke coil D and capacitors C. A filter circuit without the capacitor C_1 is called a choke input, and a filter circuit with C_1 is called a capacitor input. In the second case, as a consequence of the charging of capacitor C_1 to the peak value S, the unidirectional voltage obtained is higher by a factor of $\sqrt{2}$ than with a choke input. The choke input therefore offers the advantage of providing a potential that corresponds to the effective value and is quite independent of changes in load. Evacuated diode rectifier tubes are practically lag-free because of the mobile electrons and for this reason can also be used as high frequency detectors for demodulation. The tubes are evacuated to at least 10^{-7} Hg-S with a forepump and a mercury or oil diffusion pump. This is made somewhat clearer by the following illustration of Barkhausen. He calculated that if only one gas molecule of each billion remained after evacuation, there would still be a residue of 28 billion such air molecules (consisting of nitrogen, oxygen, etc.) per cm^3 present within the tube. Since the vacuum within the tube would deteriorate in the course of time, a chemically active material, known as getter metal[1] (usually barium, previously magnesium), is introduced into the glass envelope, and after pumping it is vaporized in a high frequency induction field by eddy current heating. Usually this technique can be recognized by a metallic silver coating on the wall of the envelope. The getter metal binds the gas residues that are liberated on heating of the various materials. Also, a conductive metal coating in the glass envelope can act to screen the tube from disturbing electric fields. It is then connected with a separate socket contact or also with the cathode, which is usually grounded in practice. Since the filaments hardly ever burn out at the low temperatures that are commonly used nowadays, the life of a tube depends to a much greater extent upon the emissivity of the barium oxide cathode. The usual tubes have an expected average life of about 3000 hours, and overdimensioned industrial or long life tubes, about 10,000 operating hours.

The maximum power to be obtained from a tube is always a thermal problem. In small transmitters the permissible cathode temperatures allow operation up to red heat. In larger tubes (20–100 kw) the heat is dissipated by flowing or boiling water.

Rectifiers for high power purposes and network frequencies are frequently designed as gas-filled tubes. After evacuation they are filled with mercury vapor or argon at a low gas pressure. In Chapter I a description was given of how permanently neutral gas molecules

[1] Recently titanium has been used instead of barium. This metal has a greater capacity for binding gas residues. According to the industry, the gas pressure in a tube utilizing this material can in the course of time be reduced from 10^{-6} to 10^{-8} mm Hg.

are split into positively charged ions and electrons by impact ionization. The positive ions move toward the negative pole, and the electrons toward the positive, but these ions quickly collide with other gas molecules, which are also ionized as a result of the impact,[1] so that a chain reaction occurs in which many current carriers are formed. Tubes of this sort have an extremely low internal resistance and, consequently, a low energy loss or heat loss. Gas-filled tubes cannot be used for high frequency because of their high lag (long ionization formation period, as a result of the ion mass, which is about 2000 times that of an electron). They are incapable of following high frequency a.c. fields.

In Chapter III we will take up gas-filled tubes in greater detail. In this chapter we will concern ourselves primarily with vacuum tubes. However, when dealing with areas of application where the two types of tubes verge upon each other, we will also refer to the gas-filled tubes. Thus, for example, a photoelectric cell can be designed as either a vacuum tube or as a tube filled with a rare gas. We will now turn to this electronic element, which has been introduced in great number in industry. Photoelectric cells are used for counting and sorting of objects, as well as for alarm and security installations, etc.

The photoelectric cell is essentially a diode, but the "vaporization" energy necessary for the emission of electrons is supplied by light energy rather than by heat. The photoelectric cell plays such an important part in electronics that we will consider it in somewhat greater detail. A photoelectric effect can also occur at the atoms or molecules of a gas, since electrons can be separated from them by light, with consequent ionization. The photoelectric effect was discovered as early as 1887 by Heinrich Hertz. He observed that a spark discharge between two metal plates could be initiated even at low voltages if they were irradiated with ultraviolet light by such means as the sparkover of a second spark coil. In this connection Hallwachs discovered (Fig. 35) that a negatively charged metal plate could be discharged by ultraviolet light; a positive charge, however, was not discharged because the electrons freed in this case were pulled back to the metal surface by the electrical force (+). Lenard was the first to draw the conclusion that when ultraviolet light impinged upon a smooth-ground, negatively charged zinc plate it instantaneously liberated electrons or, as one

[1] Collision of this sort is called "impact," though no real collision need occur, only a mutual interaction of forces.

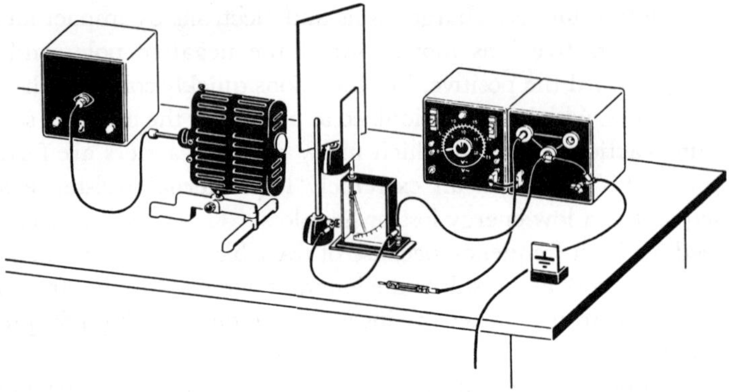

FIG. 35. Photoelectric (Hallwachs) effect. The regular occurrence of this effect is one of the bases for the assumption of the quantum structure of light. With the model photocell (by the firm Phywe of Göttingen) shown here, one experiments with the discharge of a zinc plate through irradiation with the light of a mercury vapor lamp under varying conditions. The projecting electroscope registers the effect. Also shown is the high voltage source.

would say today, the plate emitted electrons. When the experiment was conducted using light from an incandescent lamp, no matter what the intensity, nothing happened. The emission phenomenon is explained by Planck's quantum theory. According to this theory, which we touched upon in Chapter I, light occurs as a type of excited body, in the form of a light quantum with the energy $h \cdot f$. A light quantum possesses a definite mass

$$m = \frac{h \cdot f}{c^2}$$

The quanta can be imagined as minute spheres which become larger as the frequency increases. That light possesses a particle nature is also demonstrated by the "Compton effect" (Fig. 36). Photons are made to collide with electrons, and it is found that a part of the energy is transmitted to the electron, while the remaining light changes color because the energy of the remaining light quantum has been diminished. In our model the energy of blue light would correspond to a heavier sphere than the energy of red light. For most metals red light quanta do not possess sufficient energy to free electrons from the metal surface, whereas ultraviolet light quanta possess more energy and are capable of freeing them. This will be seen more clearly in Fig. 37. Metal atoms in the solid state are arranged in crystal lattices.

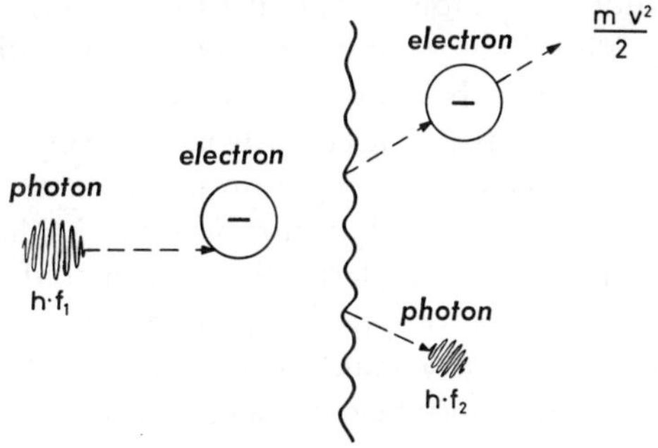

FIG. 36. Compton effect. A light quantum, e.g., from an X-ray, is shot at an electron. After the elastic collision the electron has obtained the momentum $m \cdot v$ and the energy $\frac{mv^2}{2}$, while the light quantum $h \cdot f$ has lost both in equal quantity, i.e., the light quantum continues to move at the lower frequency $h \cdot f_2 = h \cdot f_1 - \frac{mv^2}{2}$.

This is a natural consequence of the interaction among the atoms. A discussion of the bonding forces responsible for this would lead us beyond the bounds of our subject. In the drawing the quasi-free conduction electrons are depicted as separate in comparison to the electrons that move in prescribed shells about the atomic nucleus and

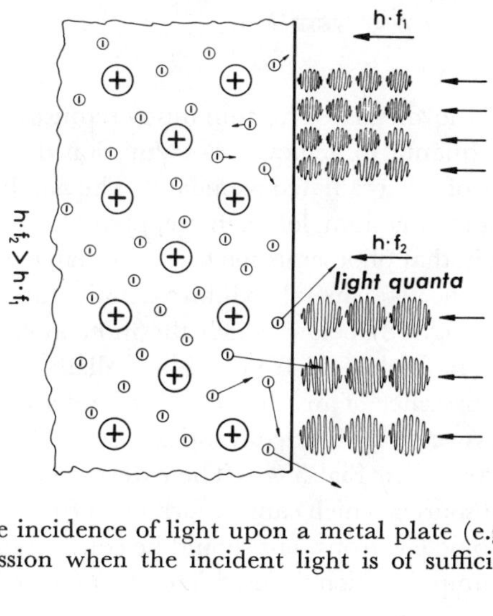

FIG. 37. The incidence of light upon a metal plate (e.g., zinc) results in electron emission when the incident light is of sufficiently high frequency.

belong to the individual atom. The quasi-free electrons, moreover, do not belong to definite atoms but are bound to the entire aggregate. If a light quantum strikes the metal and is absorbed by an electron, the kinetic energy of the electron is necessarily increased. Let us imagine that the zigzag movements resulting from the impact are intensified. This offers two possibilities. The first is that the light quantum possesses sufficient energy to accelerate the electron to the point where, if the motion is in the appropriate direction, the electron will be capable of escaping from the metal. The second possibility is that the light quantum possesses insufficient energy, which is the case with quanta from red light; this causes no photoemission. In the second case the electron is indeed raised to a higher energy level, which is the same as an increase in kinetic energy. This consequently raises the

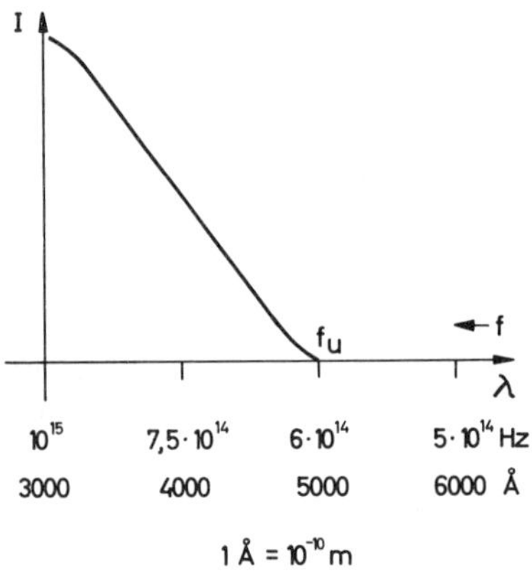

FIG. 38. Emission current of a photocell as a function of the frequency, or wavelength, of the incident light.

temperature of the metal. The minimum requisite expenditure of light energy in quanta to achieve photoemission depends essentially upon the nature of the irradiated element. The alkali metals, such as potassium, sodium or cesium, for example, present a contrast to nearly all other metals in that photoemission can be initiated even with visible light. But even the alkali metals exhibit a definite threshold frequency f_u (as shown in Fig. 38) below which the most intensive irradiation will not succeed in freeing electrons. Since alkali metals are rapidly oxidized in the presence of air, they must be kept sealed in evacuated or gas-filled glass vessels. The conventional symbol for a photocell of this sort is illustrated in Fig. 39*a*. The external circuit is connected across a current source, which cannot correctly be considered a source, as it gives rise to no electrons, but should be considered as similar to a "circulation pump," which is used merely to provide the energy necessary for movement. The photocell, which is connected in series, is almost an insulator in the unilluminated state. If a ray of light strikes the light-sensitive cathode, which consists of an alkali metal plate, electrons are liberated from the latter and escape from the

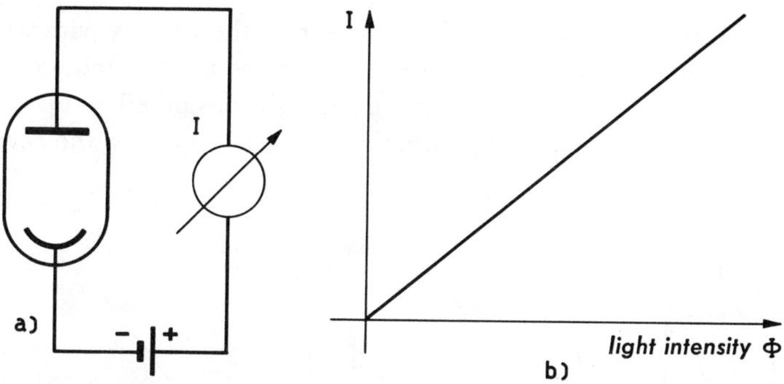

FIG. 39. (*a*) Photocell circuit. (*b*) Current as a function of the light intensity, assuming that the light is of at least the minimum frequency required to initiate electron emission.

surface, flying in the direction of the positive anode. Consequently, a current also flows through the external, closed circuit. It has been found as a basic law of photoemission that the current I increases linearly with the light intensity (radiant flux) Φ. Thus, in accordance with Fig. 39*b*, the number (not, however, the velocity) of emitted electrons is proportional to the intensity of the light. Usually the glass envelope of the photocell acts as a filter for ultraviolet light. This gives a sensitivity curve comparable to that of the human eye. Such a frequency response curve is desirable for television purposes.

Just as in radio tubes, the electrons in this case also travel to the anode even in the absence of a positive anode potential. This occurs here also as a result of a number of electrons acquiring sufficient velocity from the energy of the absorbed light quanta to overcome the coulombic force drawing them back to the atomic ensemble. In order to stop the flow of electrons, it is necessary to supply the anode with a negative potential of the order of about 1 volt. Figure 40 shows an experimental setup to measure the flow of the electrons.

The reverse voltage that will be required is purely a function of the frequency of the light employed, i.e., of the amount of energy transmitted to the electron. The experiment described above makes it possible to prove that the energy of the light quantum is used to free electrons from the metal and transmit an additional kinetic energy

$$\frac{m \cdot v^2}{2}$$

to the metal. From this it is clearly seen that the velocity with which the electrons escape from the metal increases as the frequency of the light increases. The intensity of the light has no effect upon the velocity. In Fig. 41 the experimentally found values for the reverse

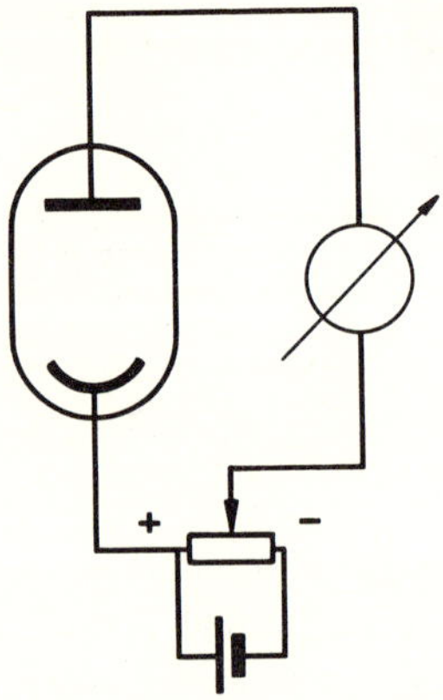

FIG. 40. Circuit to determine the relationships in the starting zone. The potentiometer is regulated until an actual current flow is established. The potential required to establish the flow is measured.

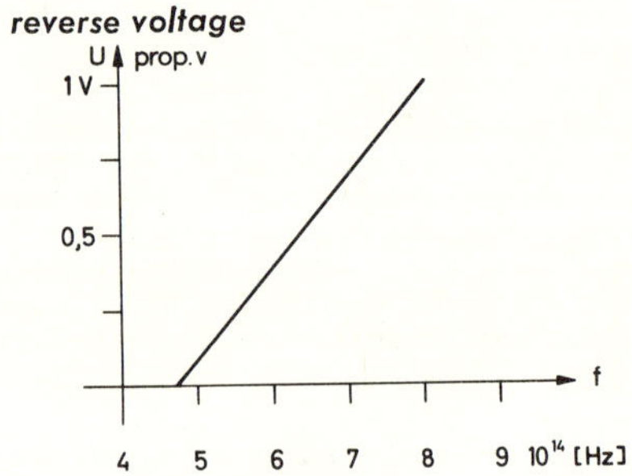

FIG. 41. Reverse voltage (negative) as a function of the light frequency.

voltage are plotted as a function of the frequency of the monochromatic light used. In order to determine the velocity of the electrons outside the metal surface, it is necessary to subtract from the kinetic energy supplied the work of separation or work function A required to overcome the coulombic attraction force. Figure 42 allows us to consider these relationships. Next to the partially occupied conduction band (hatched) there is a potential curve to the right. The ordinate (y-axis) represents the potential energy of the electron from the reversal point and the abscissa (x-axis) the perpendicular distance from the metal surface to the outer space. It should be pointed out that the potential energy is here proportional to the electrical potential at the locus of the electron. A light quantum of the magnitude $h \cdot f_3$, for example, is capable of raising the first electron e_1 to the energy level E_3. In this case the energy is greater than the requisite work factor A. The electron has an excess energy of

$$\frac{m \cdot v^2}{2} = h \cdot f_3 - A$$

If the formula is solved for v, the sought-for velocity with which the electron can escape from the metal is obtained. According to Einstein this would be the most favorable, since the observed electron was situated in the highest part of the conduction band at E_0, in the

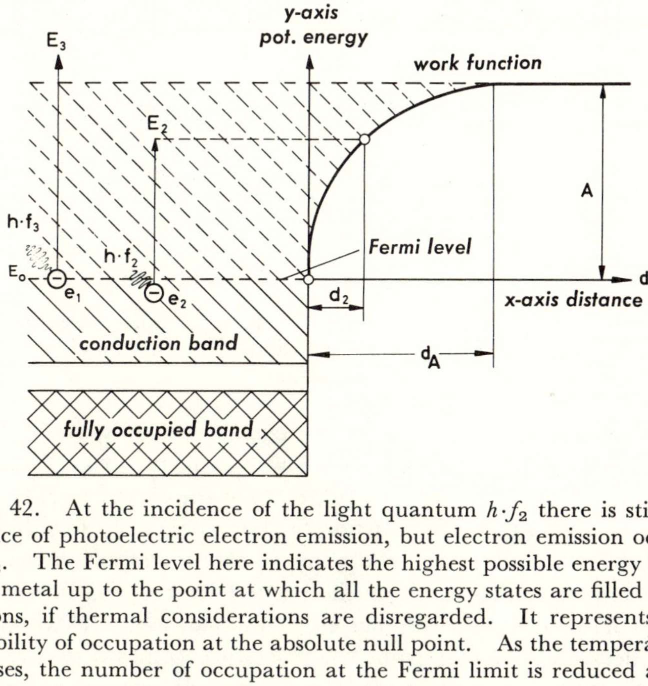

FIG. 42. At the incidence of the light quantum $h \cdot f_2$ there is still no evidence of photoelectric electron emission, but electron emission occurs at $h \cdot f_3$. The Fermi level here indicates the highest possible energy limit in the metal up to the point at which all the energy states are filled with electrons, if thermal considerations are disregarded. It represents the probability of occupation at the absolute null point. As the temperature increases, the number of occupation at the Fermi limit is reduced as its occupying electrons become thermally excited and move to higher levels.

so-called Fermi level. From this upper level the occupation density decreases with extreme speed as we move upward. At a temperature of absolute zero ($T = 0 = -273°\text{C}$) all the energy levels up to E_0 are occupied, whereas the levels above E_0 are unoccupied. As the temperature of the metal is increased the electrons absorb thermal energy, as a result of which they move upward. We have considered the situation at room temperature, at which the occupation density at E_0 has fallen to about half. In metals the Fermi level lies inside the band and in a nonconductor it lies between. To summarize, at room temperature the electrons reach higher levels than E_0. The density distribution is then spread upward. With increasing temperature actual hot cathode emission is approached (in accordance with the previous section), at which point the electrons acquire sufficient kinetic energy to escape from the surface of the metal. If from an energetic viewpoint an electron is lower in the conduction band or, in other words, if it possesses less kinetic energy, it will, even after escaping from the surface zone, at which point it can no longer be pulled back by coulombic forces, possess a correspondingly lower velocity than the

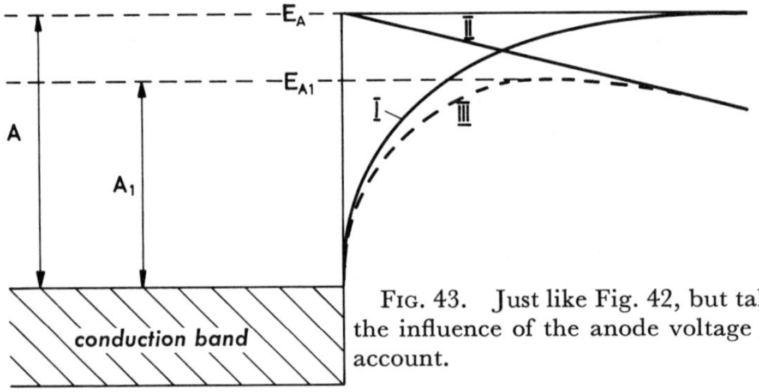

FIG. 43. Just like Fig. 42, but taking the influence of the anode voltage into account.

particle which occupied a higher position in the band. Another example: a light quantum $h \cdot f_2$ of lower frequency (e.g., "infrared") is only capable of raising the electron e_2 to the energy level E_2. This electron moves away, as shown in the potential curve, to a distance of d_2, which is still within the zone of the coulombic forces. It is consequently pulled back by the positive charge. It should, moreover, be clear that the distance d_2 corresponds to the reversal point of the electron. At that point the velocity becomes zero and all the kinetic energy is transformed into potential energy. The energy previously acquired from the light quantum is expressed as additional kinetic energy after the electron has fallen back to the metal. The increased kinetic energy results in heating of the metal in accordance with the thermodynamic logic of the situation. On the other hand, electron emission results in cooling of the metal precisely because the fastest electrons leave the metal. The average kinetic energy is consequently reduced. Normally, one creates an electric field between the anode and the cathode by means of an anode voltage. Figure 43 shows that this favors electron emission. The work function is reduced from A to A_1. As can be seen in curve I, the asymptotic potential curve leads to the upper limit E_A. This represents the minimal energy limit, the condition in which the electrons can "leave" the metal surface with a zero velocity. If a positively charged anode is placed above the cathode, the potential course of the electric field of the anode current source, represented by line II, is superimposed upon the potential course of curve I. The conduction electrons are consequently subject to the common influence of both curves. When both are added together they yield the potential course of curve III. Now we can understand why the work function is diminished. It should, of course, be added that the strength of the external electric field has a very slight effect, i.e., the work function is only slightly reduced, as a consequence of the relatively great distance between the electrodes in comparison to the atomic distances.

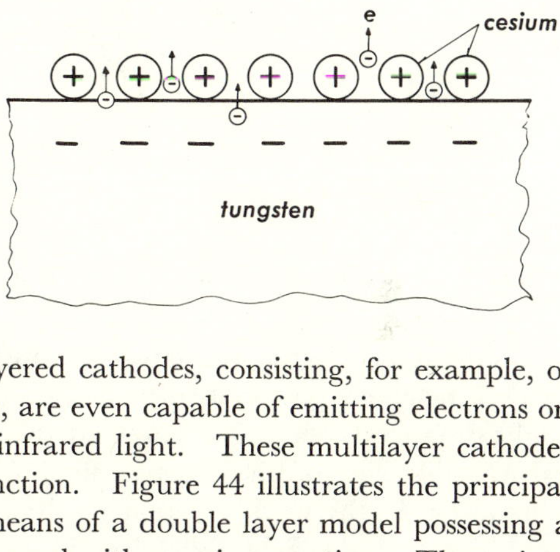

FIG. 44. Photocathode composed of a tungsten plate with a cesium coating.

Photocells with multilayered cathodes, consisting, for example, of cesium, silver, and oxygen, are even capable of emitting electrons on the incidence of red and infrared light. These multilayer cathodes have a very low work function. Figure 44 illustrates the principal manner of operation by means of a double layer model possessing a lower layer of tungsten covered with a cesium coating. The cesium is applied by vaporization in a coating so thin that it scarcely exceeds the atomic width. The cesium coating forms a sort of permeable net above the tungsten cathode. Cesium is an element that possesses only one electron in the valence shell and consequently can surrender it easily to the tungsten. It thus becomes a positive ion, while the tungsten, which has acquired an additional electron, becomes a negative ion. An electric field is thereby created between the two layers. Since the cesium outer layer is very loose and transmits light, the quasi-free electrons that are excited by the photons and attempt to free themselves from the surface of the tungsten plate do not simply fall back as before, but are pulled outward by the positive cesium atoms. This coulombic force produced by the double layer is what is responsible for the diminution of the work function. In practice the influence of the external electric field produced by the anode potential is added to this. The potential course obtained is illustrated in Fig. 45, where the potential curve IV obtained from the sum of the partial curves I, II, and III determines the work function. The internal, interatomic electric field (curve III) between the two layers, the potential drop of which lies between *a* and *b*, has a considerably

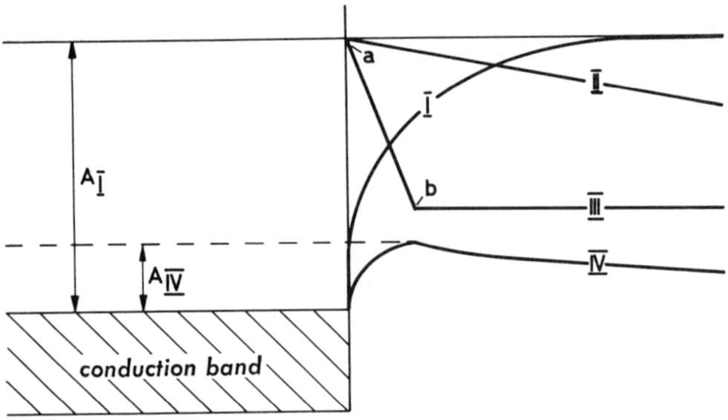

FIG. 45. The same as Fig. 43, but with the additional effect of the electric field resulting from the double layer.

greater influence upon the emission of electrons than the external field of the anode potential curve II, as a result of the extraordinarily slight distance (about 10^{-8} cm) and the, consequently, enormously greater intensity of the electric field (about 10^7 volt/cm). The diminution in the work function resulting from an external electric field is called the Schottky effect in honor of the semiconductor researcher Schottky, to whom we will have occasion to refer several times in the discussion of transistors. Since cesium vaporizes after a certain period of time has elapsed, the principal multilayer system described above had to be modified in practice. The tungsten is replaced with silver that has been oxidized with oxygen, so that in the course of the process the cesium coating is converted to cesium oxide. If composite materials containing antimony are used, blue-sensitive cathodes or photoelectric cells are obtained.

A good photoelectric cell ought to supply a relatively high current flow on the incidence of light and a low, so-called dark current (thermal noise), possibly caused by thermal emission in the tubes, in the absence of illumination. The insulator currents can be kept very small by appropriate selection of the glass and the other insulation materials. The ionization current in tubes that have not been absolutely evacuated or that have even been intentionally filled with a rare gas is quite another thing. In tubes filled with rare gas the ionization is utilized for amplification. It increases the sensitivity by a factor of from 5–10. To achieve this, however, one must be willing to accept, in addition to an augmented dark current, the increased lag resulting from the use of ionic current carriers, which is manifested by a decline in sensitivity when there are rapid changes in illumination,

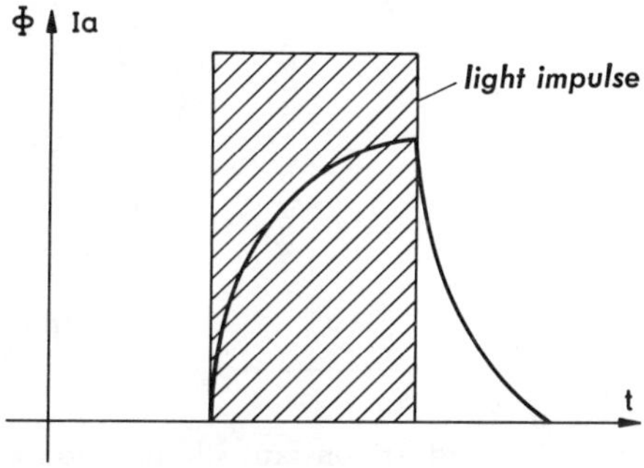

FIG. 46. Illustration of the lag of a photoelectric cell filled with a rare gas. The light impulse (hatched) has a rectilinear form and greatly distorts the photoelectric current (heavy line).

i.e., at higher control frequencies. Figure 46 illustrates the disadvantage of using gas-filled tubes for rapid light signals. The hatched section represents a light impulse. In a vacuum cell the anode current response would have the same form as the light impulse. In the gas-filled cell, on the other hand, the form of the curve is distorted. The ascending portion of the curve represents a complete distortion of the form of the impulse. This is a result of the ionization period. Even after the light ray has disappeared a current continues to flow during the deionization period. For rapid alteration of light, as for example in television, where the greatest linearity is demanded of photoelectric cells, only deep vacuum photoelectric cells can be used. The usual frequency response curves show a decrease in sensitivity of about 6 db from 8000 to 10,000 cps and a very sharp increase above 10,000 cps. In the design of sound film amplifiers this is taken into account by a corresponding frequency correction. Figure 47*a* depicts the anode current course at constant illumination as a function of the anode potential, curve *a* representing a vacuum tube and curve *b* representing a tube filled with a rare gas. Curve *a* resembles the space charge curve of a tube (Fig. 32), where an abrupt inflection of the curve sets in as soon as all the emitted electrons are drawn away by the anode potential. This is the transition into the saturation zone. The subsequent slight increase is to be attributed, according to Schottky, to the effect of the external field, which

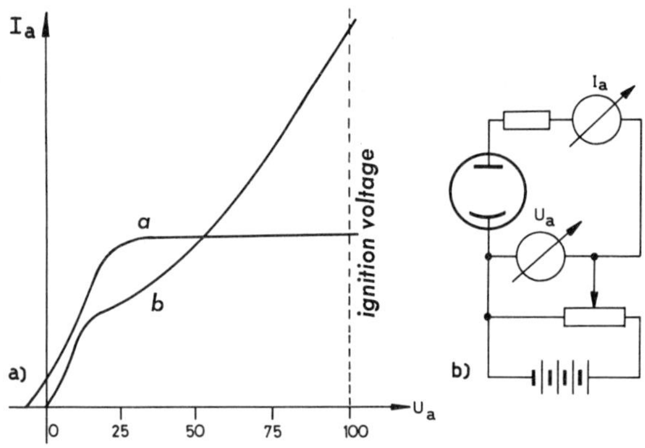

FIG. 47. (*a*) Anode current at constant illumination as a function of the anode potential for a vacuum photoelectric cell and a tube filled with a rare gas. (*b*) Measuring circuit to determine characteristics.

progressively diminishes the work function. The sensitivity of deep vacuum photoelectric cells is 20–30 μA/lumen and of gas-filled cells, 100–150 μA/lumen. Both make use of an external photoelectric effect, i.e., the electron emission is outward. There is also an internal photoelectric effect, which will be discussed in another section (with semiconductors).[1] The gas-filled photoelectric cell exhibits its greatest sensitivity just below the discharge voltage. If the anode potential is raised to the discharge voltage, the photoelectric cell becomes a glow discharge tube. Without series resistances this would result in destruction of the cell. The photoelectric cell is generally a dipole tube, but can be supplemented with other systems (television camera tube).

[1] Photoelectric elements with selenium and other such substances are described in Chapter VI.

Photoelectric cells and photomultipliers have found a very important application as light detectors in the field of spectral photometry. In this case the white light of a tungsten lamp is broken down into the spectrum colors by means of a glass or quartz prism, a diaphragm slit being employed to ensure that only a small segment of a color, i.e., so-called monochromatic light, impinges upon the photoelectric cell. The spectral width is essentially determined by the slit width, with the practical values ranging from 0.1 to several millimeters. The substance to be studied, which absorbs some portion of the light, i.e., characteristic wavelengths, is placed in the ray path. An absorption curve obtained in this manner, which can be used for quantitative, photometric analysis of a substance is shown in Fig. 232. The internal transmissivity or so-called transparency T is equal to the ratio between the intensity of the incident light and the intensity of the transmitted light I: $T = I/I_0$.

The absorption factor α is frequently used as the unit of measurement. This is equal to

$$\alpha = \frac{I_0 - I}{I_0}$$

The logarithm of the reciprocal of the transparency is called the decadic extinction:

$$E = \log \frac{1}{T}$$

The measurement is frequently made over a wide range of the spectrum, possibly from infrared to ultraviolet. Inasmuch as the absorption depends upon the thickness of the layer, calibrated vessels or standardized test tubes are used, into which prescribed concentrations of the liquid substance are introduced. Figure 231 depicts such a spectral photometer. The wavelength is adjusted by rotation or displacement of the prism. The setting is given by a scale. In addition to absorption analysis there is also an emission analysis. Here the material to be studied takes the place of the tungsten lamp. If a gas which is to be investigated is placed in an evacuated tube, the tube is turned into a gas discharge tube by the application of an external voltage. Another highly favored method of analysis is flame photometry, which can be used for definitive determination of over 60 elements. Here the material to be investigated is burned in a colorless oxyacetylene or oxyhydrogen flame. The flame gives off a bright light which can be analyzed. This method allows the determination of trace elements present in dilutions of 1 part per 10 million.

The use of photoelectric cells in such investigations is possible only for frequencies above the lower frequency limit of the work function. At low frequencies, such as those of infrared or thermal radiation, the photoelectric cell is replaced by a thermoelectric element (p. 252). The light is generally cut by a rotating shutter into light impulses (12–25 cps), which makes it possible to obtain a highly stable alternating current amplification. Photomultipliers allow measurements to be made at very low light intensities, i.e., at very small spectrum widths of the measuring light. They are used both for absorption and emission measurements. Modern spectral photometry instruments make use of automatic devices that scan the substance to be studied over the whole spectrum range and record the results on a standard sheet.

If we look back and compare electron conduction in a vacuum tube with that in a metal, we come to the conclusion that the electrons in a metal must be present in a quasi-free[1] state not bound to individual atoms. If this were not the case even the beginning of an electron flow, i.e., even a low current intensity, would require a definite minimum voltage to accomplish the work necessary to dislodge the electrons from their atoms.

Metals, however, are governed by Ohm's law $I = U/R$ or $I = U \cdot \sigma$, in which $\sigma = 1/R$ represents the conductivity. In a metal conductor the quasi-free electrons are always present. The external current source provides only for the circulation of the electrons. Any type of current source that can feed a circuit while remaining stationary can be considered an electron circulation pump. A dynamo or generator is usually thought of only in terms of the induction law $U = -d\Phi/dt$, where U represents the induced voltage and Φ the magnetic flux. This means that each change in the number of lines of magnetic force that pass through a given portion of surface creates at the edge of the portion an equal circumferential stress in opposition to the helical lines. From an "electronic" viewpoint we will consider that the magnet is stationary and that a conductor (a coil) is displaced transversely through the magnetic field. This moving conductor, however, brings its own conduction electrons with it. Their movement necessarily creates its own magnetic field, which is subjected to the deflecting forces of the stationary magnet (Lorentz force, p. 31). These forces, which are perpendicular to the motion and perpendicular to the stationary magnetic field, compel the displacement of the

[1] Quasi-free since they are still bound to the aggregate of atoms.

conduction electrons along the conductor, and thus transversely with respect to the direction of motion. This displacement is relative to the original center of the conductor; thus we are dealing with a circulation pump. Our understanding would be incomplete if we did not also consider the conductor from a relativistic standpoint as stationary and the external magnet as in motion. When a magnetic field is moved or altered it creates for a stationary observer, and thus, for example, for the electron, as Maxwell initially inferred, an electric field. (Compare this, for example, with the electric rotation field around a transformer core). This electric field accelerates the electrons. Thus, we have seen that it is also possible to consider electrical science from a standpoint that traces "cause and effect" back to the electron.

Thermoelectric elements and their "electronics" will be discussed on page 247.

In the anode battery and all voltaic elements the "electronic" relationships are even simpler.

The chemical tendency of various elements to strive toward the lowest energy level (p. 6) by chemical combination leads to the formation of negative and positive ions in an electrolyte. At the surface of the anode the excess electrons of the negative ions continue their path through the external circuit to the cathode, where they combine with and neutralize the positive ions. The quantitative relationships are taught in chemistry, and of central importance in these is the Faraday constant.

2. Tubes with grids

The space charge cloud of negative electrons formed about the heated cathode, which was mentioned at the beginning of this chapter, creates a situation such that the potential curve between the (−) cathode and (+) anode does not increase linearly in the manner of the potential curve for the dielectric between the plates of a condenser. The wall of negative potential (created by the "electron pileup" in front of the cathode), as shown in Fig. 48, is responsible for the fact that only the fastest of the electrons escaping from the cathode are capable of "scaling the wall" or passing to the anode. The stream of electrons flowing to the anode is consequently smaller than the emission current considered possible on the basis of all the electrons emitted from the cathode. Under uniform operating conditions a vacuum diode, for example, exhibits a very stable resistance. If the electron current begins to grow, it is followed by an unavoidable increase in the space charge, the result of which is to augment the arresting effect of the negative potential wall and wipe out the current increase. The electron current flowing to the anode is consequently maintained at a constant value. If, however, a so-called grid is introduced into the electric field of the previously mentioned potential wall—thus in the vicinity of the cathode—the stabilizing effect of the potential wall can be increased or diminished depending on the magnitude of the initial grid voltage (grid bias). When the grid bias is positive the effect of

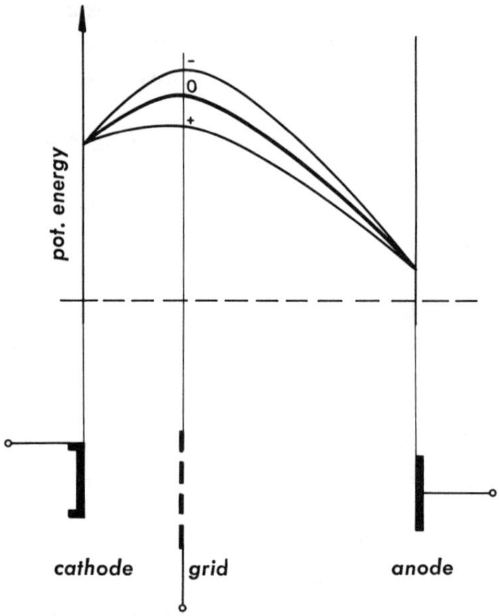

FIG. 48. The potential course between the cathode and anode of a triode.

the space charge is diminished, and more electrons pass to the anode; and when the grid bias is negative the space charge is augmented. In order to allow the electrons to reach the anode, these control electrodes are usually made in spiral or lattice shape and, consequently, are of porous design, consisting of a filament coil, wire mesh or net. These electrodes are, therefore, properly known as grids. A tube with one grid possesses a total of three electrodes and is called a triode. If it has two grids, there are four electrodes (a tetrode); there are also five-pole tubes (pentodes), six-pole tubes (hexodes), seven-pole tubes (heptodes), eight-pole tubes (octodes), and a nine-pole tube (the enneode). From a technological viewpoint a tube is a vessel, either evacuated or filled with a gas, as described in Chapter III, possessing separately insulated and vacuum-sealed leads to the electrodes. The electrode used as a cathode should emit electrons especially easily, i.e., at low temperature; whereas the anode and grid should be selected for poor electron emission. In many circuits the tubes possess the same cathode voltage. It is therefore possible to save space and combine different systems, e.g., a diode and a triode, into a compound tube with a common cathode. In choosing materials for the emission cathode, preference is given to metal conductors exhibiting the lowest possible work function, inasmuch as these have very low heat requirements and do not produce much vaporization. This keeps the requisite filament wattage low and increases the operating life of the tube. Previously cathodes required tungsten wires that were heated to 2300°C. They were, therefore, appropriately known as radio lamps because the light they gave off was capable of illuminating the surroundings. Later, the use of barium oxide coatings on the heating filaments made it possible to reduce the heating temperature to the dull red range, i.e., in the vicinity of 800°C. Today the usual cathode material is thoriated tungsten[1] next to the barium oxide coating. The latter is vaporized onto a heating filament. In direct heating the electrons are emitted directly from the incandescent heating filament. Because of the low heat energy these tubes are used whenever batteries are used as the power source. When alternating current supplied by input tubes is used, indirectly heated tubes are necessary to avoid "hum." Here the heating filament is used only for heat generation and is electrically completely separate from the cathode. When the filament is used only to supply heat, it is not considered an electrode because it would be conceivably possible to obtain equally effective heat from the sun's rays (concentrated by a lens). In indirect heating, a coating of barium oxide is applied in the form of a paste to a nickel tube that encloses the tungsten filament, from which it is insulated by an intermediate layer of ceramic material (Fig. 49). Nickel-cobalt alloys (Kovar) are favored as lead wires in order to prevent subsequent cracking of the glass envelope. The thermal expansion of this alloy is very close to that of glass.

The anode is frequently constructed in the form of a cylindrical metal sheet within which the heating filament is situated along the

[1] This type of cathode gives the same electron emission as a pure tungsten cathode and requires about one-third the wattage. The operating temperature is about 1600°C.

axis of rotation; it may be made of tantalum or some other material that exhibits a high work function in comparison to the cathode material. In transmitter tubes graphite anodes are also used. This material, like tantalum, has the advantageous property of adsorbing gas residues and moreover reflects heat well. In the interests of better cooling it is often provided with fins. It should be added that the electrons flowing through the vacuum bombard the anode, which is initially still cool, at a velocity corresponding to the

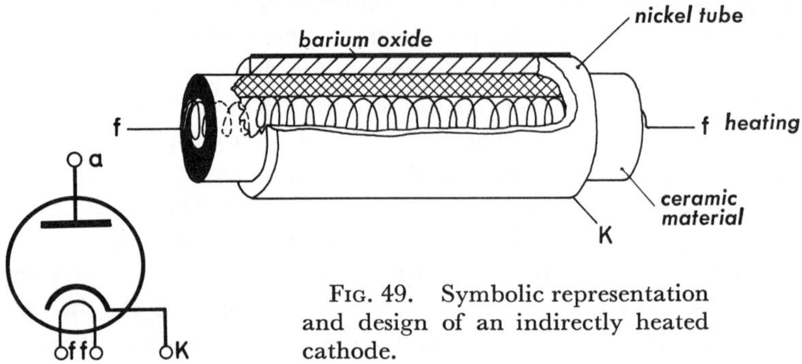

FIG. 49. Symbolic representation and design of an indirectly heated cathode.

voltage. There they generally give up their kinetic energy to the electrons within the metal of the anode, the velocity of which is thereby increased. Since the kinetic energy of the electrons, which is augmented by elastic collisions with cathode electrons, determines the heat content of the metal, the temperature of the anode rises correspondingly. The power supplied to the anode is the product of the intensity of the electron current and the voltage between the cathode and anode.

With this general view of the metallurgy of most types of tubes, we can return to the manner of operation of grid electrodes. Let us first consider the triode. When the control grid is positive with respect to the cathode, the electrons are accelerated; when it is negative, the electrons are slowed down in their flight toward the anode and can even be repelled if the voltage reaches a certain magnitude. The anode current is thus diminished.

Since the grid is much closer to the cathode than the anode, it requires a considerably lower voltage than the anode to cancel the effect of the positive anode potential. The anode current can be made to disappear completely with the use of a negative grid bias, the precise magnitude of which will depend upon the design of the tube.

The circuit at the right of Fig. 50 enables us to obtain the so-called characteristic curves or characteristics. The I_a/U_g characteristic shown at the left plots the anode current I_a as a function of the grid bias U_g for various fixed anode voltages. This gives a family of curves in which the anode voltage U_a appears as a so-called parameter. As the anode voltage is increased the curves are displaced to the left, i.e., toward greater negative grid bias U_g. The left side of the coordinate system is shown since it is primarily the negative voltages which are of interest. When dealing with transistor technology—Chapter VI—we will have occasion to consider other quadrants of

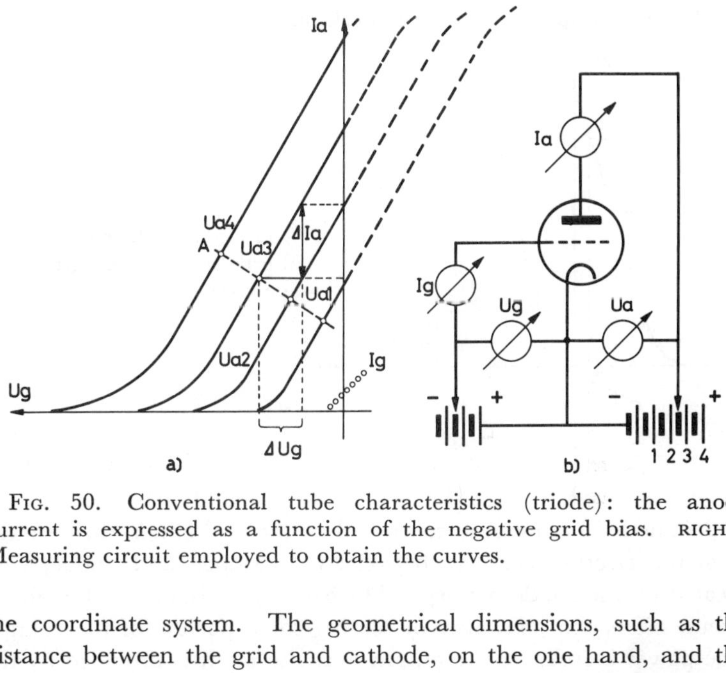

FIG. 50. Conventional tube characteristics (triode): the anode current is expressed as a function of the negative grid bias. RIGHT: Measuring circuit employed to obtain the curves.

the coordinate system. The geometrical dimensions, such as the distance between the grid and cathode, on the one hand, and the anode and cathode, on the other, determine the characteristic features. The ability of the field generated by the anode potential U_a to "penetrate" the grid is expressed by the so-called grid penetration factor D. It is, in other words, the measure of how much weaker the effect of the anode potential U_a is than the grid potential U_g upon the anode current I_a. The grid penetration factor can also be approximated by the ratio of the partial capacitances C_a/C_g, where C_a represents the anode-cathode capacitance and C_g the grid-cathode capacitance.

Later we will discuss other characteristic values that describe the behavior of a tube. It is precisely the degree of control obtained by the use of the grid that opens the way to the numerous and varied applications of tubes, as, for example, low and high frequency amplifiers, oscillators, etc. If an a.c. current potential is superimposed upon the grid bias, the anode current is controlled in a corresponding rhythm (see Fig. 51).

Generally we speak of amplification when an input signal appears at the output side with amplified power (in contrast to the behavior of a transformer). Our consideration of the amplification effect would be incomplete if we did not establish that the electrons passing

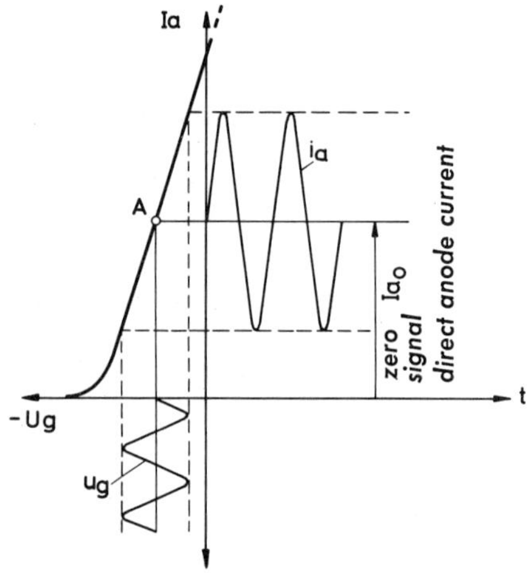

FIG. 51. Tube control. An alternating current grid voltage gives rise to an alternating anode current.

to the anode in a quantity controlled by the grid obtain their energy from the electric d.c. field between the cathode and anode, e.g., created by the anode battery. The external resistance in the anode circuit offers the possibility of reusing some portion of the energy. The power increase of the output over the input is provided by the anode current source. Another requirement of a linear amplifier is that the output signal be proportional to the input signal. A tube operates as a so-called A-amplifier if the grid bias is chosen in such a manner that the average working point, as shown in Fig. 52*a*, is situated in the center of the linear section of the characteristic. The anode current can then be controlled in such a manner that it is essentially proportional to the changes in the grid voltage. Figure 51 illustrates the amplification of a sinusoidal a.c. current. In class B amplifiers the working point is located in the vicinity of the lower inflection of the curve (Fig. 52*b*). When this manner of operation is selected the zero signal direct anode current I_{a0} is reduced. Compared to the previous method, this method allows considerably greater control with the same heat and power losses, which are determined

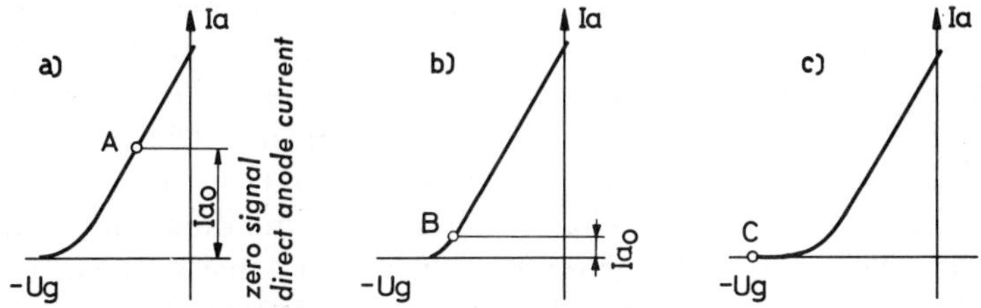

FIG. 52. Various working points resulting from the choice of negative grid bias. (*a*) A-operation, suited for simple amplifiers; (*b*) B-operation, suited for push-pull amplifiers; (*c*) Co-operation, suited for high frequency power amplifiers.

by the construction of the tube. This means that the efficiency is enormously increased. However, since a certain distortion is to be expected, especially in the low frequency range as a result of the bend in the curve in the starting zone, special measures, such as negative

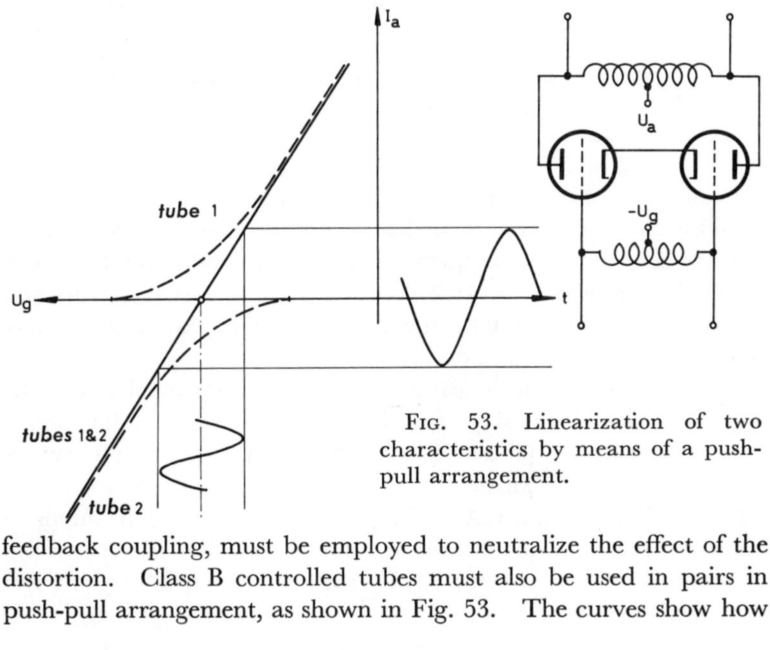

FIG. 53. Linearization of two characteristics by means of a push-pull arrangement.

feedback coupling, must be employed to neutralize the effect of the distortion. Class B controlled tubes must also be used in pairs in push-pull arrangement, as shown in Fig. 53. The curves show how thoroughly the asymmetries are eliminated. The class C amplifier, which cannot be used as a low frequency amplifier and is used exclusively as high frequency amplifier or special purpose switch, exhibits even more favorable relationships from the point of view of efficiency. The working point, as Fig. 52*c* shows, lies all the way to the left, and the class C amplifier, therefore, exhibits the highest attainable efficiency. An oscillatory circuit in a high frequency amplifier stage or sending stage operates in intermittent pulses in a manner somewhat analogous to a rocker drive, to use a mechanical example. The current no longer flows for the full duration of a half cycle. The control voltage is relatively high and may even enter the positive grid current zone without causing concern. Because of the intermittent grid current the class C amplifier requires a power control. In the class A amplifier, where there is a lossless control,

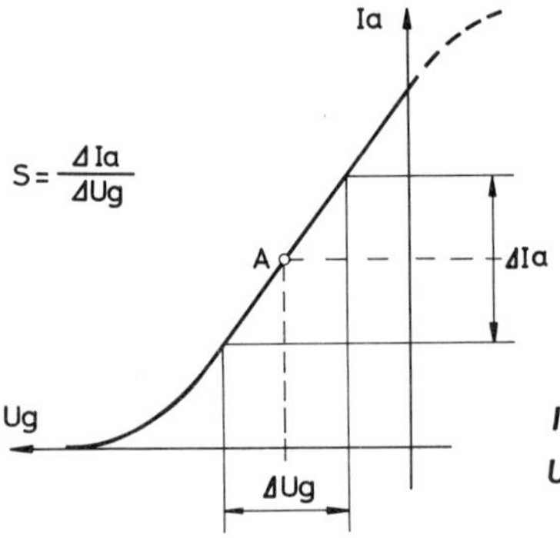

FIG. 54. Determination of slope S from the characteristic.

I_a = anode current in mA
U_g = grid voltage

the control must not be pushed to the point where the grid potential would become positive. A further requirement for lossless control is that the control voltage frequency must not descend into the range of the centimeter waves, because then it becomes necessary, as we will see in Chapter V, to take the transit time of the electrons into consideration.

A tube is described technically by the characteristics or characteristic values. The so-called slope S of the current increase as a function of the grid voltage is plotted graphically as in Fig. 54 and can be computed from the formula $S = \Delta I_a/\Delta U_g$, where I_a represents the anode or plate current mA and U_g the grid voltage. The anode or plate voltage U_a is maintained constant here. The slope is not a constant magnitude over the whole characteristic but is constant over the linear portion of the curve.

The internal resistance R_i (Fig. 55) is computed from the ratio of the change in plate voltage divided by the change in plate current in accordance with Ohm's law:

$$R_i = \frac{\Delta U_a}{\Delta I_a}$$

In this case the grid voltage U_g is kept constant. It is possible to plot a I_a/U_a characteristic for every fixed grid voltage. As the grid

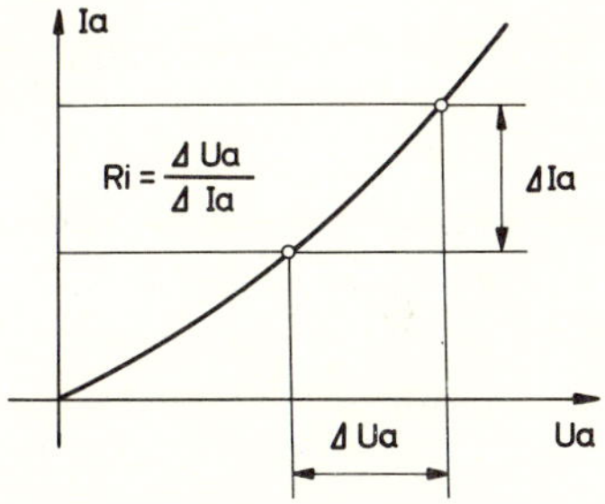

FIG. 55. Determination of the internal resistance R_i from the characteristic.

voltage U_g becomes increasingly negative the characteristics are displaced toward the right, i.e., increasingly higher plate voltage is required to create the same current. The internal conductance G_i is the reciprocal of the internal resistance $1/R_i$. Since there is a grid current at positive grid voltages, the cathode current in this mode of operation is equal to the sum of the plate current and grid current.

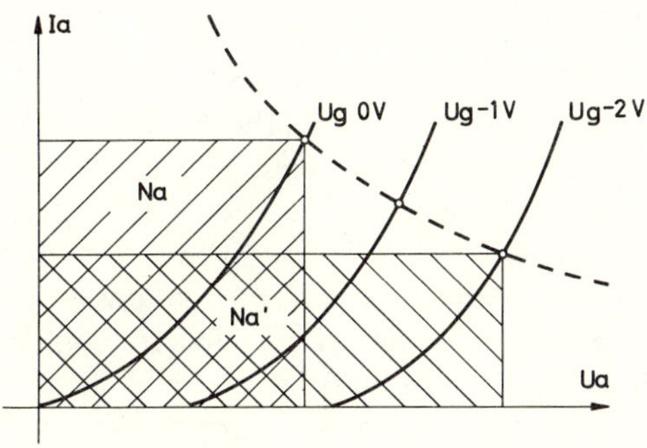

FIG. 56. The maximum permissible current load. The limit is determined by the power hyperbola.

There is a definite battery current that a tube is capable of handling, and this depends upon the construction of the tube. This value is known as the maximum current dissipation and is given in watts. It follows from this fact that for any anode voltage there is a definite maximum permissible current. In Fig. 56 the product of the "side lengths" I_a and U_a, and thus the enclosed rectangular area, gives the magnitude of the current dissipation. The permissible maximum values result in a hyperbolic characteristic ($U_a \cdot I_a$ = a constant). The operating conditions must consequently lie inside the power hyperbola if the tube is not to be ruined. The voltage amplification factor V, which represents the maximum degree of amplification, is attained only when the load resistance approaches infinity. It is equal to the ratio of the amplified output amplitude to the control amplitude of the grid:

$$V = \frac{\Delta U_a}{\Delta U_g}$$

The practical values for triodes lie between 10 and 50. The current amplification factor in class A operation, in which there should be only a negligible grid current (max. 1 microamp.) with a good vacuum, approaches infinity. The grid penetration factor, with which we are already familiar, can be computed from the formula:

$$D = \frac{\Delta U_g}{\Delta U_a}$$

to obtain which the plate current I_a is held constant. The grid penetration factor is usually expressed in percentages, and the value then multiplied by 100. A comparison with the formula shows that the grid penetration factor is identical with the magnitude $1/V$.

In the course of devising the previously cited definition Barkhausen arrived at the following important formula:

$$D \cdot S \cdot R = 1$$

With this formula it is always possible to obtain the third characteristic value for a tube if the other two magnitudes are known. The above described characteristics relate to static conditions. Under normal conditions a tube operates with a load resistance R_a, at which the plate current fluctuations appear as potential fluctuations.

If great amplification is important, as, for example, in a microphone preamplifier stage, a high load resistance must be selected.

For special purposes it is sometimes desirable in practice to obtain tubes with a variable curvature, i.e., with a so-called exponential characteristic curve. The first experiments consisted in the use of an axially conical grid. It is better for the distance between grid windings to be increased as illustrated in Fig. 57. This results in a sort of parallel arrangement of tubes with different slopes. Tubes possessing exponential curves and, therefore, variable amplification, are used in automation; they are also used in radio engineering for automatic volume control, and in electronics for a constant voltage regulation. In low frequency amplifiers tubes of this sort are only suited for use in the initial stages, where the control amplitudes are

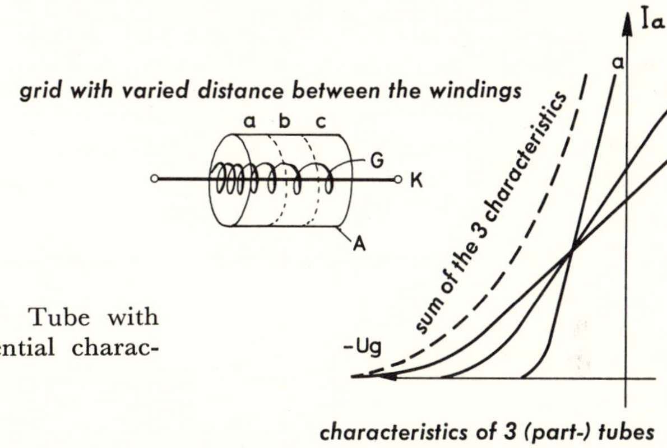

FIG. 57. Tube with an exponential characteristic.

still sufficiently low in relation to the curvature of the curve. There are various shortcomings attached to the triode. These are primarily:

1. The plate feedback, the effect of which is obtained from the formula

$$I_a = S \cdot U_g + \frac{1}{R_i} \cdot U_a$$

2. The relatively high grid-plate capacitance, as a result of which the high frequency oscillations cause a displacement current to flow back from the anode to the grid.

Moreover, the slope attainable in practice has been found to deviate considerably from the statically computed characteristic and to follow a much flatter course. The operating slope or so-called dynamic slope S_a is computed from the formula

$$S_a = \frac{R_i \cdot S}{R_i + R_a}$$

A particularly unfavorable limitation is that the plate current-voltage characteristic can only be partially utilized. The I_a/U_a characteristic climbs slowly, as shown in Fig. 58, and the first part of the working range is thereby wasted. These drawbacks can be eliminated by the introduction of a new grid with a constant, positive bias. Thus, we come to the four-pole tube or so-called screen grid tube (Fig. 59*a*).

In this tetrode the electrons, as a result of the positive screen grid voltage, are accelerated independently of the instantaneous plate voltage. The fluctuations in the plate voltage consequently have little effect upon the plate current resulting from the electron flow. When the distance between plate and screen grid is small, which is

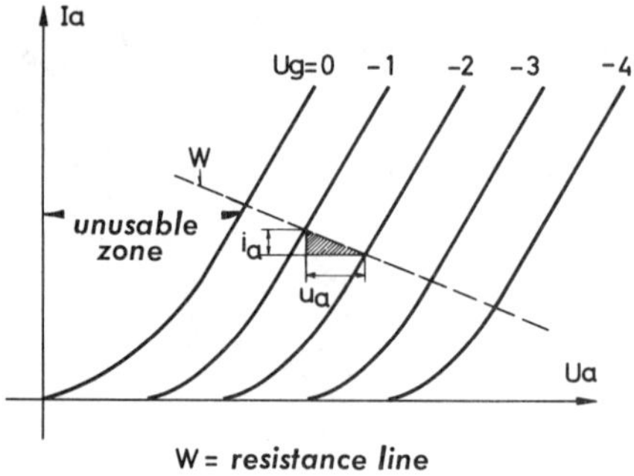

FIG. 58. I_a/U_a characteristic of a triode.

the case in radio tubes, another disadvantageous phenomenon is observed. The electrons accelerated by the positive screen grid strike the plate with such force that they liberate secondary electrons, which return toward the positive screen grid. These secondary electrons are lost to the plate current. If the number of secondary electrons

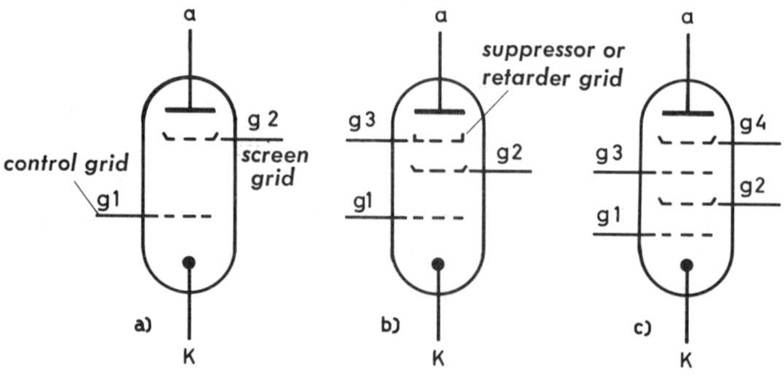

FIG. 59. Multigrid tubes. (*a*) Tetrode (4 electrodes); (*b*) pentode (5 electrodes); (*c*) hexode (6 electrodes).

were greater than the number of primary electrons, the plate current would run in the reverse direction. However, the reversal point of the characteristic is reached before this (Fig. 60). As the plate voltage increases the plate current drops; this corresponds to a negative resistance, which leads to a damping reduction of, for example, a connected oscillatory circuit and turns the whole arrangement into an oscillator. The non-linear zone caused by the secondary electrons disappears once more as soon as the plate voltage becomes greater than the screen grid voltage. The dip in the curve can be eliminated

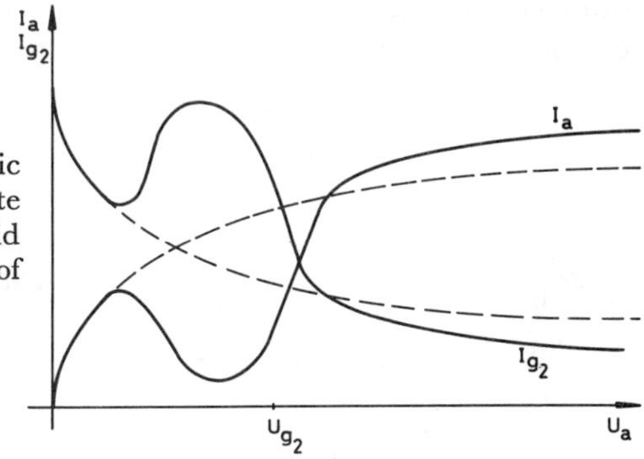

FIG. 60. Characteristic of a tetrode. The plate current I_a and screen grid current I_{g2} as functions of the plate voltage.

by a negative suppressor field in front of the anode, created by an additional suppressor grid. Thus, we come to the five-pole tube or so-called pentode shown in Fig. 59*b*. As with the tetrode the amplification factor of the pentode is one degree of magnitude greater than

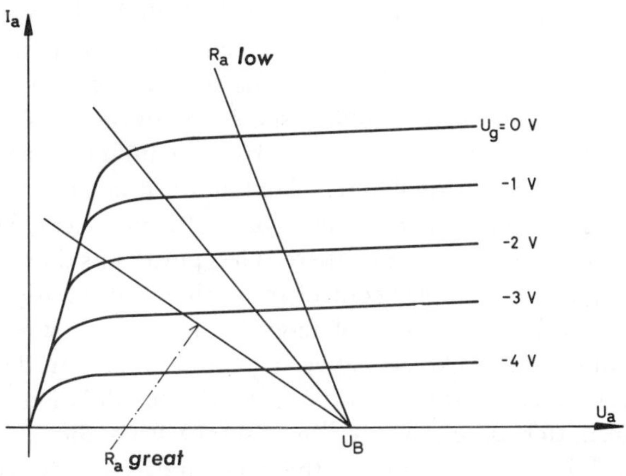

FIG. 61. I_a/U_a family of characteristics for a pentode with the resistance lines for various operating resistances.

that of the triode. Because of the higher internal resistance the plate feedback is considerably slighter than in the triode, which should be immediately clear from a consideration of the designated formula. Figure 61 shows the favorable course of the I_a/U_a characteristics for pentodes. As a result of the steep ascent the unusable portion is considerably smaller than with the triode. The transition to a path parallel to the abscissa shows that the plate current is independent of the plate voltage. In the I_a/U_g set of curves the curves are situated very close to each other, as Fig. 62 shows, and the static characteristics

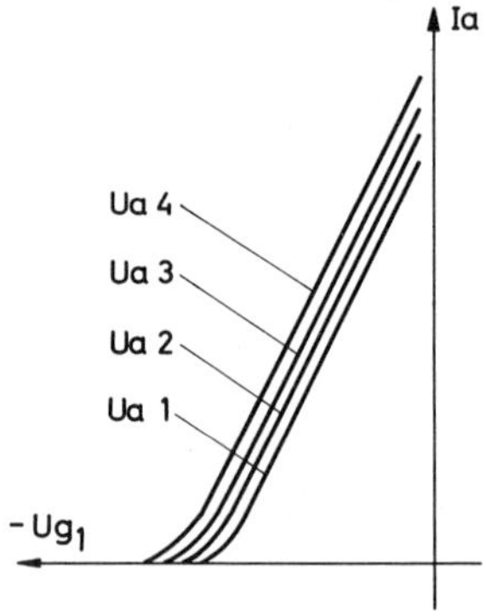

FIG. 62. The closely adjacent, typical I_a/U_g characteristics of a pentode at various plate voltages U_a.

are nearly in agreement with the dynamic characteristics. The screening effect of the screen grid results, as in the tetrode, in a low grid-plate capacitance, as a consequence of which the disadvantageous reverse current is eliminated. The screen grid current is lower by about one degree of magnitude than the plate current, but is lost as consumed power. Final amplifier stages in transmitters, as is well known, receive their control voltages from preliminary stages, which are in turn fed by an oscillator. Pentodes are favored for smaller final stages, and tetrodes for larger ones. In both cases, because of the higher amplification factor, there is less power loss than in triodes, and consequently a simpler control transmitter can be used. Quite apart from this, the lower plate-grid capacitance simplifies the neutralization. The choice between pentode and tetrode depends upon various factors aside from the method of modulation.

In transmitter tubes, where the distance from the plate to the screen grid is unusually great, the comparatively slow secondary electrons—those accelerated with no more than about 6 electron volts velocity—are unable to reach the screen grid. As in the previously described triodes, space charge clouds are created around the plate and drive the electrons back toward the cathode. It is very important to select a plate material possessing very poor secondary emission properties. The tetrode may also compete with the pentode if the plate is designed in an appropriate form. In the beam tetrode the secondary electrons are deflected away from the screen grid and continue their path toward the plate. Another solution of the problem is to place the screen grid very close behind the control grid, so that the grid wires of the two electrodes lie one in back of the other.[1] This arrangement reduces the requisite screen grid potential and eliminates the concern over disturbances from secondary electrons. We come to the hexode when four grids are placed between the cathode and the plate as shown in Fig. 59*c*. The grids 1 and 3 are used to control the plate current, while grids 2 and 4 are used as screen grids. Accordingly, the potential of grids 1 and 3 is usually negative with respect to the cathode, while the potential of grids 2 and 4 is positive and may be equal to about half the plate voltage. The hexode may be thought of as if it were a series connection of a triode with control grid 1 and a tetrode with control grid 3. The plate current is controlled by both grids. These tubes are primarily used in mixing stages. If two grids of such a tube are fed two different frequencies one obtains in the output both the sum and difference frequencies in addition to the two original frequencies, as a result of the multiplicative mixing. The first grid controls the electron current in the direction of the screen grid g_2; the space charge created in front of this is known as a virtual cathode. The second control grid g_3 is a current divider control, which either allows the electrons to continue to the plate or drives them back to the first screen grid. We are, therefore, concerned with two independent slopes,

$$S_1 = \frac{\Delta I_a}{\Delta U_{g1}} \quad \text{and} \quad S_2 = \frac{\Delta I_a}{\Delta U_{g3}}$$

the second of which is the smaller. This grid is supplied with the higher amplitudes of an oscillator. If the small receiving signal is fed to the first grid we obtain at the output the well-known intermediate frequency that can be computed from the difference of the two oscillations. Since primary electrons strike the plate, secondary electrons may be emitted, which in turn pass to the screen grid g_3. These secondary electrons have an effect upon the operation of the hexode. This disadvantage can be avoided by the introduction of a suppressor grid between the screen grid and the plate, and this takes us to the so-called heptode. Here also two different control voltages

[1] When the screen grid lies in the shadow of the control grid, much fewer of the electrons passing from the cathode to the plate collide with the screen grid. The screen grid current is consequently smaller.

are used to regulate the tube. The suppressor grid is coupled to the cathode. There are also special tubes that contain more electrodes. Aside from being used as combination oscillators and mixers, they are most frequently used for phase comparison purposes. In tubes of this sort each grid must be supplied with a certain minimum value in order for there to be any plate current at all, or the magnitude of the plate current will be a function of the phase position of the individual control voltages. Usually eight- and nine-electrode tubes, the so-called octodes and enneodes, are used here.

In the tetrode the secondary electrons liberated from the plate by the impact of the accelerated electrons produce a sensibly disadvantageous effect (see Fig. 60). However, in certain tubes, the so-called electron multipliers, this secondary electron emission is deliberately put to use (see Fig. 7*c*). The electrons, which in this case also are obtained by thermal emission, are controlled by one or more grids. Each primary electron that strikes the target anode liberates several secondary electrons. These secondary electrons then fly toward the real anode or plate, which possesses a positive potential with respect to the target anode. When a multiplied current of secondary electrons is obtained we are dealing with a current amplifier. There is an increase in efficiency if each primary electron that strikes the target anode dislodges a great number of secondary electrons from the anode material. Copper-beryllium forms a suitable basis for target electrodes of this sort. The practical multiplication factor lies between 5 and 8. It makes no difference here whether heat or light rays cause the emission of the primary electrons from the cathode. Since the target electrode acts like an anode for the primary electrons and as a cathode for the next "target anode," it is called a dynode in reference to its double nature. The dynode current is negative with respect to the plate current, since it gives up more electrons than it receives. The secondary electrons are released very rapidly (10^{-10} sec), so that these multiplier tubes are very well suited for use at up to the highest frequencies. The upper frequency limit is determined by the differences in electron transit time. It is possible to build broadband amplifiers with a thermal cathode.

Secondary electron multipliers are, however, more often built with photoelectric cathodes. In modern photoelectric cells with secondary electron multipliers, which will operate quite satisfactorily even at candlelight and at great distances (sensitivity 100 amp/lumen), the photocathode is vaporized onto the inside of the bottom of the tube.

Cesium-antimony is used as the light-sensitive material. In order to ensure the emission in large quantity of secondary electrons, the target anode is activated with silver-magnesium. For great amplification, a large number, e.g., up to more than 15, target electrodes are arranged in a cascade, so that the electrons are directed from one dynode to the next. The use of 10 stages results in an amplification factor of over 10^6. Aside from the correct choice of the form for the metal foil, deflection is obtained by the use of an electric field; previously a magnetic field was used. The potential is increased from dynode to dynode by about 100 volts. A stabilized voltage divider supplies the individual voltages to the individual target electrodes. In photoelectric cell electron multipliers the light quantities can be measured out in magnitudes as low as individual light quanta. Since the ratio of working voltage to disturbance voltage is more favorable in cascade connection than with multigrid amplifier tubes, the former system is preferred in television camera tubes. Figure 116 in Chapter IV shows an electron multiplier incorporated into a television camera tube. A common radio tube can be used for oscillation, rectification, or amplification; but here also, somewhat analogously with optics or microscopy, where for physical reasons there is a definite limit to the degree of magnification, the lowest voltage or current limit is physically fixed. The signal we would like to amplify or rectify can be drowned in noise. The sources of the noise are to be found in the quantized energy values that we discussed in Chapter I. Particularly significant noise sources are the shot effect and the thermal resistance noise. There are also other sources. The shot effect was studied by Schottky, who found that the source of the noise was the individual electrons that appeared in the form of energy quanta. A stream of electrons toward an anode is somewhat comparable to the dropping of small metal balls. Studies of the diode led to the following finding: the square average of the noise current in the saturation zone (therefore, at high anode potential) is given by the formula:

$$i^2 = 2 \cdot e \cdot I \cdot \Delta f$$

$$\text{or} \quad i = 5.66 \cdot 10^{-10} \sqrt{I \cdot \Delta f}$$

This is at the same time a formula from which it is possible to determine the elementary charge e. In the formula I represents the anode direct current, e the electron charge ($1.6 \cdot 10^{-10}$ amp/sec), Δf the frequency range to be transmitted in which the noise is measured (effective band width in cps).

The comparison with falling metal balls in the place of the unit charges as cause of the background noise might lead to confusion. We must, therefore, keep in mind the effect of induction described in Chapter I, in accordance with which the flight of the individual electrons from the cathode to the anode generates a current throughout the whole circuit, and the sum of these numerous individual events forms the total anode current.

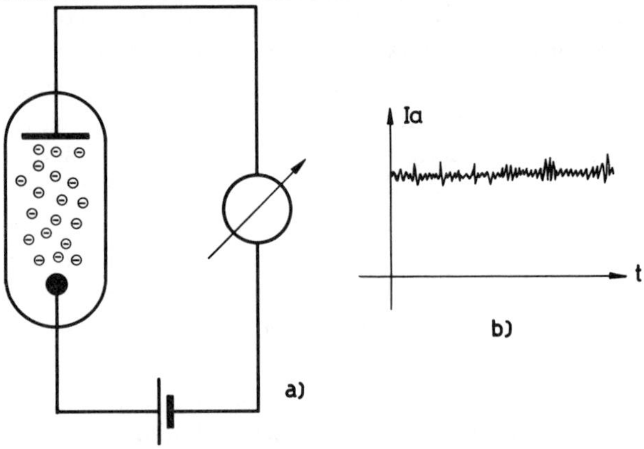

FIG. 63. (*a*) Principal measuring circuit consisting of a diode, measuring instrument (e.g., cathode-ray oscillograph) and plate current source. (*b*) Oscillogram of the plate current.

In the saturation current zone all the free electrons begin independently of each other and fly as individual projectiles discontinuously toward the anode (Fig. 63). In accordance with thermodynamic principles we arrive at the Nyquist formula:

$$i^2 = 4 \cdot kT \frac{1}{R_i} \Delta f$$

in which T is the absolute temperature, k the Boltzmann constant and Δf the band width of the connected amplifier.

In the starting zone of a diode the space charge cloud has a stabilizing effect. We obtain the formula:

$$\bar{i}^2 = 0.5 \cdot 4 \cdot kT \frac{1}{R_i} \Delta f$$

(0.5 is the so-called attenuation factor.) Thus, the tube noise increases as only half the cathode temperature. In the space charge zone, and consequently between the starting and saturation zones, where electrons are still drawn from the reservoir of the space charge cloud to the plate, the separate electron emergences from the cathode are likewise smoothed out. The experiment yields the following results:

$$\bar{i}^2 = 0.644 \cdot 4 \cdot kT \frac{1}{R_i} \Delta f$$

The tube noise in this space charge zone varies as 0.644 of the cathode temperature. When several grids are used in a tube, the unfavorable effect of the quantic nature of the electrons is enhanced by the discontinuous current carriers. This brings us to a current distribution noise, which leads to the consideration that a triode possesses less background noise than a pentode, a fact which is extremely important

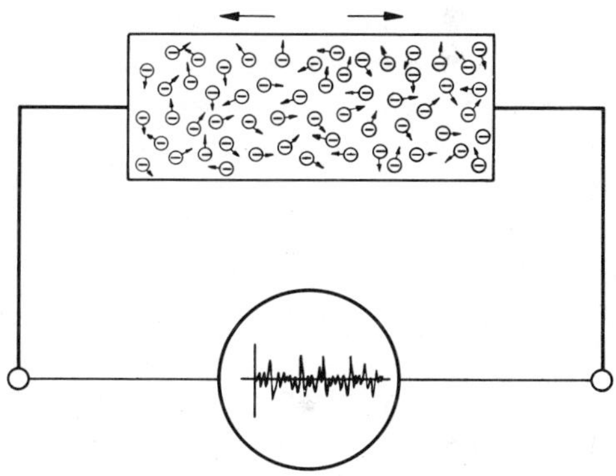

FIG. 64. Measurement of the noise potential at a resistance.

in the selection of tubes for the initial stages in a highly sensitive amplifier. Inasmuch as the electrons emerge from the barium oxide layer of the cathode as from many active small centers, somewhat analogous to the site of origin of the little gas bubbles in a bottle of soda, or, if one prefers, to little volcanoes, there is still another source of noise, the so-called flicker effect.

This can be explained physically if one assumes individual emission centers for the cathode with an exponential fadeout.

It has been found that resistances must also possess background noise at temperatures above absolute zero (Fig. 64). A noise potential must be created at the ends of such a resistance. As a consequence of the thermodynamic effect, the electricity carriers are constantly in lively motion. The Nyquist formula gives the following relationship:

$$U^2 = 4kTR\Delta f = 1.6 \cdot 10^{-20}\, R \cdot \Delta f$$

from which the effective value of the noise potential is obtained:

$$U = 1.26 \cdot 10^{-10}\sqrt{R \cdot \Delta f}$$

or of the noise energy at a resistance:

$$N = \frac{U^2}{R} = 4kT\Delta f$$

Here we are dealing with a specific internal noise energy only one-quarter of which can be made available for use. The noise of a complex resistance, e.g., consisting of coils and capacitors, is created only by the real component, i.e., by the ohmic resistance. But all sorts of real resistances must create noise, for example, an antenna with its radiation resistance. In radio and television it is customary to make use of a noise index, which indicates the factor by which the noise ratio in a tested receiver is worse than that of an ideal receiver without additional noise sources. The noise energy, or characteristic noise, appears in television sets as so-called rain, snow, or grit. It is the result of the flaring-up of many individual points in an unco-ordinated manner upon the picture screen. The sensitivity is thereby limited and is given as a function of the specific power kT. This is referred to a band width of 1 cps. The minimal signal power level obtained from the antenna must exceed the noise level. The noise level can be kept low by an appropriate selection of input circuits and tubes.

In television sets the noise indices today are between 4 and 8 kT_0, where T_0 is room temperature at 293° Kelvin or 20°C. At the same time attempts are being made to construct supersensitive receivers that are as noiseless as possible for the reception of signals from rockets that are millions of kilometers away; the lower attainable limits here are fixed by the formulas given above. Whether radio tube amplifiers with their shot effect offer the best solution for this problem will be discussed in Chapter VI.

We will conclude this chapter with the definition of an electron tube. An electron tube is a discharge tube in which the electrical current is carried primarily by free electrons. Gas-discharge tubes are also electron tubes. Mercury merely acts as a sort of catalyst.

Classification of tubes:

(1) High vacuum tubes
(2) Gas-filled tubes

III. FROM GLOW-DISCHARGE TUBES TO THE THYRATRON

Gas diodes—triodes and tetrodes

Gas-filled tubes comprise as large an area as vacuum tubes. The simplest gas diode is the so-called glow-discharge tube. It consists of a glass envelope that encloses two electrodes, usually in the form of rods, coils, or cylinders. In conventional glow-discharge tubes cold rather than hot electrodes are used. In the manufacture the glass envelope is first evacuated and then filled with a small quantity of a chemically inert gas[1] such as neon. A low gas pressure about 1/100 of atmospheric pressure is employed. The current in the gas diode, as in the gas triode, in contrast to high vacuum tubes, is carried by ions as well as by electrons. Consequently the electrical behavior of the glow-discharge tube is characterized by the fact that until a certain minimum potential difference between the electrodes is attained the tube does not light and there is almost no current flow. In Chapter I (see Fig. 13) the physical basis of the ionization phenomenon was already discussed. In review, we may say that the electric field between the two electrodes accelerates the chance electrons of the filling gas in the direction of the anode. On the way these electrons collide with neutral gas molecules. The collision may result merely in an elastic rebound, but may also lead to the formation of new ions and electrons. Since these new electrons also collide with other gas molecules, more and more current carriers are formed. The electrons pursue their path toward the anode. The positive ions strike the cathode and there result in the emission of new electrons. If the electric field is strong enough for each individual electron to liberate on the average more than one new, free electron, a self-sustaining discharge is obtained. Tubes of this sort are known as glow-discharge

[1] Inert gas = the gaseous state of a pure element, i.e., of an element whose valence shell is completely filled with the maximum permissible number of electrons.

tubes because the excitation of the atoms leads to the emission of light. The glow-discharge tube can be used as a long-lived indicator of superior performance, and also as a stroboscope,[1] in which the light is emitted as a consequence of the temporary and subsequent elevation of electrons to a higher orbit, from which they then drop back to their stable state. As explained in Chapter I, the return of the excited atoms to the fundamental state is the sole cause of the radiation, and each element emits radiation with a characteristic spectrum. There are mathematical-physical relationships which make it possible to

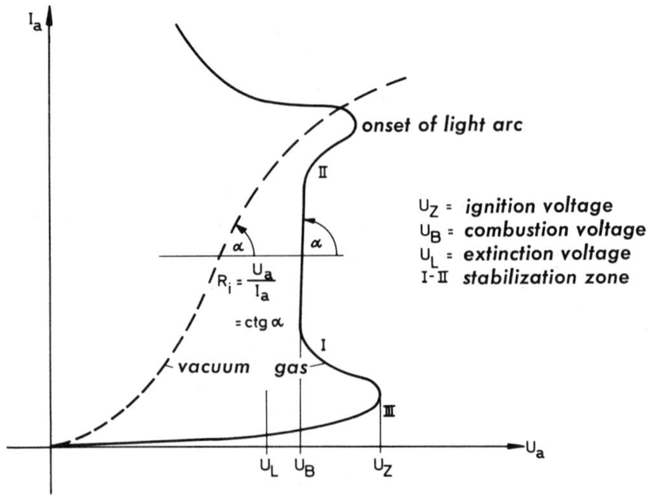

FIG. 65. I_a/U_a characteristic of a vacuum diode and a gas diode. The remarkably large range I to II, over which the combustion voltage is nearly independent of the current, makes gas discharge tubes suitable for maintaining a constant continuous voltage. Tubes specially produced for this purpose are known as stabilizer tubes.

predict the wavelength of the spectral lines for a given atomic weight. Conversely, it is possible to use the spectral lines to determine the composition of an unknown substance by spectroanalysis. In the glow-discharge tube the dominant spectral lines must lie, as the name implies, within the visible range. The color of the light varies with the type of gas atoms and, consequently, with the energy differences

[1] When rapidly rotating objects, e.g., a turning wheel or the moving valve of a motor, are illuminated, if the rotation speed or number of movements is synchronized with the periodic illuminations of the glow-discharge tube the rotating objects appear to be standing still.

between the orbit levels. Thus, the color obtained from neon is red, from mercury blue, from krypton green, from xenon (the material of electron flash bulbs) white, from nitrogen light blue, etc. If there is a difference in the surface area of the electrodes, the current is a function of the polarity. The resulting rectifying effect between a cylindrical and rod-shaped electrode is characterized by a forward/reverse maximum ratio of 200:1. Cold diodes of this type were introduced even into radio sets in the years between 1930 and 1935, but disappeared afterwards because of the voltage drop of about 80 volts and because the incomplete rectification necessitated far too much expense in filter devices. About 15 years later the ion tube rectifier reappeared. In contrast to the earlier forms, these gas-filled discharge tubes possessed a hot cathode.

In Fig. 65 the U_a/I_a characteristics of the vacuum diode and gas diode are compared. The gradual climb of the current with increased plate voltage in a vacuum tube was discussed in the preceding chapter. The efficiency of such rectifier tubes is limited. The voltage drop U in a vacuum tube depends essentially upon the current I flowing

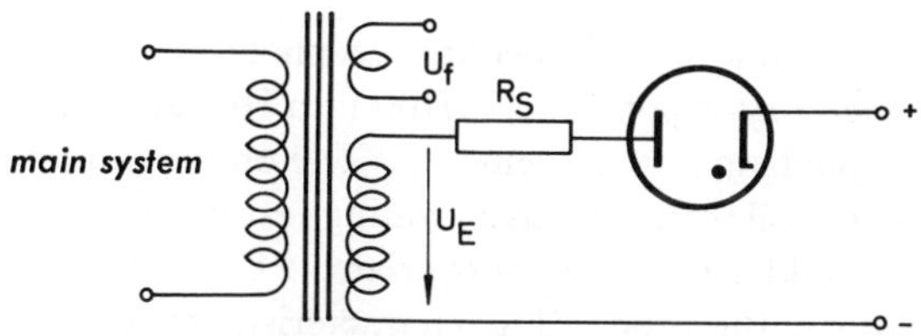

FIG. 66. Half-wave rectifier with symbolically represented gas-filled diode. Because the gas ions also participate in the current conduction, gas-filled diodes are particularly well suited for use with heavier currents.

through and the distance between the electrodes. The voltage drop can be calculated from the approximate formula

$$U = \left(\frac{I \cdot b^2 \cdot 10^6}{A \cdot 2.34}\right)^{2/3}$$

where b is the distance between electrodes and A the cathode surface. If a rectified current of 100 amp flows through such a vacuum tube in which the electrodes are 2 mm apart and the cathode surface area A is equal to 2 cm², the voltage drop will be of the magnitude of several thousand volts. Consequently, a high vacuum tube of this sort could be used at such heavy currents only with voltages higher than the voltage drop. The use of such vacuum tubes is consequently limited to low currents. On the other hand, the gas diode curve indicates an almost instantaneous, intense onset of current as soon as the ignition point is reached. The sharp inflection (III) of the curve indicates the spontaneous impact ionization. The discharge is externally manifested by the illumination of the tube. If the applied voltage exceeds the ignition voltage the anode current should continue to increase without limit. To protect the tube from being ruined it is necessary here, as in all types of gas-discharge tubes, to connect a protective resistance R_s in series with it (Fig. 66) or, if alternating

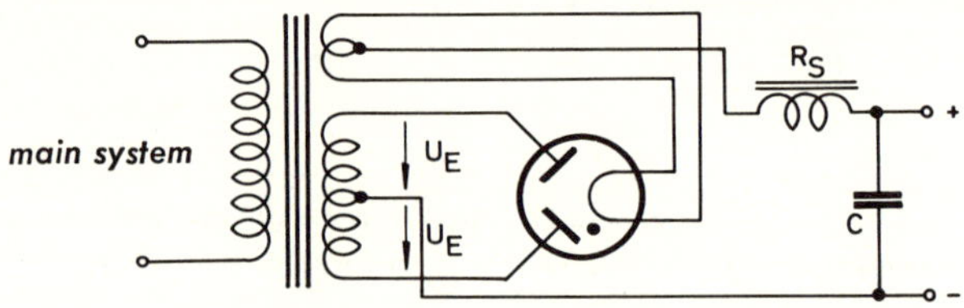

FIG. 67. Full-wave or two-way rectifier with choke coil input.

current is being used, a series impedance (a choke coil, Fig. 67). In practice most mercury-vapor tubes (1/100 mm Hg) are actually used in the choke coupled input arrangement. The ohmic inductive resistances of the choke coil then serve as a limiting resistance. A capacitor coupled input would be very inappropriate. The charging-current pulse would destroy the tube. The characteristic curve (Fig. 65) of the mercury-vapor tube shows the ratio of potential change ΔU to current change ΔI, i.e., the internal resistance

$$R_i = \Delta U / \Delta I$$

is very low. A rectifier of this sort therefore exhibits an extremely low voltage drop and low line loss or, in other words, a high efficiency. It is therefore well suited for alternating current voltages as high as 100 volts. The fixed voltage drop resulting from the combustion voltage of ~ 10 volts gives an efficiency of about 90%. Because of the low internal resistance the voltage is to a great extent independent of the load.

The vertical climb (I–II) of the gas diode curve (Fig. 65) shows that the combustion voltage can hardly be influenced at all over a wide range of current intensity. The glow-discharge tube is, therefore, suited for use as a voltage standard. In stabilizer tubes the cathode possesses a large surface area. Electron emission requires a constant current density per unit of surface from the cathode. If the current is raised beyond point I, i.e., to point II, the area of the glow covering the cathode will expand progressively.

Voltage stabilizers are constructed for regulation current loads of from several mA to about 100 mA. Since the combustion voltage is constant to within approximately $\pm 1\%$, the regulated voltage is constant to about $\pm 1\%$. Depending upon the type of gas and the pressure, the stabilization or combustion voltage ranges between 10 and 150 volts.[1] Even for the same type of tube, variations in pressure and the presence of impurities make it necessary to figure on certain tolerances. Moreover, the electrical data are subject to change with heating and in the course of time. If higher voltages of, for example, 280 volts are to be stabilized, it is necessary to make use of a four-diode system, the diodes of which are unheated in practice and receive 70 volts each; these diodes are arranged in series with each other and can be incorporated into a single envelope or may be separate. The protective resistance R_s for individual or cascade-connected stabilizer tubes can be computed from the formula

$$R_s = \frac{U_e - U_b}{I_{\max}}$$

in which U_e is the input voltage, U_b the stabilization voltage or combustion voltage (also often called the arc voltage). $I_{\max}$ is the maximum permissible current, which depends primarily upon the size of the electrodes, i.e., upon the dimensions of the tube. For stabilization purposes U_e should be set at from one and a half times to twice the value of U_b to ensure firing. Zone III (in Fig. 65) shows that a higher voltage than the combustion voltage, the so-called ignition voltage, is required to initiate the ion avalanche. Immediately after ignition is initiated the voltage in the tube drops to the combustion voltage U_b. When the tube is extinguished the sequence is reversed. The applied voltage must drop below the combustion voltage for the gas discharge to be broken off. This so-called extinction voltage is always lower than the ignition and combustion voltages. If the tube is to be fired again the voltage must once more be raised to the ignition voltage. The hot-cathode

[1] The combustion voltage depends upon the type of cathode material used, the geometry of the design and, especially, primarily upon the distance between the electrodes.

gas-discharge tube is generally more stable than the cold-cathode gas-discharge tube, and is consequently more demanding. The anode potential that is to be rectified, particularly in mercury tubes, can only be applied after heating. Also, the factory-prescribed limit to the filament voltage must be adhered to exactly if the tube is to have a long life. In practice thermal time-delay relays are employed to ensure against switching on of the anode voltage until heating is completed, though sometimes manual switches are used. Mercury drops on the electrode must be vaporized beforehand. If care is not taken, current can pass through the tube in the opposite direction (arc-back). Figures 66 and 67 show half-wave and full-wave rectifier circuits, which, with the exception of the necessary protective resistance R_s or the choke coil input, are no different from vacuum tube rectifiers. The full-wave rectifier can also be arranged in a Grätz bridge circuit,

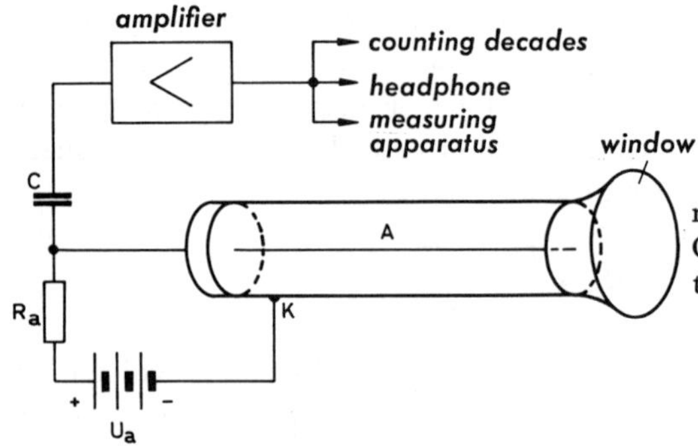

FIG. 68. Circuit arrangement of the Geiger-Müller counter tube.

which eliminates the necessity of the secondary compound winding of the transformer. The full-wave rectifier requires less filter equipment than the half-wave rectifier, and the fundamental frequency of the half-wave rectifier is 50 cps, while that of the full-wave rectifier is 100 cps.

The counter tube invented by Geiger-Müller in 1928 is a gas-discharge tube. It contains two electrodes between which is situated an electrostatic field. Figure 68 illustrates the circuit arrangement. One electrode consists of a metal cylinder, or is a conductive graphite coating on the inner wall, connected to the negative pole of a current source. The other electrode is a thin wire fixed in the center. This serves as the anode. The two electrodes form a kind of cylindrical condenser. The cylinder and the wire are connected to a socket by leads that are fused into the glass. The electric field, as is well known, is strongest in the vicinity of the wire, since that is the site of the greatest concentration of force lines per unit of surface. If, for example, there happens to be an individual, free electron present in the counter, the electron is increasingly accelerated in the direction of the inner electrode, with the result that it collides with other gas molecules and gives rise to new ionization processes. In this way it is possible for a single primary electron to give rise to a greater number, let us say *n*, secondary electrons, which move in the direction of the anode. The atoms from which these electrons are dislodged become positive ions and travel in the direction of the cathode. The newly freed secondary electrons are also capable of dislodging other secondary electrons from gas molecules in the course of their path to the anode, thus giving rise to an avalanche-like formation of charge carriers. The first free electron in the gas chamber of the Geiger-Müller tube must, however, arise from an atom. Such an atom of gas must have collided with some radiation or with a high energy particle, e.g., a beta ray, coming from outside and penetrating the wall of the tube. The emission of nuclear electrons, which occurs in radium, is called *beta radiation.* This consists of extremely fast electrons, which move with a velocity close to that of light. In the course of their path they are capable of ionizing thousands of molecules. *Roentgen* and *gamma quanta* also liberate electrons from the molecules of the gas and the wall of the counter tube by means of the photoelectric effect, the Compton effect and, possibly, the pairing effect (these phenomena will be taken up in detail later on). Thus, it is possible to count in this indirect way not only particles but also electromagnetic waves, i.e., roentgen and gamma quanta.

Alpha particles, which are positively charged ions (helium nuclei), as well as other electrically charged particles also lead to ionization and trigger a signal in the counter tube. The table on page 90 contains a list of the particles that frequently have to be counted.

Neutrons, which under normal conditions are not found in the unbound state or outside the nucleus of an atom in nature, have a zero charge and are consequently unable to effect ionization. Neutrons can pass through thick metal plates and can therefore not be kept in radiation containers. To demonstrate the presence of neutrons it is necessary to undertake nuclear reactions with them; these nuclear reactions give rise to ionizing particles or radiation. This indirect method is the only way of demonstrating the presence of neutrons. Thus, it is in principle possible to demonstrate directly or indirectly the presence of alpha, beta, or gamma rays and neutrons, as well as other particles, such as positrons[1] and mesons.[2]

Name	*Symbol*	*Charge*	*Rest mass in kg*	*Mass where* $O^{16} \triangleq 16$
α particle*	d, ${}^{4}_{2}He^{++}$	$+2e$	$6.643 \cdot 10^{-27}$	4.00276 ME
Proton	p, ${}^{1}_{1}H^{+}$	$+e$	$1.6723 \cdot 10^{-27}$	1.007593
Neutron	n	0	$1.6746 \cdot 10^{-27}$	1.008982
Electron	e^-, β^-	$-e$	$9.107 \cdot 10^{-31}$	$5.488 \cdot 10^{-4}$
Positron	e^+, β^+	$+e$	$9.107 \cdot 10^{-31}$	$5.488 \cdot 10^{-4}$
Photon (light quantum)	$h \cdot f$	0	0	$f \cdot 4.42 \cdot 10^{-24}$

$$* \; d = \frac{2p}{2n} = \text{helium nucleus}$$

A single rapidly moving electron (a beta ray) leaving a train of ions or ion track in its path is sufficient to trigger an impulse inside the counter tube. One may also say that the introduction of the quanta to be measured gives rise to a supply of current carriers.

The two typical zones I and II shown in Fig. 69 are put to use in radiation counters with Geiger-Müller tubes.

In the proportional zone the impulse level is a function of primary ionization intensity.

Here, as in a gas-filled photoelectric cell (sound film), gas amplification takes place. The gas amplification depends upon the type of gas with which the tube is filled, but it depends even more upon the relation of the voltage to the ignition voltage. To make the onset point of the self-sustaining impact ionization less critical, the tube is filled with a gas that absorbs light quanta, so that no or few free electrons are created photoelectrically. Halogen gases such as chlorine or bromine, as well as alcohol vapors, have the property of absorbing light quanta and thereby eliminating the photoelectric effect. Other requirements of the filler gas are tube operability at low external temperatures and a low operating voltage. Since these

[1] Particles of the same mass as an electron, with an equal but positive charge.

[2] Mesons are short-lived particles with a mass larger than that of the electron and smaller than that of the proton; they may be electrically positive, negative or neutral. They were first discovered in the course of disintegration of nuclei that had been bombarded with high energy radiation.

conditions reduce the degree of gas amplification, only intensively ionizing particles like alpha rays, protons and beta rays can be measured. Gamma rays cannot be recorded, i.e., they are recorded with a probability of less than 1%. Proportional counter tubes are consequently very well suited for determining the presence of the above-mentioned particles in areas where there is gamma ray disturbance that one does not wish to measure. The time resolution is given as 10^{-6} seconds. In a proportional counter tube there should, therefore, be no glow discharge. The situation is different with the Geiger-Müller tube, which operates in the linear saturation zone.

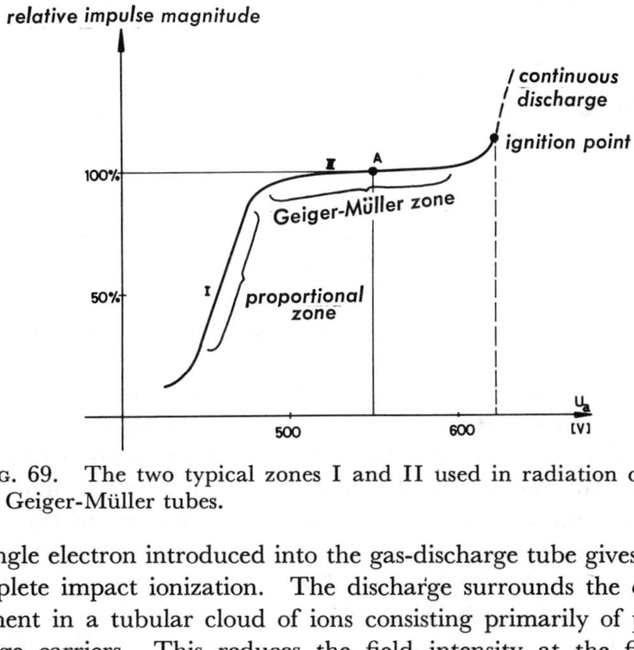

FIG. 69. The two typical zones I and II used in radiation counters with Geiger-Müller tubes.

A single electron introduced into the gas-discharge tube gives rise to complete impact ionization. The discharge surrounds the counter filament in a tubular cloud of ions consisting primarily of positive charge carriers. This reduces the field intensity at the filament electrode to such a degree that the discharge is extinguished within a thousandth to a ten-thousandth of a second. In this linear, so-called Geiger-Müller zone, the impulse level is constant and no longer a function of the intensity of the primary particles. Since the photoelectric effect, i.e., the characteristic glow discharge, participates in the amplification, the gas amplification factor obtained is over a million.

Each impact is heard as a clicking noise over a headphone. The impacts can, however, be fed into a measuring instrument, where the individual impulses are integrated by means of a condenser to a mean value (so-called dose calibration). A still better method is the use of an electronic meter, which counts the impulses continuously. These typical Geiger-Müller tubes are very sensitive; they do, however, possess a relatively low impulse frequency limit, i.e., particles occurring during the recovery period of the counter tube are not counted.

It is possible to reduce the time required for neutralization of the ion cloud, i.e., the time required for the positive ions to reach the cylindrical electrode, to about 10^{-3} to 10^{-4} seconds by the addition of appropriate quenching gases.

The operating voltage is set at about the middle of the horizontal course of the characteristic curve in order to obtain the most favorable operating conditions. The counters are usually designed for adjustable measuring ranges of 0–0.1 mr/hr and 0–3 mr/hr (milliroentgens per hour). To measure lower radiation intensities other methods, which will be discussed later, are used. The positive helium nuclei have a very great ionizing effect. Because of their size (mass number 4), however, there is a great tendency for them to be absorbed. Even in the atmosphere the mean path length of alpha rays arising from radium decay is only about 1 cm; alpha particles are, therefore, strongly absorbed by the entry window of the Geiger-Müller counter tube and pass through only with difficulty, i.e., it is possible only for high energy (fast moving) particles possessing a definite minimum energy to pass through. If low energy alpha particles are to be measured, the window must be a very thin foil. A mica window about 3/1000 mm thick is frequently used. The decisive factor is not so much the thickness of the window material but the mass number, which is why the mass is frequently indicated in milligrams/cm^2. The lower the energy and the greater the mass of the particles to be measured, the thinner and less absorbent the window must be for the rays to enter and not be absorbed on the way. Fast electron radiation (several hundred kev), because of the small size of the particles, is usually capable of penetrating thin aluminum sheets; beta rays can even penetrate sheets up to millimeter thickness. To measure gamma rays or roentgen rays, which is possible only after these radiations have first liberated electrons from the cathode, the cathode, or outer cylinder, must be made of a material possessing a high nuclear charge number. Electrons and protons are then no longer able to pass through this absorptive material, with its great mass, unless they possess a minimum velocity that lies beyond the measuring range.

X-ray tubes also belong among the diodes. Only in the early stages of development were they constructed as cold-cathode gas-discharge tubes. Later they were equipped with a hot cathode and evacuated. The direction of their development was from ion tubes to electron tubes. The principle, which is illustrated in Fig. 70, is

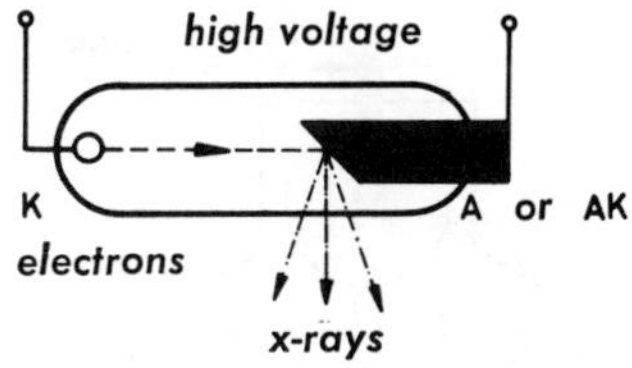

Fig. 70. X-ray tube. These tubes were originally gas-filled, but at present they are designed only as hot-cathode vacuum tubes. K—cathode; A or AK—anticathode. When the electrons strike the AK, x-rays are emitted.

the same in both cases. Under the influence of a very high anode voltage (20,000–1,000,000 volts) between the cathode and the tungsten anode, electrons are fired at very high velocity toward the anode, where they are retarded. The sudden, intense retardation of the electrons in the field of electric force lines gives rise, in accordance with electromagnetic theory, to waves in the form of the well-known x-rays. However, x-rays are also created as a consequence of the

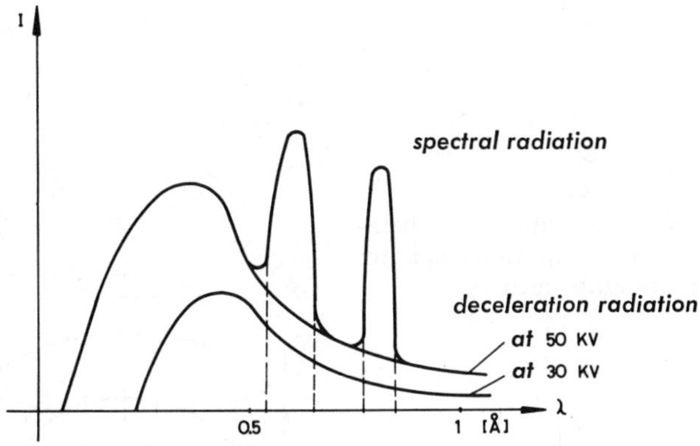

Fig. 71. Intensity distribution over the electromagnetic frequency range (x-ray range).

impact or excitation produced by the rapidly moving electrons upon the atoms of the anode material. Electrons are knocked loose from inner electron shells, and when these gaps are refilled by the dropping back of electrons, x-rays of definite spectral frequencies are emitted;

the frequencies of these emitted roentgen rays depend exclusively upon the type of material used in the anode. In contrast to the fundamental radiation, this characteristic radiation is created by the transition of the excited atoms back to the fundamental state. Both types are illustrated in Fig. 71. Fig. 72 shows a modern x-ray tube.

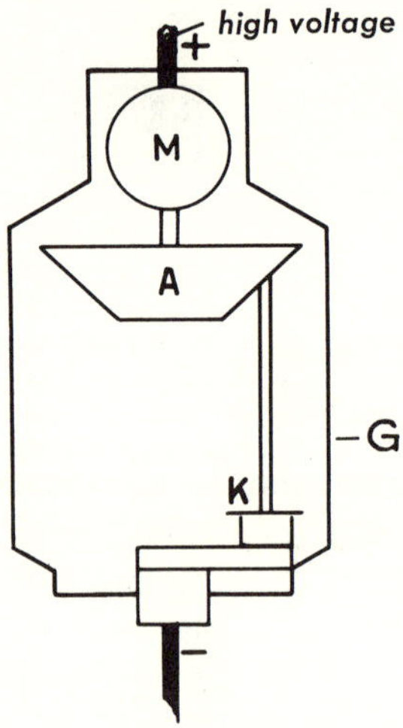

FIG. 72. Modern x-ray tube in which the anode is rotated by means of the motor M in order to effect heat dissipation. Such tubes permit the use of greater loads. A is the rotating anode, K is the cathode, and G is the glass wall of the evacuated tube.

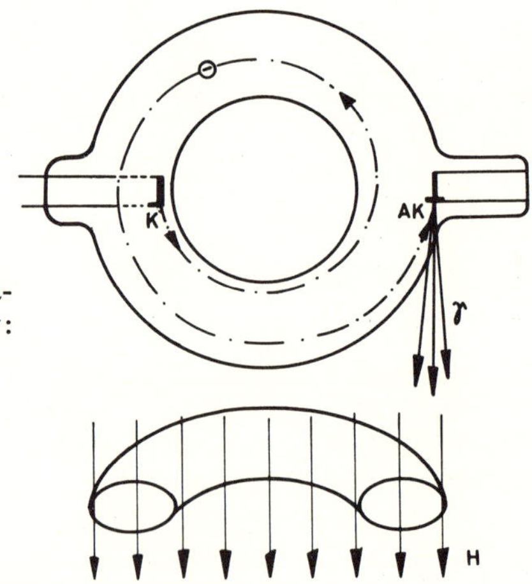

FIG. 73. Principle of the betatron. ABOVE: top view; BELOW: section of betatron tube.

Since the anode is also the point of departure for the x-rays, it is also known as the anticathode. X-ray tubes reverse the procedure of photoelectric cells and utilize electrons to give rise to light quanta.

The betatron, developed by Brown-Boveri, represents a further development of the x-ray tube (Fig. 73). In order to bring electrons to extremely high velocity, they are accelerated in the shortest possible time in a vacuum tube closed to form a toroid, in which they are forced to make about a million revolutions, being kept on a circular path by means of periodically increased magnetic fields. The annular tube is arranged like a short-circuited secondary winding around an iron plate core. A coil serving as the primary winding is supplied with 50 cps alternating current. In the first quarter of the alternating current cycle—from the zero moment to 1/200 second—the magnetic field builds up, as, however, does the coil voltage, in accordance with the sine form of the curve. This is the period utilized by the electron accelerator, and it is within this same period that the electrons complete their numerous revolutions around the inside of the toroidal tube.

The electrons must be removed from the system before the completion of this quarter cycle. At the beginning of each new alternating current cycle, or shortly after the magnetic induction has passed through the null point, the electrons must, therefore, be rapidly "injected" into the orbit of the acceleration tube, whereupon rotational acceleration sets in as a result of the electrical rotational field. The increasing velocity would be accompanied by an increase in the electron orbit if the periodically increased magnetic field did not force the electrons into smaller orbits, as a consequence of the Lorentz force which we have studied. A condition of equilibrium can be created by the appropriate arrangement. Moreover, the transformer core possesses an air gap in the center and a specially designed pole piece at the outer edge to ensure proper direction of the electrons to the center of the orbit inside the vacuum tube. With each revolution of the electrons, the kinetic energy is increased by precisely the coil voltage of the transformer, so that with a large iron core (e.g., 30 volts per winding) and about 1 million revolutions it is possible to obtain energies of the order of 30 million volts (30 Mev). After the acceleration is completed, i.e., just before the end of the $\frac{1}{4}$ cycle, the magnetic field of the core is diminished by means of auxiliary coils and the discharge of a capacitance across a thyratron, so that the orbit increases and the electrons pass to a platinum cathode. Because of the great kinetic energy of the electrons, which are then suddenly retarded in the electric fields of heavy atomic nuclei, the x-rays thereby created are of very great intensity ($h \cdot f$) and are sufficiently "hard" to pass through the thickest metal sections. However, it is also possible to change the orbit by a corresponding modulation of the auxiliary coil current in such a manner that the electrons emerge directly from the acceleration tube. In the betatron there is no continuous beam; instead there are pulselike (5 microseconds long) electron packets that are delivered repeatedly with each alternating current cycle. By means of special arrangements it is possible for the acceleration process to be completed twice in the course of every signal cycle. These shortest-wave x-rays of the order of magnitude of many Mev per individual light quantum are used in industry for nondestructive materials testing (x-ray method) of large castings, such as Pelton

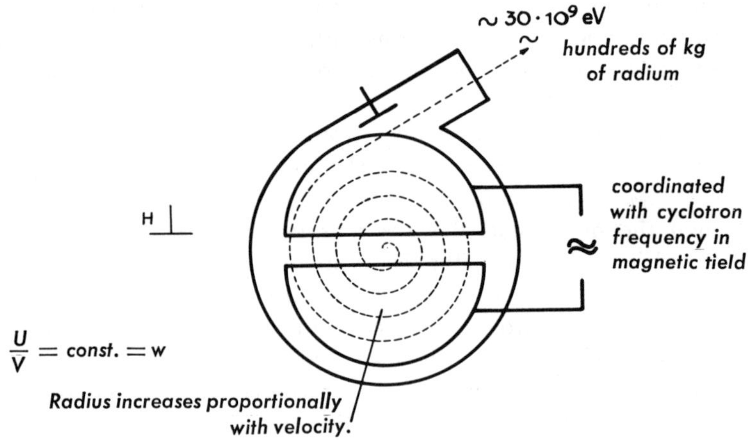

FIG. 74. The cyclotron. Acceleration of protons and particles. Acceleration by steps.

turbine wheels, and in medicine for the treatment of cancer patients.

The x-ray tube developed from an original gas-filled tube to a vacuum tube. The fast electrons required for the generation of the electromagnetic radiation are obtained in this case by means of higher anode voltage, i.e., by direct linear acceleration of the electrons emitted from the hot cathode. This type of accelerator is called a linear accelerator. The betatron and cyclotron (Fig. 74) belong to the class of circular accelerators. Electrons are not, however, the only particles that can be accelerated. It is also possible to accelerate larger particles, for example, hydrogen nuclei (protons). In the accelerator (a proton synchrotron) recently put into use by CERN in Geneva, these positive ions are accelerated to a kinetic energy of 28,000 Mev (28 Gev or 28 billion electron volts) in a, naturally, large circular orbit. This accelerator employs an evacuated, circular tube 628 m long. The arrangement can be roughly compared to that of a large betatron. The protons are accelerated in their orbit by the electric fields of twenty resonant circuits of variable frequency disposed along their path. These resonant circuits reach a maximum once in every acceleration cycle. Beginning from the injection point the frequency is raised from 2.9 Mcps to 9.5 Mcps. This frequency is twenty times higher than the revolution frequency. The peak voltage per acceleration gap is 6300 volts, which, with the indicated resonator losses, corresponds to a high frequency power of 7 kw per unit. With each revolution, energy is supplied in a cyclical sequence as a result of the acceleration intervals. The particles would fly in a straight path if the Lorentz force were not used here too to compel the particles to follow the fixed orbit of the circular tube. The maximum magnetic field strength reached in the guiding field is 14,000 gauss. This is attained by means of a ring magnet composed of 100 magnet sectors of C-shaped cross section. As in the deflection coil of a television tube, here also the current must increase linearly in accordance with the requirements of the magnetic field. The exciter coils are connected in series, and the exciter current rises to 5000 amps. The exciter voltage applied is 5400 volts. At the end of each acceleration phase, which takes about one second, the magnet core system must be demagnetized as quickly as possible.

The acceleration mechanism of a synchrotron is somewhat different from that of the betatron and can only be understood properly after the completion of Chapter V, where velocity controlled tubes are discussed. Nevertheless, the discussion of the synchrotron belongs in this section with ion tubes.

As with the injection of electrons into the betatron, the injection of the ions into the synchrotron takes place at the beginning of each acceleration cycle and occurs within a few microseconds, i.e., within a shorter period of time than is required for a single revolution in the synchrotron ring. The ions are generated by ionization of hydrogen in a ceramic vessel at low pressure (10^{-3} mm Hg) by means of a high voltage–high frequency field of 130 Mcps. The ionization is in pulses synchronous with the operating cycle of the synchrotron. The ion current (about 50 milliamps), consisting of positively charged protons, is accelerated by a negative high voltage of 25,000 volts in the completely evacuated preaccelerator after passing through a 3 mm thick nozzle opening. After being subjected to further acceleration procedures the proton beam, which has by this time become very fast, is deflected into the synchrotron ring by means of magnetic deflection fields. Pulsed deflection and focussing systems guide the ions tangentially into the central, stable acceleration orbit of the ring tube.

The highly accelerated particles—which can be considered as projectiles of nearly the velocity of light—are allowed to strike other materials. The impact phenomena or interactions make it possible to study the character of various known or new elementary particles.

Probably the simplest gas-discharge tube is the nullode. It consists simply of a glass tube filled with a rare gas. No electrodes are sealed into the glass. This tube is primarily used as a high frequency indicator for decimeter and centimeter waves. When the tube is brought into the vicinity of a strong high frequency field the great field strength produces ionization inside the tube and results in illumination of the tube.

Long, thin glow-discharge tubes with a single capacitative electrode at one end of the tube are also used for tuning centimeter-wave wave guides such as those used in radar. As the voltage is increased the light column grows longer. If a number of such small glow-discharge tubes are inserted into the wave guide and distributed along its length, they provide the simplest way of obtaining an "optical" image of the course of the standing waves. This makes it possible to control aerial tuning.

Dipole glow-discharge tubes without barrier resistance are used as rare gas lightning arresters. They are connected between electrical leads to protect them from overvoltages (e.g., lightning discharges). When the overvoltages occur the tube is ignited and, as a result of the combustion voltage, protects the instruments to which it is connected from overloading. These tubes are also used as gas-discharge voltage suppressors.

Since glow-discharge tubes are very good conductors during the process of impact ionization, i.e., they possess slight resistance, they are favored as electronic switches. Elements of this type are encountered in the wave guides of radar instruments, which connect the common parabolic antenna with the transmitter or receiver section. Two such tubes (TR and anti-TR cells), connected at the nodes and antinodes of the standing waves, ensure that during the sending impulse all the energy flows to the antenna and never to the receiver. In the unignited state, i.e., during the wait for the echo, all the incident receiving energy is drawn off to the receiver.

The gas triode or so-called thyratron represents a further development of the normal gas-filled rectifier tube. The thyratron possesses a hot cathode. A power thyratron is usually filled with mercury vapor. Only low power gas triodes are filled with gases like hydrogen. Figure 75 compares the I_a/U_g characteristics of the vacuum triode and gas triode (thyratron) at a determined plate voltage. In the vacuum tube the anode current varies continuously with the grid bias voltage. The thyratron curve shows that the thyratron suddenly becomes conductive and remains so as soon as the original negative biasing

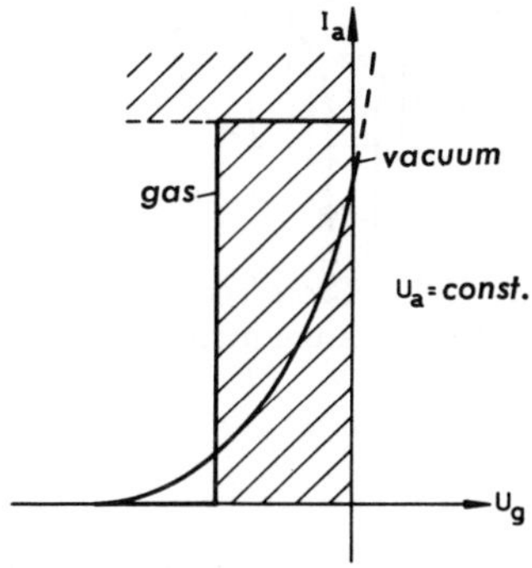

FIG. 75. Comparison of the I_a/U_g characteristics of a vacuum triode and a gas-discharge triode.

voltage is brought to a certain more positive value. With the use of a grid it is possible to "adjust" the ignition voltage. The tube does not operate as a continuous system like the vacuum tube, but as a switch or relay. From a physical viewpoint no electron current flows —as long as the grid bias is sufficiently negative—and consequently no ionization takes place; there is also no ion current. If the negative grid bias is shifted to the ignition point, an electron current begins to flow until the chain reaction of self-sustaining impact ionization is initiated; this leads to an arc discharge. Figure 76 shows a cross section of the electrode system of a thyratron. The grid consists of a perforated partition between the cathode and the disc-shaped anode, which is in turn connected, primarily for insulation purposes, with a sleeve contact introduced into the envelope. If the grid were to enclose only the cathode or only the anode, the unenclosed electrons would be greatly affected by the electric field created by charges on the walls of the vessel. Since the condition at the walls varies from moment to moment, this would lead to considerable fluctuation in the operating conditions. The grid is therefore designed in such a manner that it encloses both the cathode and anode. Figure 77

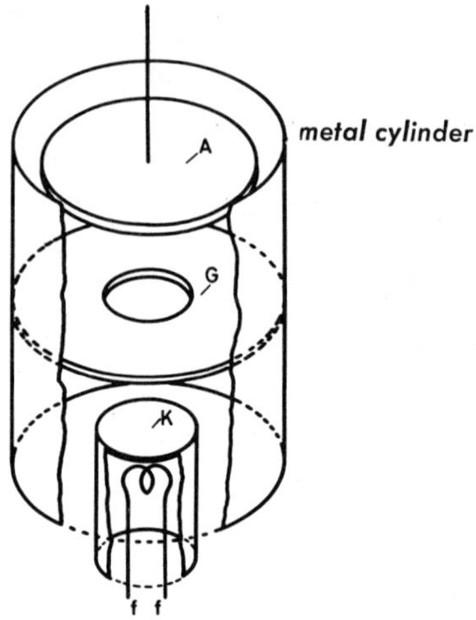

FIG. 76. Cross section of the electrode arrangement of a thyratron.

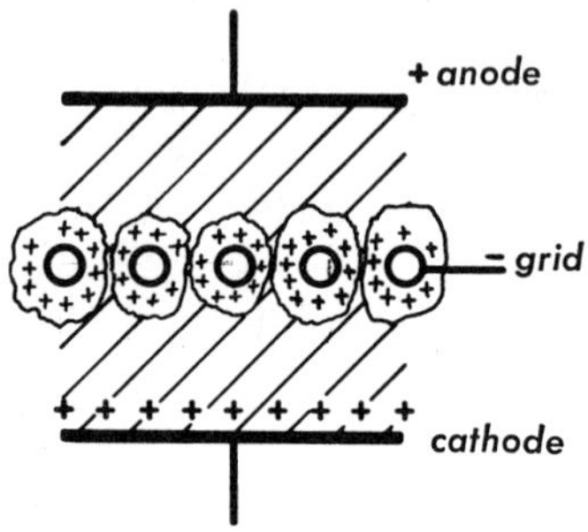

FIG. 77. Ion concentration in a thyratron. The grid is surrounded by positive ions. During the deionization period these are destroyed, i.e., neutralized.

shows how a layer of positive ions is formed in front of the cathode (cathode trap) in the ignited tube. The negative grid bars trap the positive ions. Despite the negative potential it is therefore possible for a considerable grid current to flow.

The point of onset or so-called ignition point of the current flow depends, on the one hand, upon the magnitude of the negative grid bias voltage and, on the other, upon the level of the applied anode

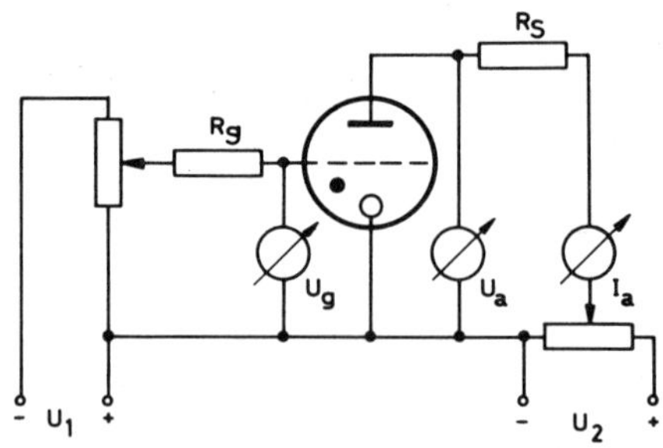

FIG. 78. Measuring circuit for plotting the characteristic curve of a thyratron. If a positive potential U_a is applied between the anode and cathode, the grid (U_g) can be used to set a point in time at which the passage of current through the tube will begin.

voltage. The U_a/U_g characteristic can be plotted with a measuring circuit shown in Fig. 78. For each of the different grid bias voltages a different anode voltage is required for ignition of the tube. If the grid bias voltage is negative, the anode voltage required for ignition is correspondingly higher. If there is a fixed negative grid bias voltage while the tube is in the unignited state, continuous increase of the anode potential at a certain point will result in ionization and the tube will become conductive. At the same time, the grid will, as mentioned before, lose its effectiveness. The tube can be extinguished only when the anode voltage drops below the extinction voltage, which is only slightly below the combustion voltage. With a unidirectional voltage the anode voltage must be switched off, and with

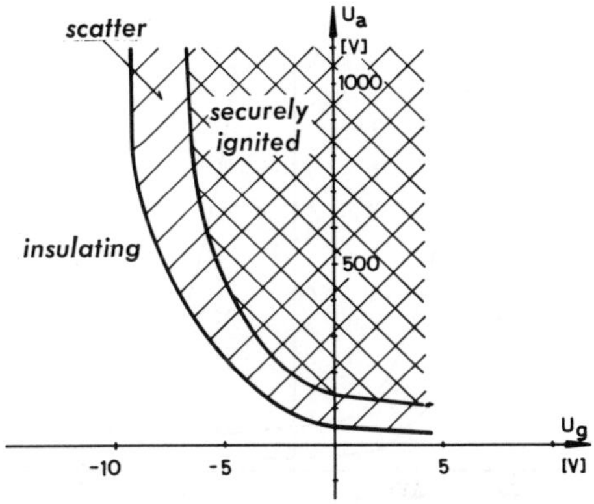

FIG. 79. Thyratron characteristic with scatter.

an a.c. voltage one must wait for the half-wave to drop below the extinction voltage. Figure 79 shows the so-called ignition characteristic or control curve. From this curve it may be seen that the more negative the grid bias voltage, the higher the anode voltage must be to ensure secure ignition. The ignition zone is crosshatched. The large tolerances of gas-discharge tubes are usually expressed as a scatter zone between two adjacent curves. The obligatory value lies somewhere between. The strong temperature dependence is also taken into account because the gas pressure doubles with each 10°C increase in temperature. The tube ignites more easily at higher temperature. Thyratrons are manufactured for various grid bias voltages or, in other words, with the most varied control ranges, depending upon the purpose for which they are to be used. In certain types, the grid bias voltages required to secure the beginning of ignition are negative for all the practically occurring anode voltages;

in other types grid bias voltages required for ignition fall into the positive range (Fig. 80). We speak of negative or positive control curves. The thyratron, the anode of which is usually supplied with an a.c. voltage, is used as a rectifier.

In contrast to the vacuum diode rectifier the anode current in the thyratron does not start at the beginning of the positive half-wave, but only after a definite ignition voltage controlled by the chosen grid bias voltage is reached. The mercury vapor thyratron[1] is primarily used when greater power and, specifically, greater voltages of several hundred to thousand volts are desired and for currents up to about 20 amps. An important characteristic of the thyratron is the so-called ionization time and deionization time. This is the period of time

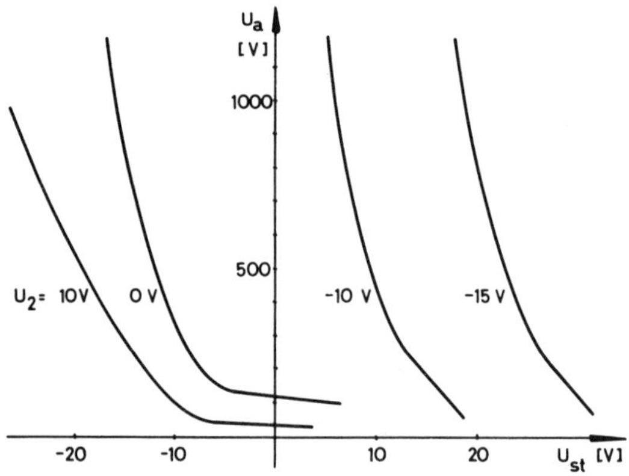

FIG. 80. Negative and positive control curves for various types of thyratron.

required for the build-up of ionization and the period after extinction before the grid recovers control of the electrical processes inside the tube, since the ions do not disappear immediately after the discharge has ceased. The deionization time depends upon the gas pressure, the kind of gas used, etc. In mercury-vapor tubes deionization takes about 1 millisecond. The operating frequency depends upon the deionization time. In the usual mercury tubes, which have a small internal voltage drop (a combustion voltage of about 10 volts), the operating frequency is below 500 cps. Smaller tubes filled with rare gas possess deionization times of only a few microseconds and are suited for operation at frequencies up to 150 Kcps. Here also the lag arising from the ionization and deionization time is not so great as to prevent the use of these tubes for high frequency demodulation.

[1] Recently, in addition to the use of mercury, a tube filled with a mixture of mercury and a rare gas has been found advantageous (BBC).

The ionization time can be related to the molecular weight of the filling gas. The heavier the molecules, the more slowly they move and the longer it takes for enough ions to form for the arc to occur. Mercury vapor requires a considerably longer ionization period than hydrogen. Here it should be mentioned that mercury is a conductor like all other metals when in an atomic ensemble, i.e., when it is in a solid or liquid aggregate state, and must therefore be vaporized beforehand by the filament voltage during a heating-up period of several minutes to prevent disruptive discharges. Rare-gas thyratrons with hot cathodes are less troublesome, in that the heating-up period is shorter. In both types of thyratrons the heating voltage must be held to within $\pm 5\%$. If the voltage is too low the oxide layer of the cathode will not be sufficiently emissive. It will be overloaded and will be ruined by sputtering. If the heating voltage is too high, the cathode material is vaporized and collects upon the grid. This can lead to grid emission and reduce the effectiveness of control.

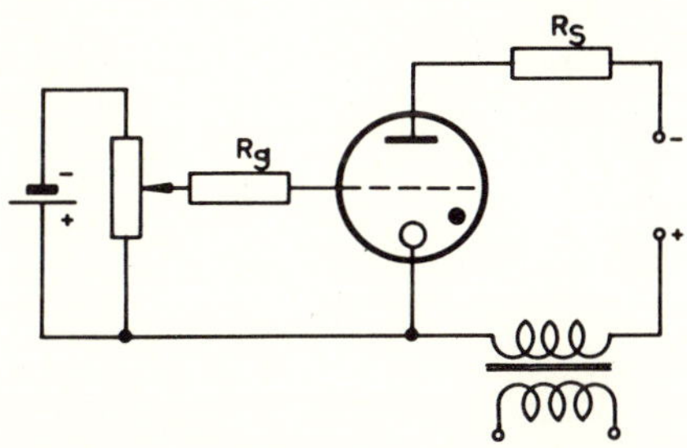

FIG. 81. A simple thyratron control (a.c. operation). R_g—resistance limiting the grid current. R_s—protective resistance in the anode circuit. The current onset can be controlled by variation of the grid bias voltage. The mean value of the current flowing through the tube varies proportionally.

The life of the tube is determined by the scatter range, the combustion voltage being taken as a criterion. All types of gas-discharge tubes deteriorate as a result of absorption of the gas filling by the negative electrodes (ion bombardment). Mercury-vapor tubes are less subject to this type of deterioration because they possess a sufficiently great supply of mercury (or mercury vapor).

The factory-prescribed limit for the operating conditions should not be exceeded with either mercury-vapor or rare-gas thyratrons. Both types belong to the single group of gas-filled tubes, though a further subdivision into hot-cathode and cold-cathode tubes is possible.

Gas-filled tubes are represented symbolically in circuit diagrams by hatching or, more recently, by a dot—to indicate the gas filling. If in the circuit shown in Fig. 81 the characteristic curve of Fig. 80 is taken as a basis and the grid bias voltage is set at zero or preferably at a positive value, as in the diode a current flows during almost the entire positive half-wave. Control starts at the point at which the ignition voltage is reached. The magnitude of the current depends essentially upon the external load or limiting resistance. If the grid bias voltage is set at -10 volts, the current does not begin to flow until the positive half-wave reaches several hundred volts. Since a complete wave train in phase angle measure corresponds to a value of

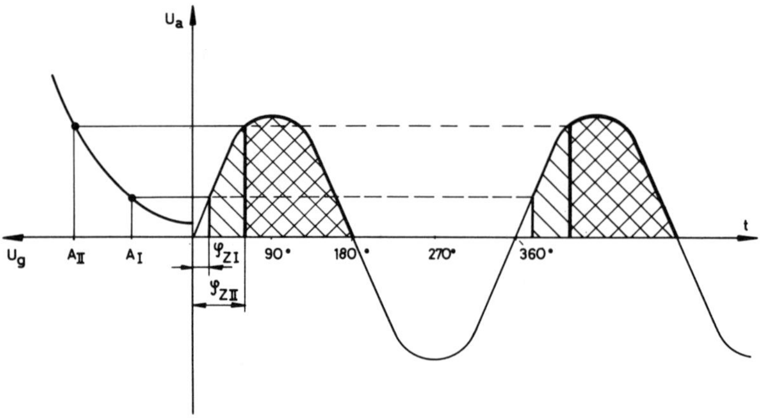

FIG. 82. Control of the start of ignition by selection of the grid bias voltage. Higher (—) grid bias voltages result in a delay in ignition. The ignition angle ω_z then becomes greater.

360°, the circuit of Fig. 81 offers the possibility of control between about 0° and 90°. Figure 82 illustrates the relationships at two different operating points I and II. At working point I ignition begins sooner and the mean rectified current, shown by the hatched area, is greater than at working point II. The phase angle mentioned here is technically known as the ignition angle. The anode current is smoothed out with the usual filtering devices consisting of choke coils and capacitors. Thus, one obtains the average rectified voltage, which can be varied as desired by variation of the grid bias voltage. In practice, two thyratrons are usually coupled in a push-pull arrangement in order to make use of both half-waves. Thyratrons are suited for efficient, sensitive control circuits. In contrast to a variable resistor, the use of a thyratron results in only slight losses.

The ignition time can be controlled by other means than an adjustable continuous grid potential. A plain method is the use of square-wave control pulses (represented in Fig. 83 as double lines), which can be supplied at various points of time $\pm \Delta t$. In this arrangement the start of ignition can even be made to occur after the half-wave maximum, i.e., it can be delayed up to 180°. Rectifiers of this type can be controlled down to the 0 value. The dotted line U_{gz} in Fig. 83, which lies below the positive half-wave U_a, represents the negative minimum grid bias voltage at which the tube will still ignite

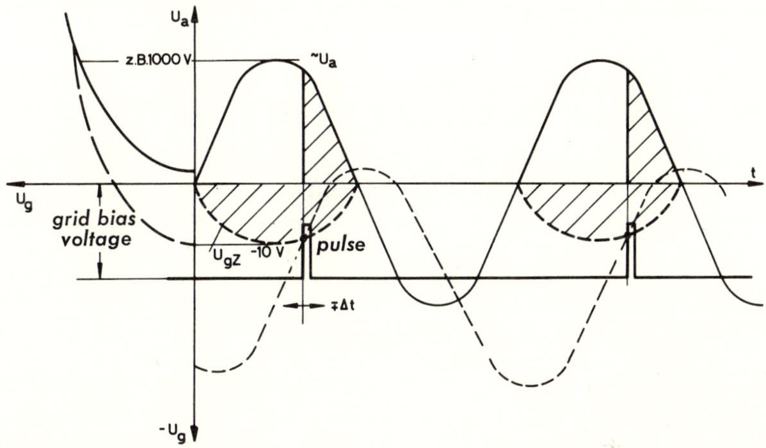

FIG. 83. Control of a thyratron by phase-displaced pulses or by phase-displaced a.c. voltage.

at a given instantaneous anode voltage. The ignition zone is hatched there. Here also, inside the half-wave, we see the typical characteristic of the thyratron in that for different anode voltages the grid bias voltage must also vary for the arc discharge to start. The dotted line, the so-called ignition characteristic, can be obtained graphically or by computation directly from the control characteristic. If a pulse occurs in the hatched area, ignition follows immediately. Instead of a pulse it is possible to use a continuous a.c. control voltage (dotted line) for the grid. By shifting the phase position in relation to the anode voltage, by means, for example, of a potentiometer and capacitor in accordance with Fig. 84, it is possible to set all the ignition angles between 0° and 180°. At 180° the rectifier output is exactly 0 and is at a maximum at 0°. The disadvantage of this sine-shaped a.c. voltage control as opposed to pulse control is that the ignition characteristic cannot be sharply defined. The ignition start is therefore blurred and no longer clear. Pulse generation is a circuit technique and will not be discussed further within this volume. The pulse

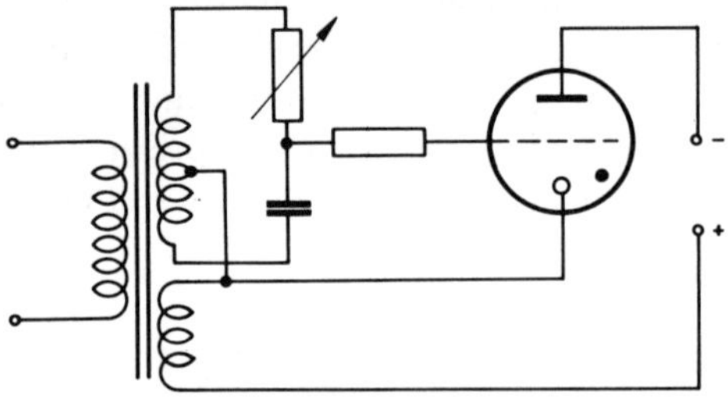

FIG. 84. Circuit for phase control of alternating grid potential.

width, however, must fall within the magnitude of the ionization time. In contrast to the push-pull arrangement, Fig. 85 illustrates the so-called antiparallel connection, in which two thyratrons are coupled together in such a manner that the controlled output is an a.c. voltage rather than a unidirectional voltage. Each tube takes over one half-wave and a more or less truncated alternating current curve,

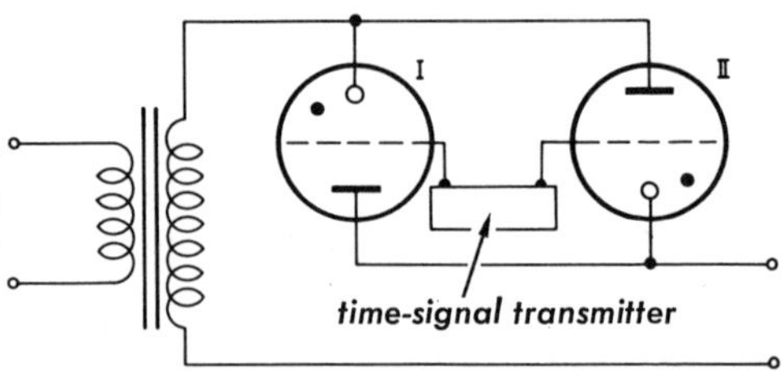

FIG. 85. Antiparallel arrangement for controlling alternating current output.

depending upon the magnitude of the control applied, is obtained on the output side, as shown in Fig. 86. The average a.c. voltage or power is electronically regulated in accordance with the magnitude of the section cut off from the half-wave. In industry the thyratron has found many different types of uses. It is commonly used for varying the brightness of fluorescent tubes. Because of the requisite ignition voltage it is impossible to reduce the brightness by voltage reduction, and it is therefore necessary to regulate it by shortening the length of the intermittent combustion periods. Thyratrons are also used to regulate the speed of drive motors. Closely related to the thyratron is the so-called ignitron shown in Fig. 87. It is also a mercury-vapor-filled discharge tube; however, it contains a cold cathode. The primary emission arises from a liquid mercury cathode at which, after ignition has been initiated, a high temperature focal spot is created as a result of the impact of positive mercury ions upon the liquid

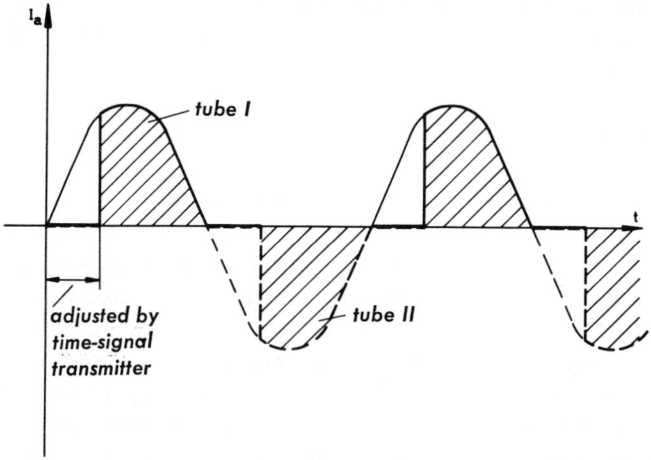

FIG. 86. "Truncated" alternating current obtained at the output from the antiparallel arrangement.

surface, as in the classical multiarm mercury rectifier. All the ignitron components are enclosed in a metal vessel. The anode, which consists of a large graphite plate, is fastened to the upper end of the vessel, from which it is insulated. The ignitron possesses instead of a grid an ignition pin made of a semiconductor material, e.g., silicon

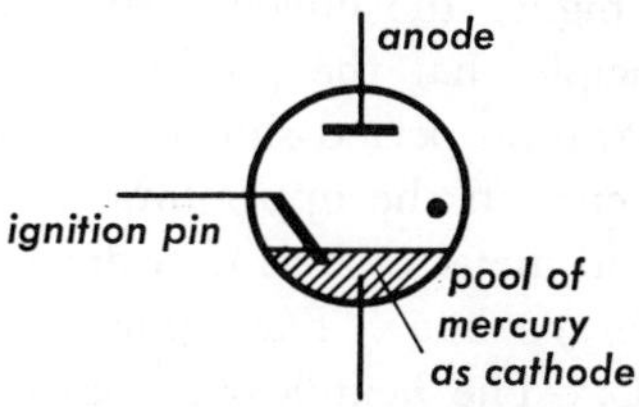

FIG. 87. Basic diagram of the ignitron, which is very well suited for industrial applications. The control function of the grid in the thyratron is taken over by the ignition pin in the ignitron.

carbide, which is immersed in the liquid mercury cathode. If a positive potential is applied to the ignition electrode an arc is created between it and the mercury cathode. This initiates the main discharge between the anode and cathode. Electrons are emitted from the mercury surface. On the way to the anode these electrons collide with molecules of mercury vapor and give rise to ions. The positive ions in turn move to the surface of the mercury cathode, where they cause heating, i.e., they lead to the formation of a combustion basin (extremely hot spot), which then takes over the task of primary emission. It is necessary for the anode to possess a positive charge for major discharge between cathode and anode to be initiated. Once the discharge is triggered it is maintained until the anode voltage drops below the combustion voltage of about 10 volts. The ignitron is relatively simple to control because ionization never begins until the ignition electrode has received about 150 volts (at several amps). In the absence of the ignition pulse the tube acts as an insulator. The ignition pulse is applied at the end of each half-wave. If the voltage to the ignition electrode is interrupted in the course of operation, the current flow is maintained until the last ignited half-wave falls below the combustion voltage. The mercury cathode is considerably more rugged than the hot cathode of the thyratron; it can take peak currents of up to 2000 amps. The average current values lie in the vicinity of several hundred amps. The characteristic gives, in addition to the maximum operating voltage (to prevent arc-backs), which is usually between 500 and 2000 volts, the permissible anode current as a function of the duration of current flow. For a long life the tube should not be thermally overloaded. In spot and seam welding, extremely high peak currents of short duration are required. The waiting periods between the current pulses are, however, long. Consequently, a favored practice is to express the duration of the current flow as a percentage of the so-called integration time. For large ignitrons the integration period is about 10 seconds. From thermal considerations it follows that the shorter the circuit duration, the higher the current can be. In general it can be taken as a principle that the product of the average anode current and the integration period should not be exceeded. The permissible anode current sets the upper limit here. Ignitrons are housed in double-walled metal casings in order to dissipate the heat losses by efficient water cooling. The appropriate operating temperature is about 40°C. The heat loss to be dissipated is approximately 10 to 15 volts times the current intensity, the latter value being extremely high. If a thyratron is connected in series with the ignition electrode of an ignitron, infinitely variable control of the ignitron can be obtained easily by adjustment of the ignition delay (ignition angle adjustment) in the grid of the thyratron.

Two types of ignitron are distinguished: the so-called welding ignitron, which is suited for high currents and low voltages; and the rectifier ignitron, which is suited for high voltages and high average current values.

In addition to being used as controlled rectifiers, ignitrons are also frequently used in antiparallel arrangement as switches for high alternating currents. They are used primarily to replace mechanically controlled switches and are particularly well suited for high time-on percentages. The excitron is a variation of the ignitron. It possesses an auxiliary anode in addition to the mercury cathode, the anode and the ignition electrode. The ignition electrode is used only to set the tube in operation initially; afterward the hot combustion focus on the surface of the cathode, which is the source of emission, is maintained by the auxiliary anode. The auxiliary anode is therefore similar to the grid of a thyratron. Also related to the ignitron is the senditron.

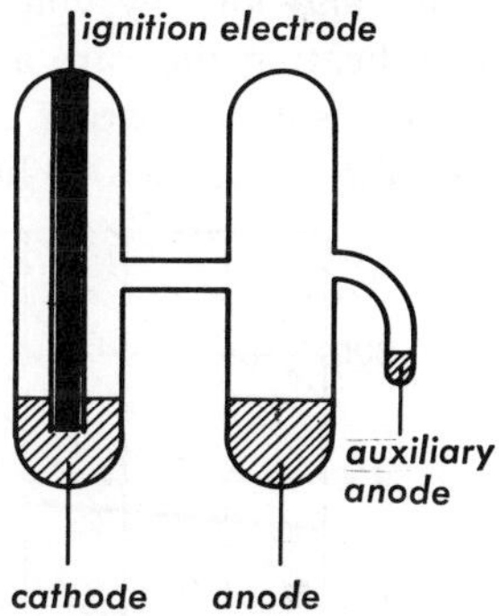

FIG. 88. Basic diagram of a senditron.

This is illustrated schematically in Fig. 88. It consists of an elaborately shaped glass tube in which the anode as well as the cathode is liquid mercury. An ignition electrode, which is completely enclosed in a nonconductive coating, is immersed in the liquid cathode. The ignition is capacitive and results from the development of a strong electric field between the ignition electrode and the cathode. As compared to the ignitron the control current is very low and the voltage is high, amounting to about 10,000 volts. The auxiliary anode, which is placed to the right of the main anode, is also composed of mercury. A high voltage is applied to this auxiliary anode so that the tube will ignite immediately and remain ignited if the main current, i.e., the external load, drops below a certain minimum. By lesser or greater delay of the ignition within the positive half-wave it is possible to readjust the mean current value and switch it on again. The senditron is used in industrial electronics as a switch for heavy currents of the order of several thousand amperes. In welding engineering there is a demand for heavy currents with short switching duration. Ignitrons and senditrons regulated by electronic time switches can satisfy these requirements.

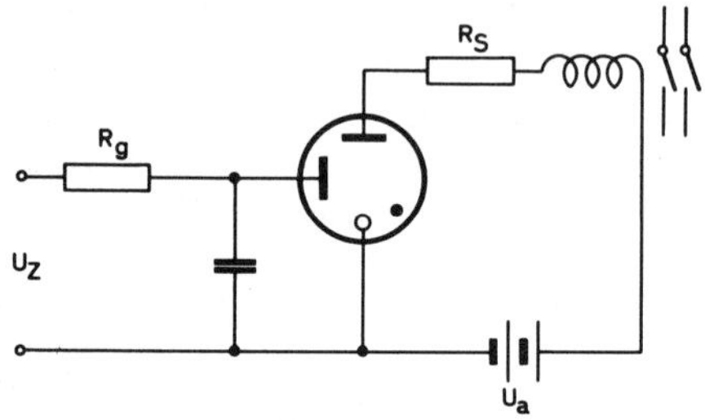

FIG. 89. Circuit arrangement with a gas-discharge triode (frequently called a relay tube).

Gas triodes are not only designed for use in heavy current engineering, but also for low voltage and power uses. Since they do not require heating they are always ready for use and do not result in power losses from heating. They are particularly well suited for use as long life relays (30,000 hours) and are used in warning and alarm

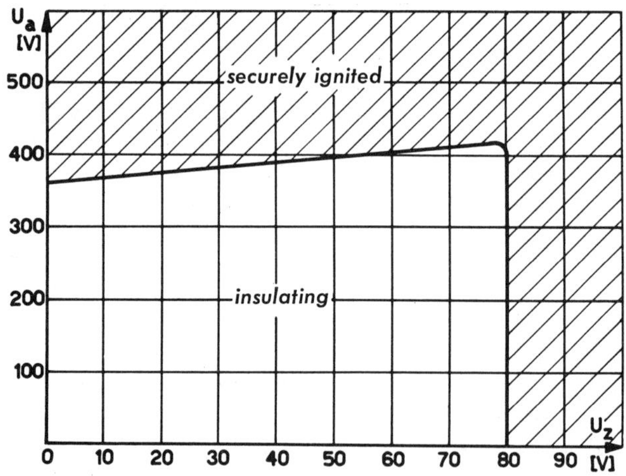

FIG. 90. Characteristic operation of a cold-cathode gas-discharge triode.

units and metering relays. These so-called cold-cathode relay tubes are filled with rare gas. The cathode is made out of pure metal, most often molybdenum. Figure 89 gives a basic circuit diagram with a gas triode. The anode is placed at a distance from the cathode, and, therefore, as shown in Fig. 90, must be supplied with an ignition voltage of over 400 volts to become conductive. The ignition or starter electrode, which corresponds to the grid of a thyratron, is placed very close to the cathode. At 80 volts the electric field between the ignition electrode and the cathode is sufficiently strong to initiate the self-induced impact ionization, which, in turn, gives rise to the main discharge. The anode, which possesses a large surface area, is capable of carrying an anode current of about 30–100 mA (depending upon the type of thyratron used) without overheating. This considerable current flow can be initiated by an ignition voltage of only 80 volts and an ignition current as low as a few μamps. To

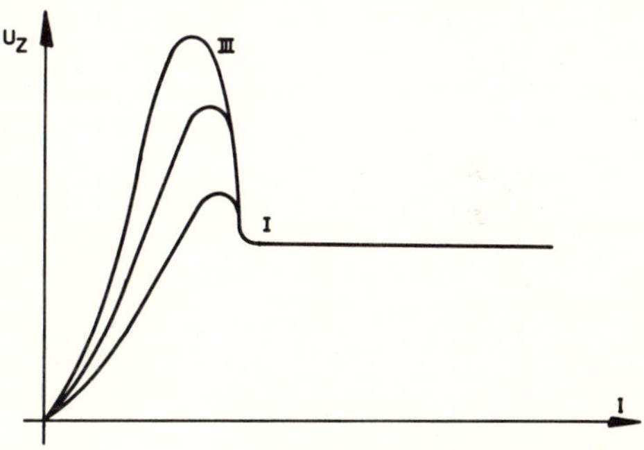

FIG. 91 (RIGHT). Current-voltage characteristic (ignition voltage for three different preliminary ionizations).

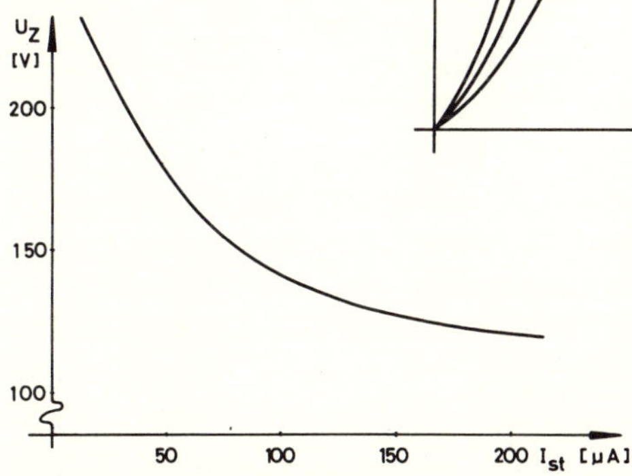

FIG. 92 (LEFT). Control characteristic of a gas triode. The ignition voltage as a function of the starting current.

ensure that ignition is initiated only by the auxiliary anode, the anode voltage is kept 30% lower than the anode ignition voltage. The supply voltage lies between the ignition voltage and the combustion voltage. If the control zone is ignited a preliminary ionization process sets in, which, depending upon its intensity, can bring the ignition voltage down almost to the combustion voltage. In terms of Fig. 65, this would mean that the lower hump in the curve between points III and I is increasingly levelled as the control or starting current is increased. Figure 91 illustrates these relationships. In practice the control current lies between 30–150 μamps (Fig. 92). If such a current is not available, as in Fig. 89, a capacitor that is charged with a slight current (10^{-6} to 10^{-12} amps) and that will discharge in pulse fashion as the starting ignition is attained is placed between the starter and the cathode. Such a tube operates as an electronic fine relay. When connected into the anode circuit it is capable of controlling high ohmic mechanical relays, meters, etc. The control voltage can also be supplied in the form of a short pulse. Here, the ionization time must be taken into account. If the cathode is designed with a photoemissive layer, it can be used to control a relay on the incidence of a light ray. Generally, however, separate circuit elements consisting of a photoelectric cell and a gas triode are preferred. Another type of relay tube in which the anode is coated onto the glass envelope is ignited with the use of a capacitor in connection with a high frequency or some other alternating current field. As long as the control signal is not continuously present the tube is extinguished in a.c. operation at the end of each half-wave or when the voltage drops below the extinction voltage of about 80 volts. In d.c. operation the tube works as a holding contact in the relay; the anode voltage must be momentarily interrupted. Here the deionization period must also be taken into account.

It is all right to connect several tubes in series, but as a rule they may not be connected parallel because of the combustion voltage scatter. When several tubes with the same stabilization range but different combustion voltage are connected in series, the overall combustion voltage is the sum of the individual combustion voltages.

Cerberus Männedorf has developed a new cold-cathode tube for low control voltages, the so-called glow-thyratron (*Glimm-Thyratron*), which is shown in Fig. 93. Cold-cathode tubes with pure metal cathodes have since their introduction proved themselves as very long-lived and reliable circuit elements. For certain purposes, however, the high voltage required for control is a disadvantage.

The new glow-thyratron eliminates this disadvantage. It can be controlled with voltages as low as about 5 volts and, moreover, possesses all the special advantages of cold-cathode tubes, such as freedom from temperature-dependence, visibility of the switching condition, small tolerances, stability, and ability to withstand overloading.

In the relay tube a control discharge must be ignited to initiate the main discharge, whereas in the glow-thyratron the control discharge is maintained in continuous combustion as an auxiliary discharge and its controlling effect upon the main discharge is regulated by a grid.

The control mechanism of the glow thyratron is illustrated in Fig. 93. Between the auxiliary cathode HK and the (main) cathode K an auxiliary discharge, in which the cathode acts as the anode, is maintained continuously. Electrons from the auxiliary discharge pass through an opening in the cathode into the space between the cathode and the grid G. This space is controlled by a composite field consisting of the actual field of the grid and the anode field that penetrates the grid aperture. If the grid is sufficiently negative with respect to the cathode, the electrons, as may be seen in Fig. 93, enter a retarding

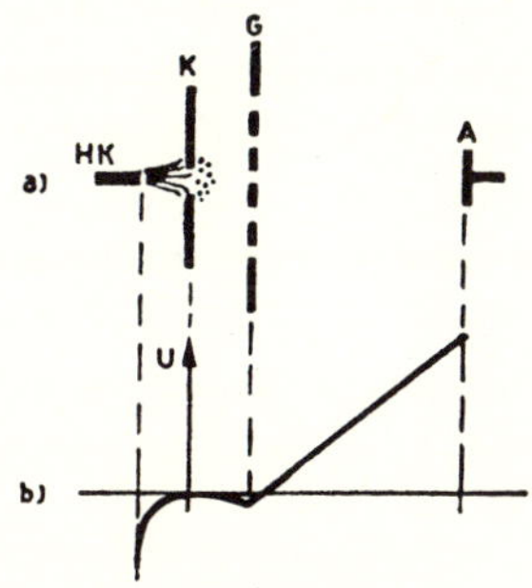

FIG. 93. (*a*) Schematic diagram of the arrangement of the electrodes of a glow-thyratron; HK is the auxiliary cathode, K the cathode, G the control grid, and A the anode. (*b*) Potential distribution along the axis of the tube system with a negative grid.

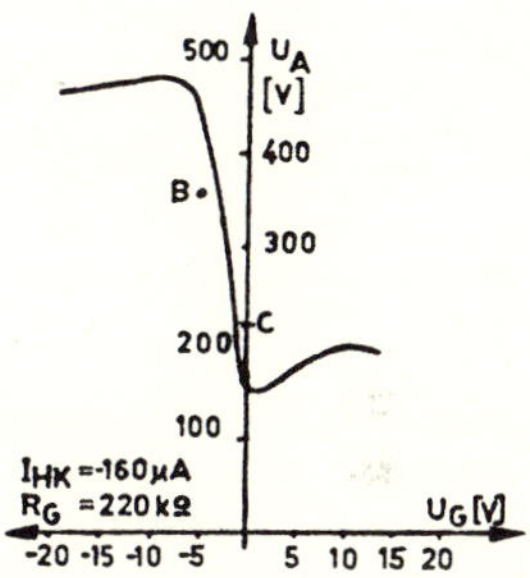

FIG. 94. Control characteristic of a glow-thyratron. B and C are control points of the border characteristics. To the left and below the characteristic, the tube is blocked. To the right and above the characteristic, the tube is conducting.

field, which delays them and drives them back to the cathode. If the grid is less negative, the electrons from the auxiliary discharge enter an acceleratory field, which accelerates them through the grid to the anode A, whereupon they ionize the gas in the space grid. This gives rise to a discharge between the auxiliary cathode and the anode, which in turn ignites the discharge area between the cathode and anode.

The control characteristic in Fig. 94 shows the anode voltage necessary to ignite the tube as a function of the grid voltage. The left side of the characteristic shows the tube action suppressed and the right ignited. Points B and C are control points for the limiting characteristics.

When the tube is ignited the grid (like the grid of a thyratron or the starter of a relay tube) loses its controlling effect. To extinguish the tube the anode potential must drop below the combustion voltage. If the power source supplies an a.c. voltage or half-waves, this happens automatically toward the end of each positive half-wave. The combustion voltage (voltage drop of the ignited tube) is determined by the type of discharge and cathode. In a glow-thyratron, which utilizes a glow discharge, this is 115 volts for a molybdenum cathode.

A further development consists of gas-discharge tubes with additional supplementary electrodes. The most familiar of these are the tetrodes, which contain a so-called auxiliary anode in addition to the starter electrode. A tube of this type is illustrated in Fig. 95. However, thyratrons containing a control screen as the fourth electrode are also produced for work with heavy currents. Here we may also speak of a gas tetrode; in this case the screen does not have the same function as in the vacuum tetrode. Figure 80 depicts the ignition characteristics of a thyratron, in which the screen voltage U_2 appears

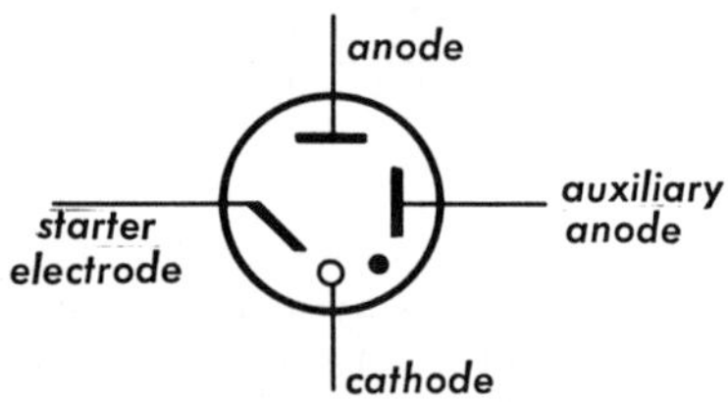

FIG. 95. Cold-cathode tetrode.

as a parameter. The ignition characteristics can be shifted to the right or left by the application of a voltage U_2 to the screen. It is therefore possible while using the same type tube to shift to a certain degree the control characteristic into the positive or negative zone. An interesting gas-discharge tube is the so-called coincidence thyratron, which is designed for use in counting chains. Its two grids are wound together in such a manner that the anode current cannot flow unless both grids are unblocked. There are other special purpose tubes containing still more additional electrodes. In certain numerical counting tubes a number of electrodes are arranged at intervals so that with each new impulse another stage is ignited or the first stage leaps across to the next. A simple decade counter is obtained when ten counting electrodes are introduced into a single envelope. The individual glow-electrodes can be designed in the form of numbers or areas in a round number system somewhat similar to that on the face of a clock.

The so-called numeral-indicator glow tube (multicathode glow tubes) are used for many purposes. These possess a common anode and ten cathodes in the form of the numerals 0–9 placed closely one above the other. When a particular cathode is switched on, the corresponding numeral is illuminated.

A new type of gas-filled counter tube with a maximum counting frequency of over 1 Mc/s has been developed by H. von Gubelberg (Elesta). The electronic counting mechanism is illustrated in Fig. 96. The counting tube EZ 10 contains a disc-shaped anode and twenty identical cold cathodes arranged in a circle. Every second cathode acts as a main cathode with a number from 0–9. These main cathodes are led separately out of the glass envelope and connected across a 6800-ohm output resistance to the negative pole of the current source. The remaining ten cathodes, the intervening auxiliary cathodes, are connected across a 20–50 kilo-ohm resistance to the negative pole of

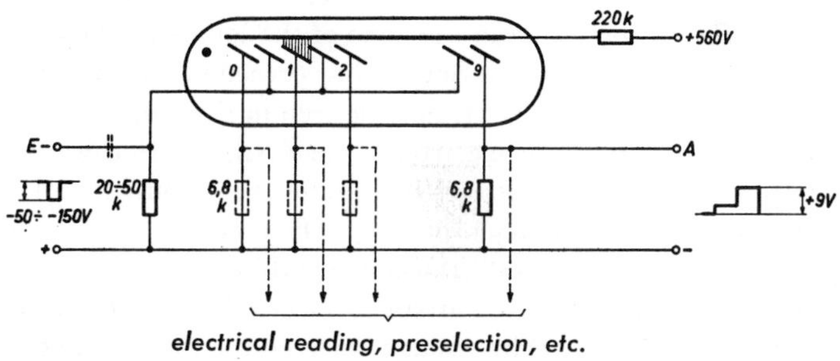

FIG. 96. Rare-gas counting tube.

the anode current source. The anode voltage passes across a 220 kilo-ohm resistance to the common disc anode. If the tube is to count, negative input pulses must be impressed upon the parallel connected auxiliary cathodes. This increases the intensity of the electric field between them and the anode. Let us begin by assuming for the purpose of discussion—as shown in Fig. 96 by the hatching—that the glow discharge has already occurred at main cathode 1. The blue illumination can be seen from outside through a hole in the anode plate, and it is designated as number 1. Since the anode voltage is 560 volts and the combustion voltage of the gas discharge between the main cathode and the anode is 280 volts, the voltage drop at the anode resistance is figured as 280 volts. The auxiliary cathodes which, in accordance with the design, are placed before and after each main cathode, are so arranged that they are in the immediate area of the discharge and—as a result of the probe effect—receive a small ion current.

This type of voltage divider creates a bias of about +20 volts at the input resistance, as a result of which the electric field between the anode and this auxiliary electrode is reduced in intensity, thus preventing the discharge from spreading to the next auxiliary cathode. Ignition of the neighboring main cathode is also impossible because this would require a higher voltage than the combustion voltage, i.e., the so-called ignition voltage. The glow voltage at cathode 1 can therefore be considered as stable. Counting occurs when the tube is supplied with rectangular negative pulses of the order of magnitude of 40–150 volts. The pulses must be of greater magnitude than the positive bias of the auxiliary cathodes. During the pulse the strength of the electric field between the auxiliary cathodes and the anode is increased. A discharge must therefore be initiated between an auxiliary cathode and the common anode. Both of the two neighboring auxiliary cathodes lie in the zone of the ionization field of the ignited main cathode 1, so that neither has precedence over the other in this connection. However, as may be seen in Fig. 96, as a result of the oblique placement of the electrodes, the auxiliary electrode on the right projects directly into the existing discharge. Consequently, the ignition of the right auxiliary electrode is favored.[1] An increase in the voltage drop at the anode resistance stabilizes the electric field between the parallel connected auxiliary electrodes and the anode to the combustion voltage. The voltage between the main cathode and the anode then drops below the combustion voltage, with the result that the main cathode 1 is extinguished. The presence of the negative pulse causes the discharge to pass from cathode 1 to the auxiliary electrode on its right. This is the first stage of the counting process. The second occurs on cessation of the pulse. As a result of the voltage drop at the input resistance a positive potential reappears at the auxiliary cathode as before. The adjacent main cathode number 2 on the right extends into the illuminated area of the above auxiliary electrode. The main cathode possesses no positive bias, so that its potential is higher and, consequently, the discharge spreads to it. Thus the reception and disappearance of pulses results in counting in a left to right or clockwise direction. As a result of the circular arrangement the number 9 is followed by a return to the number 0, and the output pulse of about 7 volts can be used to control the following group of ten. Since the ten main cathodes are attached to

[1] Auxiliary electrodes of this type are frequently called lure electrodes in the literature.

separate leads outside the glass envelope it is possible to obtain a preselection or a resetting to zero by means of a brief application of a negative potential of about 120 volts to the desired cathode. Switching off all the cathodes also causes a resetting to zero. This can be done by a push button. Counters of this type, which do not require complicated or delicate circuitry, are favored for use with photoelectric cells, i.e., with photoelectric scanning devices.

In vacuum tubes, on the other hand, the current is carried by electrons and the illumination must be produced indirectly by incorporating fluorescent materials that are excited by the impact of the electrons. The trochotron is a counter tube in which an electron beam is guided along the path of a cycloid or trochoid. It contains ten counting electrodes coated with a luminescent material. By varying the voltage relationships it is possible to make the beam excite the luminescent material in stages, thus making counting possible.

In electronics ion tubes and electron tubes are frequently found in competition for the same applications, but in some cases they supplement each other. Thus, there are, for example, power amplifiers with gas-discharge rectifier tubes, glow signal lamps and high vacuum amplifier tubes.

IV. CATHODE-RAY TUBES AND OTHER SPECIAL HIGH VACUUM TUBES

From the previous chapters it can be seen that the anode current in vacuum tubes can be influenced in the most diverse ways. Figure 97*a* illustrates a conventional grid control, where, for example, the electron current can be increased by means of a positive voltage or diminished by means of a negative voltage.

In the cases illustrated by Figs. 97*b* and 97*c*, the electron current can be deflected from its trajectory to the anode by means of a

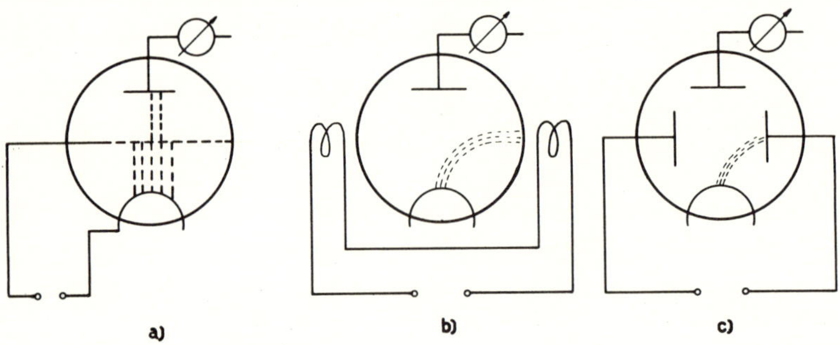

FIG. 97. (*a*) Conventional grid control. (*b*) Static grid control. (*c*) Magnetic grid control.

magnetic field. The anode current in this case can be varied by variation of the magnetic field. Similar effects are obtained by the use of electromagnetic control as shown in Fig. 97*c*. These methods of control can be put to use in electronics. If the anode is designed as a hollow metal tube (hollow cylinder) axially oriented in the direction of the electron path rather than as an intercepting plate, the electrons pass through this anode until they are absorbed somewhere, possibly by a fluorescent screen or converted into some other form of energy (light). An arrangement consisting of a cathode and a cylindrical anode axially oriented in the direction of the beam is known as an electron gun. Such an arrangement is used in a cathode-ray tube, and all three methods of influencing the anode current are used in a color television picture tube. We will first take up the simpler type in greater depth.

In the cathode-ray tube (also known as an electron beam tube, an oscillograph tube, or a Braun tube after its inventor Ferdinand Braun [1880]) electrical energy is converted into light. Figure 98 illustrates the principle of such a tube. The apparatus is enclosed in a glass vessel which has been evacuated to about 10^{-6} mm Hg. Out of

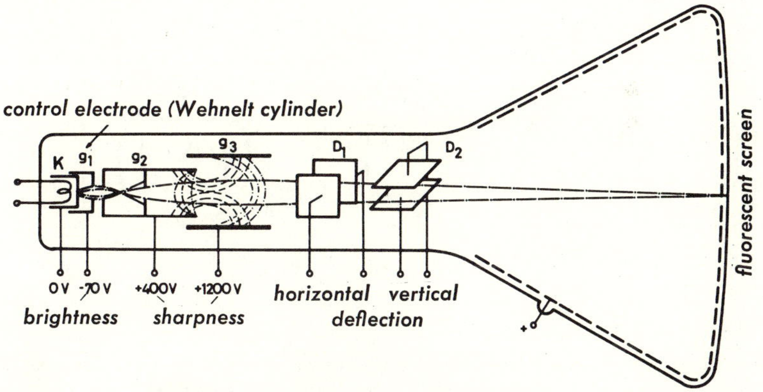

FIG. 98. Electrode arrangement in a cathode-ray tube. Focus and control of the electron beam is by electrostatic fields.

considerations of strength (danger of implosion) the front end of the tube is arched outward—as in a bridge—because a diameter of 30 cm, for example, must resist an air pressure of 680 kg. Electrons are emitted thermally at the cathode K. The electrodes g_2 and g_3, which have the form of hollow cylinders, accelerate the electrons with their positive potential and focus them into a sharply defined beam in the direction of the axis of the tube. Because the effect is produced by a stationary electric field this is known as electrostatic focussing. The field formed by the electrodes acts as a sort of lens for the electrons, and the area electronics that is concerned with the movement of electrons under the influence of electric and magnetic fields is generally referred to as electron optics. There is an actual analogy in geometric optics. When light falls upon a lens the rays at the edge are more strongly refracted than those closer to the center. The ray that follows the axis remains completely unaffected. Lenses for "electron beams" consist of axially symmetrical, electric or magnetic fields whose strength increases progressively as the radial distance from the axis is increased. As may be seen in Fig. 98, the focal point determined by the grid is projected upon the screen by this lens. The electrons focussed in a beam are projected against the opposite inner wall of the expanded face of the tube. This is coated with a luminescent material that lights up when struck by an electron beam. The color of the light can be varied by appropriate choice of materials. The preferred color is usually green—or blue for photographic purposes.

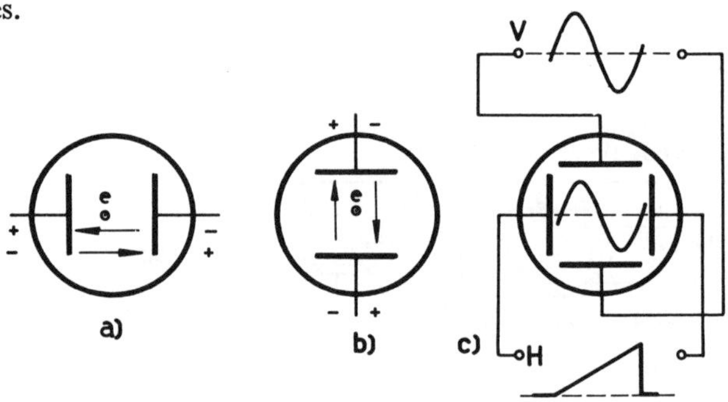

FIG. 99. (*a*) Deflection of the electrons traveling in a path perpendicular to the picture plane by means of horizontal deflection plates. To obtain a time-proportional deflection, the voltage must vary in proportion to the time. (*b*) As above, but with vertical deflection plates. (*c*) The horizontal system is supplied with a saw-tooth, so-called sweep voltage and the vertical system is supplied with the voltage to be tested.

The bright, white image seen on the television tube is usually based upon the combination of two silver-activated luminescent components (zinc sulfide for blue and zinc-cadmium-sulfide for yellow).

The intensity of the electron beam and, consequently, the brightness of the spot as well can be adjusted, depending on the magnitude of the negative potential with respect to the cathode, with the help of a control grid g_1 (Wehnelt cylinder) consisting of a perforated metal sheet (diaphragm). If the grid g_1 is strongly negative with respect to the cathode K, the electrons cannot pass through the exit aperture. The electron beam or the light spot on the screen is suppressed. Moreover, between the electron gun and the fluorescent screen there are situated two pairs of plates mutually dephased by 180° for horizontal (D_1—from the English expression "deflection factor" in volt/inch) and vertical (D_2) deflection. The position of the spot on the picture screen (Figs. 99*a* and 99*b*) can be varied at will by means of the two pairs of deflection plates. To prevent a zero error it is necessary to set the mean voltage at the deflection plates equal to the anode voltage. If one of the two plates has a positive potential with respect to the other, the electrons ($-$) are drawn in that direction or repelled by the negative plate. As long as the electron is passing through the capacitor it is subjected to the constant force $K = e \cdot E$ and since $K = m \cdot b$ (where m = constant) it receives the acceleration b perpendicular to the axis. The cathode ray that is deflected describes a trajectory parabola in the capacitor field. When it leaves this field it travels in a rectilinear path once more. If the horizontal deflection plates, for example, receive a variable voltage that guides the electron beam slowly and uniformly from left to right, let us say, across the picture screen and then quickly back to the starting point, it is possible to use this saw-tooth or sweep voltage as a linear time scale. If the voltage to be measured is fed to the vertical deflection system and the saw-tooth voltage, which is time-dependent, i.e., its frequency can be adjusted, is supplied to the horizontal deflection plates, the point of light describes a time functional curve (an oscillogram) such as is generally used in electrical engineering for alternating current processes. If in addition the time of one horizontal sweep corresponds to the period or to an integral multiple of the period of the alternating current or the oscillatory process under investigation, the curve will always begin at the same point as before and an apparently still picture will be obtained (Fig. 99*c*). If the period of the alternating current is not in agreement with the deflection frequency, the curve picture will wander forward or backward in accordance with the relative positions of the sweep and alternating current frequencies. Certain circuit arrangements make it possible to synchronize the saw-tooth or sweep frequency with the alternating current being tested. This ensures a still picture. In order to obtain a simultaneous image of two processes such as the current and voltage curves, so-called two-beam or multibeam oscillograph tubes are used. These combine several beam systems in the same tube. Under certain conditions, however, it is possible to obtain at least apparently simultaneous pictures of two different processes with one single-beam cathode-ray tube by making use of an electronic switch by means of which the beam is periodically switched back and forth between two separate inputs.

The tubes are commonly protected from disturbances from foreign magnetic fields by mu-metal cylindrical shields.

A cathode-ray oscillograph is considered good if the deflection sensitivity, upper frequency limit, brightness, and accuracy of measurement (absence of spot persistence or distortion) are high. Today cathode-ray tubes themselves without supplementary amplifiers possess a deflection sensitivity of 3–4 volts/cm beam deflection. In accordance with the general formula, this sensitivity can be attained only if the deflection plate distance d at the site of entrance of the electron beam is small, i.e., if the field strength E is relatively great ($E = U/d$). Moreover, the electrode length l along which the electron beam is influenced should be as long as possible (Fig. 100*a*). In a pointer instrument the lever principle applies, and the longer the pointer, the greater the amplitude of movement. The free, rectilinear

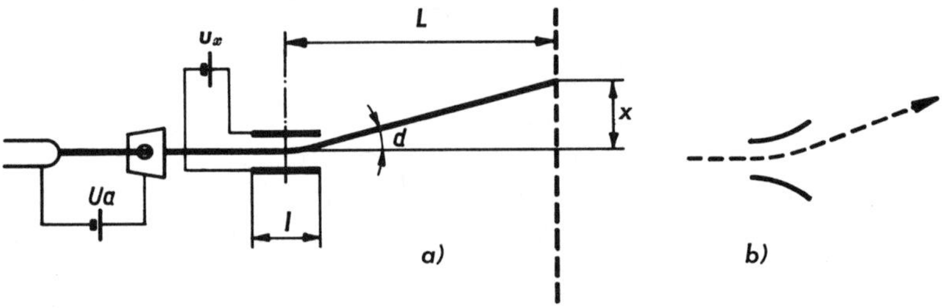

FIG. 100. (*a*) Ray geometry of the deflection formula. (*b*) Bending the deflection electrode upwards permits a greater direction outward of the beam.

electron beam between the deflection plates and the fluorescent screen can be compared to such a pointer. The deflection plates must therefore be placed very close to the electron gun. Frequently, however, two sets of deflection plates are desired (television). In that event the second pair must necessarily be placed somewhat further away, and the deflection sensitivity for the second pair is therefore less than for the first pair. In order to obtain a wide scale of deflection the angle between the deflection plates, which were previously placed parallel to each other, can be opened up as shown in Fig. 100*b* to prevent them from cutting off the electron beam on its passage through. The upper frequency limit is determined by the relationship of the magnitude of the anode voltage (velocity of the beam) to the capacitance of the deflection plates, which act as capacitors. Today the rise time from 0 to full deflection is about 20 nanosec (nanoseconds = billionths of a second). For all practical purposes these tubes operate with almost no lag when they are not connected to other circuit elements such as amplifiers.

The factors determining the brightness, aside from the excitation, are primarily the intensity of the beam or the beam current (usually of the order of magnitude of μA) and the magnitude of the anode voltage, since the product of these is equal to the energy per unit of time. The intensity of the beam current is frequently controlled for the bright and dark picture points as in television, but the acceleration is practically constant. For a bright point of light the acceleration energy must be high. Consequently, a high anode voltage must be aimed at. However, the brightness can also be improved by certain tricks. Thus, it is possible to place a thin aluminum foil behind the fluorescent screen and cause all the reflected fluorogenic light rays within the tube to be reflected back toward the observer. Naturally, the electrons must possess a sufficiently high kinetic energy to get through the aluminum layer without being absorbed by it.

The measurement accuracy depends upon, among other things, the sharpness of the point of light. The individual electrons within the beam repel each other because they possess charges of the same sign. This effect is diminished if the electrons have a high traveling speed, and thus a high kinetic energy, which is another reason that a high voltage is required. A high anode voltage, however, reduces the deflection sensitivity since it is a conventional practice to place the anode or the so-called electron gun in front of the deflection system. Consequently we have to resort to another trick, which allows the use of a lower anode voltage and a high deflection sensitivity, and at the same time eliminates the previously mentioned obstacles. The electrons are passed through the deflection electrodes at lower velocity and then pass through another electrode at a high voltage; this extra electrode acts as a secondary accelerator for the electrons (Fig. 101). Most frequently this is accomplished by providing the inside of the tube with a conductive graphite coating. This layer, which coats the inner circumference of the tube, is broken into two separate ring-shaped sections, one at the front of the tube and one at the back. The section closer to the fluorescent screen is connected to the secondary acceleration potential. The back section, close to the electron gun, receives the potential of the anode voltage. These "conical-surface electrodes" along the glass wall possess different sharply separated voltages, which necessarily results in the formation of an electronic lens, which easily bends the beam once again toward the center. This secondary acceleration reduces the deflection sensitivity by about 10 to 20%. To avoid steep potential leaps in the secondary acceleration system, it has for some time been found preferable to coat the inner wall of the tube with a spirally shaped resistance layer, which makes possible a continuous voltage increase. As Fig. 101 shows, the secondary acceleration voltage is about six times as high as the anode voltage. The use of this method makes it possible to obtain great brightness in addition to a high deflection sensitivity. The upper frequency limit is very high, and consequently this design has proved very successful. Not only is it possible to obtain cathode-ray

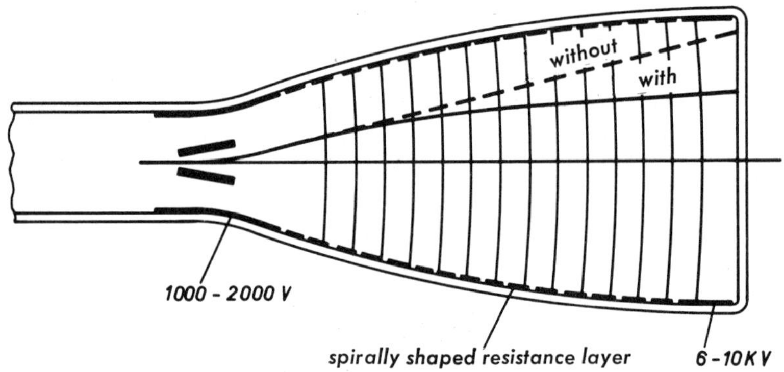

FIG. 101. Electron beam without and with secondary acceleration.

tubes with different color displays, but they can also be obtained with different afterglow times. A long afterglow time is necessary if the phenomenon to be studied is a nonrecurring event, an oscillogram, or if one desires to retain the display of a slowly recurring event for a certain period of time. There are various types of such tubes, and the afterglow time varies between 1 and 15 seconds. The times during which the display of a plotted curve will still appear visible and can be interpreted in terms of a standard depends upon the illumination of the room and the dark adaptation of the eye, in addition to the electrical data, the screen load and the excitation time. The last of these decreases as the frequency is increased.

The oscillograph is not only used for purposes of measurement; it has also been successfully introduced into industrial electronics. Of the great number of examples we will consider only ultrasonic testing.

Here we are concerned with a pulse echo process. An ultrasound (1–15 Mc/s) focussed in the form of a beam is directed across an electromagnetic transducer, a so-called sound head,[1] into the material to be tested. The ultrasound is not generated continuously but in pulse packets of a transmission duration of perhaps 1/2 microsecond. These pulse packets are repeated every 1/2000 second; they are reflected like light in a mirror at the boundaries of the material, such as back walls, and also by such cracks, pores, and faults as may exist in the material. A second acoustical transducer, or the same one, if the appropriate circuit is employed, receives the reflected pulses and converts them once more into electrical impulses. These signals, which are delayed in comparison to the transmitted signals, are fed to the vertical deflection plates of a cathode-ray tube, where they produce blips on the zero line of the screen. The distance from echo blips to the transmission blips on the left, which appeared initially on the scale, is a measure of the traveling time of the echo, for the trip forth and back, and is consequently proportional to the distance of the fault or the back wall of the material. By means of a mechanical or electronic metric calibration on the cathode-ray tube, it is possible to determine immediately the exact position of the fault. The size of the blips provides evidence of the size of the probable fault. In supersonic echo sounding, which is used to sound ocean depths, the same process is employed. In radar the mechanical oscillations are replaced by a high frequency energy, also in the form of pulses, which are directed against solid objects, e.g., an aircraft. A fraction of the reflection energy, which is received during the pauses between impulses, is amplified in the radar receiver and the signal fed to a cathode-ray tube. Here also the distance measurement radar transmitter–object is a pure time measurement, which can be represented by the nearly lag-free cathode-ray tube. The use of certain circuit techniques makes it possible to suppress the beam current during the flyback. In the previously cited examples most of the measuring processes are periodically repeated.

Cathode-ray tubes have been put to use in clock technology. The signals of a quartz clock are compared with the tick of the clock being tested. In this way the slight differences in speed can be checked.

Other applications have been found in machine construction, where the cathode-ray tube has been used for such purposes as the representation of undesirable machine oscillations. In these cases the tube is

[1] See piezoelectric transducers, p. 256.

connected to a transducer, e.g., a piezoelectric crystal, which first converts the mechanical oscillations into electrical energy.

Continually recurring vibrations of structures are oscillographed in a similar manner. In machine construction and in chemistry the properties of fuels and explosives, i.e., the compression ratio curves, are controlled in a similar manner. The heavy current engineer makes use of it to investigate such phenomena as short circuit shocks and surge pressure. The areas of application in technology are extremely numerous.

In the medical clinic doctors study the activity of a patient's heart action with a cathode-ray tube. In this case the curve is known as an electrocardiogram; however, muscle and brain currents can be represented in a similar manner.

The cathode-ray oscillograph has also proved very useful as a measuring and testing device in the adjustment and repair of television and radio sets.

Cathode-ray oscilloscopes are also used nowadays in tuning internal combustion engines, inasmuch as they provide a visual representation of the relations of injection, compression, and piston position.

In acoustics the concern is with the course of complex oscillations, which, for example, vary within the course of a short build-up time.[1] Instantaneous photographs can only obtain a photographic image of non-recurrent or unsynchronized events. For oscillographic representations of non-recurrent phenomena the anode voltage must be at least 20,000 volts as a result of the required brightness or, possibly, writing speed. The cathode-beam current of the usual tubes is about 0.2 milliamp. The beam electrons can be further accelerated afterwards by the use of an additional positive electrode. Tubes employing such electrodes are known as secondary acceleration tubes. So-called trigger-switches permit recording to begin when the external voltage appears. For time calibration the beam brightness can be periodically interrupted. In this case it does not matter from the viewpoint of the physics involved whether the voltage is regulated at the Wehnelt cylinder or at the cathode. Cathode-ray tubes have consequently found favor in electrical measuring techniques because they can be used to make visible very rapidly elapsing phenomena. The higher the acceleration voltage, the faster the electrons travel toward the screen and the less lag in the operation of the cathode-ray tube.

[1] See the book *Frequenzanalyse akustischer Einschwingungsvorgänge* (*Frequency Analysis of Acoustic Build-up Phenomena*) by W. A. Günther, Juris-Verlag, Zurich.

These tubes are very well suited for reproduction of high a.c. frequencies, e.g., up to many Mc/s. Rapid movements of the spot—which naturally have no connection with the electron velocity—create for the eye, because of its own lag, the illusion of a line or continuous curve. In the television tube, where the point of illumination moves in lines from left to right and from top to bottom over the whole picture screen, the illusion of an illuminated picture surface is created. If the intensity of the electron beam is modulated as a function of the signal received, the actual television picture is obtained. Since such a picture is composed of a rapid succession of bright points (10 million per second),[1] the highest possible acceleration voltage (16–20 kilovolts) must be used. In television tubes just before the electrons reach the fluorescent screen they must pass through a thin conductive metal layer possessing a high positive voltage. The high velocity of the electrons allows them to accomplish this without difficulty. In these tubes the inner reflector has five purposes: (1) It reflects back toward the viewer the light radiated back into the tube. This increases the amount of light transmitted to the viewer. (2) The metal layer, which is attached to a fused-in metal contact that leads to the outside and is connected to the positive pole of the high voltage, generally covers the whole inner surface of the glass as well as the picture screen and consequently also acts as a protective shield. (3) In addition to g_1, g_2, g_3, it is possible to introduce another correction lens whose last electrode is connected with the high voltage of the metal layer. The electrons are thereby enabled after leaving the last electrode to move through a space containing no electric field. (4) The metal coating, placed directly in front of the fluorescent layer, protects the latter from the damaging effect of gas ions. Since the gas ions are heavier than the electrons and therefore possess a lower velocity, they are unable to get through the metal layer and are reflected back into the tube as neutral gas atoms once more. (5) The metal layer dissipates the excess negative electron charge created on the fluorescent screen by the electron beam. The cathode emits negative ions, which have an unfavorable effect as they always strike the same point on the fluorescent screen. After a short period of operation this gives rise to a dark spot. An ion trap is used to prevent this from happening.

[1] For 625 lines a square picture format would be obtained with 625 × 625 picture points, which would give, if the picture was changed 25 times per second, a total of 625 × 625 × 25 picture points. When this is multiplied by 4/3 we arrive at the usual square television picture format.

The ion trap is based upon the fact that an oblique electric field in the electron gun will deflect equally in an upward direction both the electrons and negative ions. A small magnet (ion trap) subsequently bends the easily moved electron beam back to its original direction, but the heavier ions are less affected and end by striking against the wall of the tube.

The desired deflection of the cathode ray in the course of its path to the fluorescent screen can, as was mentioned before, be obtained either electrostatically or magnetically. In the first case the coulombic force is operative, and in the second the Lorentz force: $K = B \cdot e \cdot v \sin \alpha$, where B is the magnetic induction, e the electron charge, v the velocity of the electrons and α the angle between the direction of motion of the electron and the flux lines of the magnetic field. The deflection x of the spot on the picture screen can be computed for the electrostatic or magnetic field (Fig. 24). In the first case the following formula applies:

$$x = \frac{1}{2} \cdot l \cdot \frac{U_x}{d \cdot U_v} \cdot L$$

and for the second case the formula:

$$x \approx 0.3 \frac{B_x}{\sqrt{U_v}} \cdot L$$

where l represents the length of the deflection capacitor plates in the direction of the beam, or the active length of the deflection coil cores in the electron path; L is the distance between the fluorescent screen and the center of the deflection system (center of the capacitor plate or center of the coil core); U_x is the deflection voltage between the plates; d is the distance between the plates; B_x is the magnetic induction in the active air gap inside the electron path; U_v is the velocity voltage with which the electrons leave the electron gun and traverse the deflection system.

The factor 0.3 allows for magnetic scatter, i.e., that the electrons are to travel through a transverse field that is not homogeneous or sharply defined. The lengths are expressed in cm, the voltages in volts, and the magnetic induction B in gauss.

In television tubes magnetic deflection is preferred because it makes possible a greater deflection angle (up to 110°) or a relatively short tube, although electrostatic deflection allows a control with no power loss. The very concentrated magnetic effect upon the electron beam occurs directly behind the electrode system, the coil with the soft, magnetic core (made of the ceramic magnetic material Ferrox cube) being placed at the neck of the tube. The electrons are also deflected in an inhomogeneous magnetic field, so that they will resume their rectilinear flight after leaving the field. To ensure a linear deflection the current used in the magnetic deflection method must be saw-toothed.

To obtain a television picture in a cathode-ray tube there are three variable magnitudes aside from the various fixed voltages. The variable magnitudes are: (1) The voltage between the Wehnelt cylinder and the cathode, which varies the light intensity between black and white. (2) A synchronous, controlled saw-tooth current from the transmitter for horizontal deflection. The frequency for 625 lines and 25 frames per second is 15,625 cps. (3) A synchronous, controlled saw-tooth current from the transmitter for vertical deflection. In the usual interlaced method a frequency of 2×25 cps $= 50$ cps is required. As a result of the combination of the two saw-toothed effects $(2 + 3)$ upon the electron beam, the electron beam follows a line-by-line scanning course over the picture surface, as in reading. The number of lines on the picture screen depends upon the ratio of the two deflection frequencies. During the brief flyback, i.e., at the end of each line, as at the end of the image, the beam current is suppressed by a high negative voltage between the control grid and the cathode, since in both the pickup camera as in the television reproduction only the line of the forward sweep is necessary. The flyback or return sweep would only interfere.

If the electron beam in a Braun tube is not deflected, i.e., if it remains at the same point on the fluorescent screen, this point on the screen will be destroyed or, as they commonly say, burned out. The kinetic energy of the electrons can be increased by increasing the power of the beam, until it leads to a local melting of solid bodies. This electron bombardment effect is put to use in electron-beam welding. The object to be welded is used as the anode in a vacuum chamber. The electrons are fired from a powerful electron gun. By appropriate movement of the piece to be welded the welding seam is guided through the hotspot of the electron beam. In these applications acceleration voltages far over 30 kilovolts and currents of up to a half ampere are used. The entire beam power of many kilowatts is focussed at the beam diameter (the hotspot) of about 1 mm. This area can also be considered a part of industrial electronics.

Figure 102 illustrates the principle of large-picture television projection with an eidophor. A glass plate G placed in a vacuum chamber is covered with a transparent, oily liquid layer F that is practically an insulator and is about 1/10 mm thick. The layer is struck from above at an oblique angle by an electron beam (15 kilovolts, 10 μamp), which is deflected across the layer in a linear fashion in accordance with the television norm, and this gives rise to a charge pattern that varies with the intensity of the video signal.

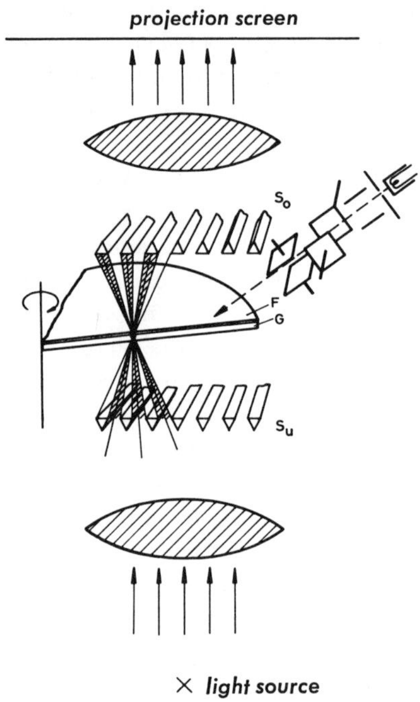

FIG. 102. Principle of the eidophor with two slit-grating diaphragms S_o and S_u.

A high frequency (15 Mc/s) is imposed upon the writing motion of the electron beam; the amplitude of the motion is modulated by the television signal. Along the lines a fine rippling—a raster—is created in the surface of the oil layer. The selected frequency of the a.c. field gives picture elements (raster elements) of the magnitude of about 1/10 mm in accordance with the line width. (The size of the picture on the oil layer nowadays is about 72 × 54 mm.)

The electrostatic forces arising from the surface charges lead to periodic rippling of the liquid layer in accordance with the magnitude of the television signal voltage. At the "dark points" in the receiver picture no electrons have been received and the surface is smooth, while at the "bright points" the oil layer has been struck by many electrons in the rhythm of the raster, as a result of which the surface of the oily layer is intensely rippled. The electrical conductivity and viscosity of the oily liquid are selected in such a manner that the charges and, consequently, the surface deformation, disappear at the conclusion of a complete image change and are ready for reception until the next scanning. In the projection arrangement this liquid film is projected upon the picture screen under special conditions by an intense xenar lamp (several kilowatts) with a concave reflector and an optical system. To ensure that no light reaches the screen when the liquid surface is smooth, two staggered slit-grating diaphragm systems are placed in the path of the beam. In the usual, so-called bright-field projection such as that obtained from the normal dia-projector, a transparent relief pattern such as that of the oil layer would not give a visible picture. Consequently, a so-called dark-field projection system is employed.

In this optical system the light source is first projected upon the (initially smooth) glass plate. In the path of the beam to the plate there is a slit diaphragm S_u with slit-shaped apertures. By means of lenses the diaphragm apertures through which the light passes are projected upon the metal crosspieces of a second slit diaphragm S_o. As a result of this arrangement no light will fall upon the projection screen even when the picture field on the glass plate is completely illuminated. It is different when the oil layer possesses a wave pattern. The projection light is laterally deflected from its normal path by the periodic undulations of the surface and is projected upon the picture screen by means of an objective. The projection image is produced by the refraction of the individual picture elements, with an optical lens system being used to secure perfect correlation of the light points upon the projection screen with the deformed starting points on the surface of the oil layer. The light yield of the eidophor is greater than that of a movie projector. With a maximum amplitude rippling of the liquid layer, i.e., at the maximum light intensity per unit of surface of the received image, practically 33% of the light is obtained upon the projection screen. The contrast, i.e., the difference between the light and dark values, is also great and is about 1:100. The liquid layer, which has to be constantly renewed, is cooled by rotating the glass plate slowly and dissipating the heat. The eidophor, which was developed for the movie theater, possesses a pumping system and must be continuously evacuated during operation. In principle the eidophor is a cathode-ray tube in which a liquid layer—as the light control—replaces the "fluorescent screen."

The eidophor process has been further refined. Now only one grating diaphragm is required instead of two. The diaphragm consists of six individual, reflecting strips arranged in steps as shown in Fig. 103. This reflecting bar system takes the place of the two separate units previously used. The light rays pass through the reflecting-strip diaphragm to the smooth concave mirror. This concave mirror is covered with the transparent oil coating that ripples in accordance with the modulation of the electron beam. The light rays pass

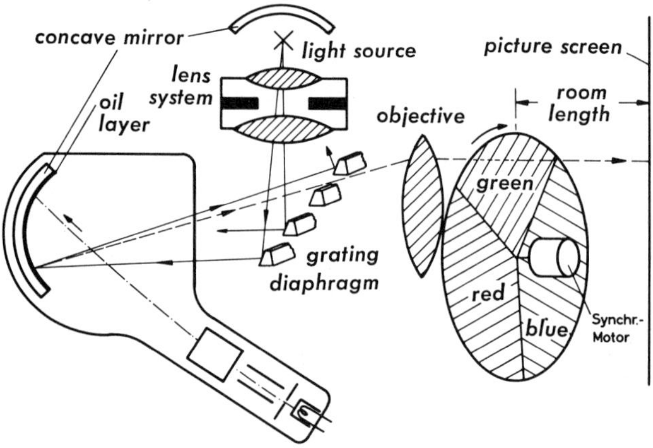

FIG. 103. Modern eidophor with one grating diaphragm.

through the oil coating and strike the concave mirror, which reflects them back to the grating. When the surface of the oil coating over the concave mirror is smooth, i.e., when there is no modulation, the light rays are projected by means of the reflecting crosspieces of the grating diaphragm back to the light source. In this case no light reaches the picture screen. When the modulation signal is present the light rays are bent, pass through the apertures between the grating bars and are deflected to the picture screen. The ripple amplitudes, which correspond to the intensity of the modulation signal, determine the amount of brightness upon the picture screen. With the right arrangement of the oil layer and mirror system, this system requires only half the charge density of the earlier one. This system is also suited for large-picture projection of color television. In front of the objective (or possibly even in front of the projector picture aperture) in the beam path—of both the television camera and the eidophor—there are three or six rotating sectors composed of color filter discs (color sequence process). The color filter wheels rotate synchronously in the transmitter and receiver, and the color filters are also always in the same relative position. If, for example, the interval during which the red filter is in position is singled out, the entire picture must be scanned one whole time from top to bottom and projected. When the lower right end of the picture surface is reached the second filter becomes effective, and the scanning and projection process begins again with the new color. Then the process is repeated all over again with the third color. By means of the combination of the three primary colors, red, blue, and green (the color triangle) all of the colors of the spectrum can be obtained. The sensation of continuous color reproduction is produced by the succession of images in the three primary colors one after another, which is to say, in sequence. The slowness of the human eye creates for the spectator the illusion of a completely colored image. As in black and white television the interlacing process is used, but to avoid flickering it is necessary to use a scanning velocity three times as high, so that all three colors are transmitted within the same picture period of 1/25 of a second.

In America the television industry has agreed upon the use of the so-called simultaneous transmission system, which does not require mechanical rotation with color filter wheels. Home television receivers contain so-called tri-color tubes, which combine three different, individually controlled electron guns within one tube, in which the electrons are directed against a common picture screen composed of fluorescent materials emitting three different colors. This area of applied electronics could only be touched upon here.

The use of electron-beam tubes has spread over so wide an area that whole books could be devoted to this subject alone. Therefore, we cannot attempt to do more than touch upon this area just long enough to obtain an overall view.

Indicator tubes, such as those which have found favor in electronic measuring instruments (bridge circuits) or as overmodulation indicators in audio band instruments or tuning indicators (magic eye) in radio sets, are also simple small cathode-ray tubes. Figure 104 illustrates the principle of one of the common arrangements. The electrons from the hot-cathode K travel to the positive fluorescent screen L, which serves as the anode. The current distribution grid is used to distribute the electrons evenly over the screen. In the path of the electrons are deflection electrodes A, which deflect the electron in accordance with their voltages and thus control a shadow sector, i.e., an unilluminated zone upon the fluorescent screen. Frequently two separate deflection systems are included in the same tube. In this way it is possible to control two sectors, each of which may be given a different sensitivity (see Fig. 104*c*). It is possible, for example, to indicate the tolerance range of the minimum and maximum modulation voltage in a frequency band instrument. The beam responds even to low voltages because the same tube contains a triode system that amplifies the control signal. The amplifier anode is

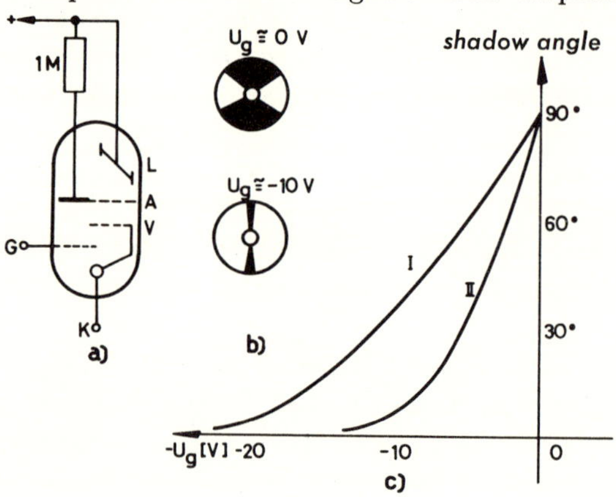

FIG. 104. (*a*) Indicator tubes with an incorporated triode amplifier. (*b*) Light pattern of such a tube. (*c*) Sensitivity curve for the two systems in the same tube.

connected within the tube to the deflection system, so that an indicator tube of this type is actually a combination of intensifier tube and cathode-ray tube.

The illumination of the screen of a cathode-ray tube is based upon the mechanism of luminescence. There are two different types of luminescence: fluorescence and phosphorescence. Fluorescence occurs only during the impact of the exciting electron beams and disappears immediately if the excitation disappears. Phosphorescence is characterized by an afterglow that remains for some time after the disappearance of the excitation, lasting, under certain circumstances, even for years. In this case the complete emission process does not occur immediately after the absorption of the electrons. There are intermediate processes that are responsible for the phenomenon. The mechanism is illustrated schematically in Fig. 105. The figure shows the energy level arrangement of zinc sulfide, a commonly used phosphor for the light screen. The ordinate represents the sum of the potential energies of the electrons in relation to the nucleus. The zinc sulfide is doped with very small quantities of copper atoms (foreign atoms). It is evident from the illustration that we are concerned with an insulator, since the filled band V, i.e., the one where every possible energy level is occupied by an electron, is connected with a vacant gap. The hatched, empty conduction band L is separated from the valence band by an energy of several electron volts. At the sites of the foreign atoms the energy level scheme is modified.

Copper with its single electron in the outermost shell has one electron less than zinc, which has two outer shell electrons. This

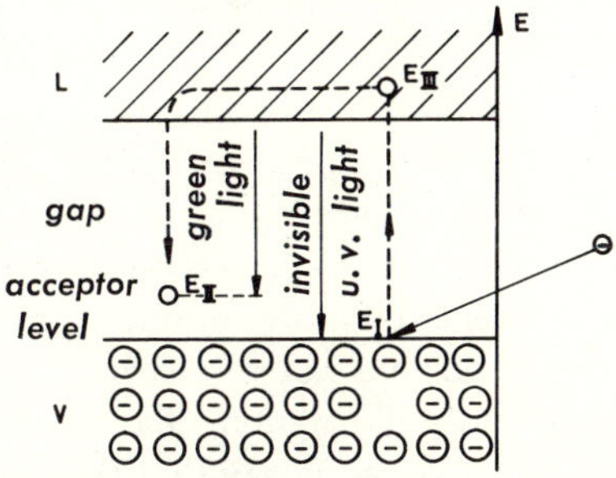

FIG. 105. The physical phenomena involving the valence and conduction bands in luminescence.

level, which is energetically changed by the presence of the copper, is known as an acceptor level. In Fig. 105 this is marked by the point E_{II}. Normally, the energy level E_{II} is empty.

If the luminescent material is bombarded by an electron, the electron transmits energy by means of the impact to the bound electrons of the zinc sulfide molecule. As a result of the energy increase, the electrons of the valence band V are raised to the level of the empty band E_{III}. For the purpose of simplification let us consider a single electron. In the conduction band the electron is mobile. But the hole created in the valence band is also mobile and travels from one atom to another. If this hole approaches a copper atom and if there is an electron in the level E_{II}, the electron is easily captured by the hole because of the slight energy difference, with the result that the hole is neutralized; this is known as recombination. The E_{II} level is once more empty. This unoccupied position becomes a point of attraction for the electrons that happen to pass by in the conduction band above. Occasionally one of the quasi-free electrons that has gotten into the conduction band E_{III} and has accidentally come to be located directly over the empty E_{II} level drops into the empty level. The radiation associated with this situation in the case of zinc sulfide is greenish. This is in contrast to the radiation associated with the energy difference between the conduction and valence bands, which lies in the ultraviolet range. On the other hand, the energy change resulting from the dropping of an electron from E_{II} to the valence band is so slight that only infrared radiation is emitted, with consequent heating of the material.

In many cases a certain lag in the luminescent effect is desired, particularly in tubes where one desires to retain the image over a certain period of time. Such is the case, for example, in radar tubes, where the scanning of the image to be transmitted is quite slow. For this purpose luminescent materials possessing an afterglow of several seconds are used. Consequently, cathode-ray tubes are manufactured with the most diverse afterglow times. Here the return of the excited atoms to the fundamental state is also effected in stages.

If a zinc atom in zinc sulfide is replaced by an aluminum atom, which is known to possess one electron more than zinc in its outermost shell, the foreign aluminum atom gives us an extra electron. This extra electron is not very strongly bound to the atomic aggregate and leads to the formation of a so-called donor level. Metaphorically speaking, we might say that there is no more room for this additional electron in the filled band, so that it must occupy a position in the next empty band (the conduction band). The loss of this electron makes the aluminum atom incorporated into the zinc sulfide ensemble a positive ion, which consequently exercises a small attractive force upon this free electron, with the result that the electron at the site of the foreign aluminum atom must occupy an energy level right below the empty band. This situation is illustrated in Fig. 106 as the donor level E_d. The copper or aluminum impurities deliberately added to the luminescent material are incorporated into the atomic ensemble. The electrons of the upper, donor level E_d are eagerly absorbed by the lower, acceptor level E_a, with the result that the donor level is frequently unoccupied. This makes it easy for the quasi-free electrons that have gotten into the conduction band to drop into the empty donor sites E_d. They are therefore known as trapping levels. Since the energy change resulting from such a drop is slight, there is only a slight infrared radiation emitted, which is likewise absorbed in the form of heat energy by the surrounding atomic lattice. Consequently, more electrons can pass from the energy level E_d into the empty band. This may take several seconds. These traps form waiting positions for the electrons. The length of the afterglow on a luminescent screen is a function of the length of these waiting periods, which occur in accordance with a statistical law. If the electron returns to the empty band, as soon as it arrives over an acceptor level E_a it drops into the available vacancy, with the result that radiation is emitted in the form of the visible greenish light mentioned before. It is possible to use other materials and obtain different energy levels, such as those yielding blue light.

Cathode-ray tubes belong to the group of high vacuum tubes that convert electrical energy into visible light. X-ray tubes, which also

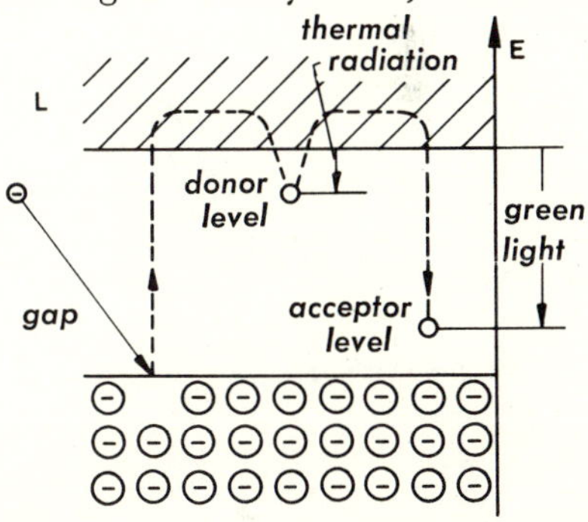

FIG. 106. Mechanism of "afterglow." The donor energy level, also known as the donor level (E_d), is located in the upper portion of the empty gap, and the acceptor energy level (E_a) is located in the lower part.

convert electrical energy into light—invisible light—were discussed on p. 93. But fluorescent lamps can also be considered as part of this group, though they are ion tubes rather than vacuum tubes and are no longer considered electronic elements in the usual sense as a result of their use as light sources.

Another form of luminescence is electroluminescence. The Destriau effect, which has been named after the discoverer of electroluminescence, is exhibited by the glowing of certain substances when they are exposed to a powerful a.c. electric field. A capacitor with a zinc sulfide phosphor dielectric that emits a bright green light on exposure to an a.c. field acts as a form of electroluminescent cell. At least one of the plates of the capacitor must consist of an atomically thin, i.e., transparent, metal coating, so that the light can be seen externally. In accordance with this principle, large plate-type light sources for glare-free ceiling illumination have already been manufactured. In electronics attempts are also being made to exploit this transformation mechanism. Electroluminescent semiconductor materials containing trace impurities possess various spectral emission bands, and it has been found that by changing the frequency of the applied voltage it is possible to alter the frequency of the light emitted. Color control of this type leads us to anticipate the use of new methods in color television tubes. The energy necessary for generation of the light is obtained from the loss-incurring displacement current flowing through the surface of the dielectric. In the cathode-ray tube the energy for the generation of the light on the screen is obtained from the incident electrons.

Luminescent coatings are encountered in various tubes. The decadic counting tube is a small cathode-ray tube with a square-wave

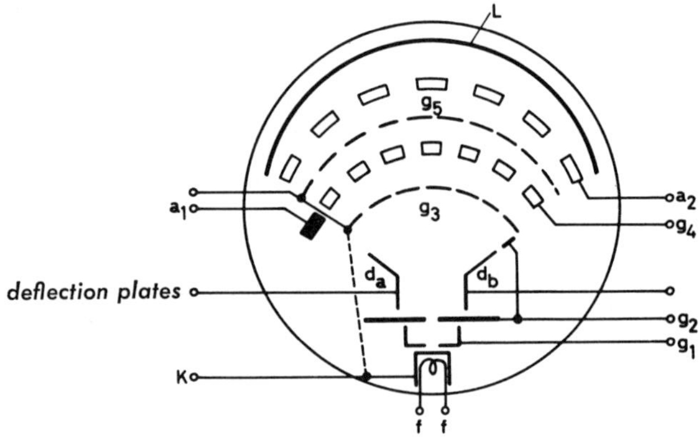

FIG. 107. The decadic counting tube is also a small cathode-ray tube.

electron beam that can be used for counting electrical impulses. The result can be seen on the screen of the tube. They are used in electronic counting and calculating machines. The counting tube illustrated in Fig. 107 is a decimal tube in which the band-shaped electron beam can assume ten stable positions that can be controlled by voltage impulses. In each of the ten positions the beam generates upon a luminescent screen L on the inside of the glass a line of light, which indicates one of the numbers 0–9 arranged upon a shield. Each positive control pulse of about 15 volts with a steep leading edge causes the beam to move one number further, until on the tenth pulse, i.e., when the number 9 is passed, the beam swings back to 0 and passes on an additional control pulse to the input of the next tube, which handles the following decade. A number of tubes, depending upon the number of decimal places desired, can be connected in cascade when necessary. Counting tubes of this type are capable of registering up to 100,000 pulses per second. The last tube of the cascade operates at a lower counting speed in proportion to the number of tubes used, so that it is possible to use a mechanical counter as the last tube. If no further impulses are received, the instrument maintains the operating condition of the last position as in a storage system, so that counting tubes are very well suited for digital computers. If the counting is to take place within a definite period of time, as for example in frequency measurement, the counting pulses must be switched off after the prescribed measuring time has elapsed. An effective method is to use quartz-controlled oscillators with connected frequency dividers, which make it possible to record oscillations of exactly 1 cps. The oscillations are used to open and close electronic counters. The operating principle of the counting tube is illustrated in Fig. 107. The electron beam, as a narrow focussed emerging band from the electron gun, first passes through the acceleration grid g_2, then the deflection system consisting of the two deflection plates d_a and d_b. The counting pulses are fed to the electrode d_a of the counting tube, and d_b is connected to the anode a_2. Both are connected to the anode voltage across a working resistance. The slit electrode g_4 with its ten vertical cut-out sections ensures that the illumination on the luminescent screen appears only at the characteristic sites. In a stable position the anode a_2 receives most of the electron beam. The remainder falls through the slit upon the luminescent screen L, where it generates the spot of green light on one of the marked numbers of the vessel. In this position the current flowing through the slit electrode g_4 is small, since the crosspieces are narrower than those of the anode a_2, and are less frequently struck by the focussed electron beam. If the voltage of the deflection plate d_a were progressively increased, the beam would be deflected to the left if the voltage at deflection plate d_b were not simultaneously increased in the same way, thus leading to a countereffect. These opposed effects are in fact employed, with the result that the beam is deflected so that it strikes the middle crosspiece of the slit-electrode g_4, and as the current deviation is increased the electrode takes over an increasing portion of the electron current—at the mid-position it takes over most of it. At the same time the current flowing through the anode a_2 decreases proportionately. The voltage drop at the common operating resistance with d_b is diminished, and the voltage increases. When the control voltage increases in this manner the beam does not leave its stable position. However, if a steep positive pulse is supplied the beam is deflected to the left because as a result of the capacitance C (consisting of the electrode and circuit capacitances) the previously compensated voltage increase cannot occur at d_b with the same velocity. In order to prevent the beam from dropping back into the previous position during the pulse drop-time, the pulse rise must be steep and the drop must be very gradual (Fig. 108). If the

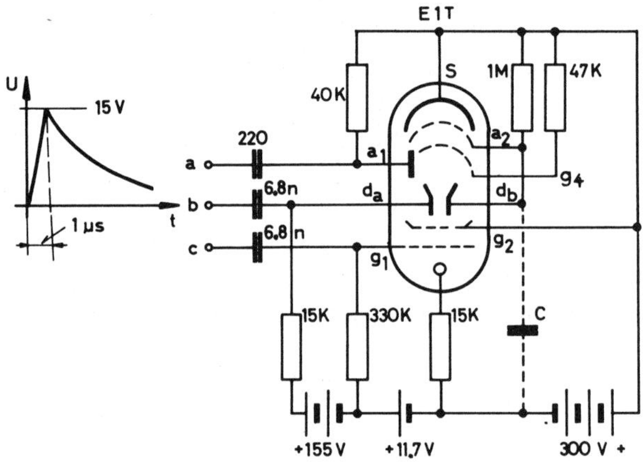

FIG. 108. Operating circuit of a counting tube. The control pulse (left) is fed to the input b. The tube is shown here in a simplified version. Right next to the anode a_1 are two round slit-electrodes (g_4). In practice there is also a retarding grid g_3 in front of g_4 and another (g_5) in front of the slit anode (a_2).

counting is to be without error the pulse form must correspond to the form prescribed by the manufacturer. To prevent other disturbances two retarding grids g_3 and g_5 are employed. The locally limited anode a_1 (switching-electrode) is placed close by the side of g_4 and receives the electron beam only on the tenth impulse. The luminescent screen, as in the indicator tube, possesses a conductive backing that is connected to the anode voltage source. This tube-saving counter is suited for use in counting circuits and calculating machines. The tube operates in principle as an addition circuit, but can also be used for subtraction if the counting is carried out backwards. If the addition process is repeated up to the number of the multiplying factor, these tubes can be used for multiplication, and the corresponding reverse operation enables them to be used for division.

The high counting speed makes it possible to carry out multiplications, for example, involving two five-place figures in less than one second if a pulse frequency of 100,000 cps is used.

Image converters represent a special type of electronic, high vacuum tube. They are capable of changing optical images of one spectral range into those of another. There are image converters that change infrared images, which are invisible to the human eye, into images in visible light. Other image converters, so-called x-ray intensifier tubes, convert ultraviolet light or roentgen rays into visible light. Figure 109 illustrates the principle of an image converter. An image is projected upon the photocathode F by means of an objective. Electrons are emitted from the cathode in accordance with the intensity of the image points. The electrons are then projected upon

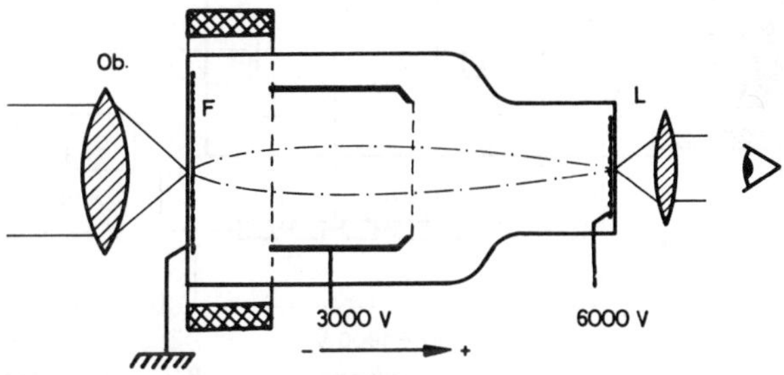

FIG. 109. Schematic view of the image converter tube. F: photocathode, L: luminescent screen.

the luminescent screen L by means of an electron-optical lens system, usually of an electrostatic nature. If desired, the image can also be magnified by the appropriate choice of an electrical field. The electron bombardment of the luminescent screen gives rise to a visible picture. If the cathode is sensitized to infrared light the converter can be used to observe objects illuminated by infrared (e.g., by infrared emitters) in an apparently dark area. With the help of an image converter it is also possible to see through a layer of mist, which, as is commonly known, absorbs less of the infrared portion of the spectrum. Heated motors, projectiles, etc., emit invisible infrared light, and infrared converters have also been used by the military to observe even at night flying projectiles, heated weapons or passing armored vehicles hundreds of meters away.

If the cathode is sensitized to x-ray radiation, the converter can be used to convert an x-ray image into a visible image. If the electrons are also accelerated by means of a high voltage, the brightness of the picture can be increased by a factor of up to a thousand.

A modern high efficiency x-ray intensifier tube consists of an evacuated glass envelope at one inner end of which is a fluorescent x-ray screen. The x-rays that strike the screen cause it to become fluorescent. Beyond this screen there is a photocathode. The fluorescent light produced on the screen dislodges at each point of the photocathode a number of electrons that is proportional to the intensity of the fluorescent light. These electrons are accelerated in an electric field, as in the image converter, and create on the opposite fluorescent viewing screen an image that is smaller than the original

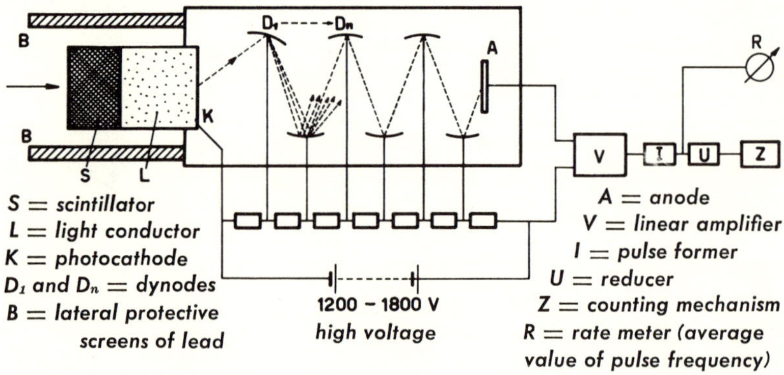

FIG. 110. Scintillation multiplier with cathode follower (impedance converter). In the photocell the light flashes are transformed into electrical impulses and amplified in the electron multiplier in order that they may be counted.

but whose luminous density has been intensified about a thousandfold. If this image is viewed through a magnifying glass, the original magnitude is restored. This sort of optical arrangement enables us to obtain a better information flow to the human eye. In addition to this there is the energetic "lumen" gain. As a result of the acceleration of the electrons in the electric field, the electrons produce a stronger fluorescent light when they impinge upon the viewing screen. Viewing devices of this type make possible longer exposure of the patient to radiation with little danger, as a result of the low intensity of the roentgen radiation required.

The so-called scintillation counters also belong among the electronic special tubes (Fig. 110). Like the Geiger-Müller tubes these tubes can be used to detect rapidly moving, charged particles and photons, such as, for example, radiation quanta (such as those arising from x-ray machines). At the front end of the vessel is a completely transparent crystal which emits flashes of light at the points of contact as soon as it is struck by the particles of radiation quanta.[1] Behind the crystal is a photocathode with an electron multiplier. This instrument is characterized by a particularly high sensitivity (quantum yield) and makes possible quantitative as well as qualitative measurements.

The physical phenomena occurring within the scintillation counter are as follows. A photon or a gamma particle strikes some atom of the scintillation material, for which sodium iodide is usually favored as the luminescent crystal. The photon liberates an electron from the K-shell of the atom, and this electron, which has been dislodged by the photoelectric effect and is now free, is reabsorbed by the scintillation crystal, i.e., it is absorbed into the shell of one of the atoms of the crystal, where it emits a flash of light (the scintillation). After about 10^{-15} sec the original atom, which had lost the electron and which had consequently been ionized, returns to its fundamental state, perhaps as a result of an outer electron dropping into the empty gap. This jump to the inner orbit has a very high energy, and roentgen radiation is emitted in accordance with the relationship $E = h \cdot f$. This in turn leads to further absorption by the scintillator, which in turn dislodges more photoelectrons. This process leads to a new scintillation flash in the same way as before. In this manner is set in motion an almost continuous alternating process of absorption and emission that lasts until all of the original gamma quanta have been converted into visible light. Depending on the amount of energy of the roentgen or gamma quanta, there arises (see Fig. 111) a photoelectric effect, a Compton effect, or a pairing of electrons and positrons. All of the light quanta created as a result of this process appear as visible fluorescent flashes and are capable of liberating electrons from the photocathode of the multiplier tube. Let us now consider one such photoelectron emitted from the photocathode. The electron is attracted in the direction of the first dynode D_1, which possesses a positive potential of 100 volts; on striking this first electrode the electron liberates more electrons. In this way it is possible for one electron to dislodge 5–6 new electrons. These new electrons travel in the direction of the next dynode, which possesses a potential 100 volts higher. Again each of the electrons impinging upon the electrode

[1] Examine the radioactive dial of a pocket watch with a magnifying glass.

dislodges several secondary electrons. All of these electrons then travel to the next dynode, which has a potential 100 volts higher. This electron multiplying process is repeated with from 5 to 10 target electrodes. By means of this method the photoelectric current generated at the cathode is increased by a factor of one million. This type of electronic tube is known as an electron multiplier. To ensure a high cathode efficiency the photocathode must be sensitized for the emission spectrum of the scintillation crystal. The spectral line for these luminescence crystals usually lies at around 4000 Å. The entire process takes place in an evacuated chamber.

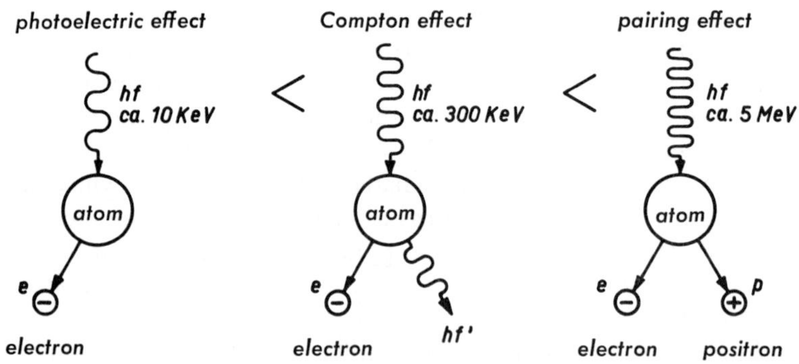

FIG. 111. Depending on the energy or frequency of the primary irradiation, there arises a photoelectric effect, a Compton effect, or a pairing effect. If, for instance, the energy of the primary photons (gamma rays) is 100 megavolts, the absorption is reflected only in the pairing of electrons and positrons, which in a magnetic field, because of their opposite polarities, fly apart in opposite directions.

The impulse frequency here is very high in comparison to that of the Geiger-Müller gas-discharge tube, since in the present case all the current carriers are electrons. The time resolution is better than 10^{-8} sec. This limit is set only by the non-instantaneous light decrease time or the half-value time. Radiation intensities of less than 0.1 milliroentgens/hr can also be detected.

Despite the internal amplification obtained by the use of the electron multiplier, it is still necessary to connect the output of the scintillation counter to the input of a linear amplifier. Then, depending upon the stringency of the measurement requirements, either the mean impulse frequency is obtained from an indicator or, for higher requirements, the impulses are all fed sequentially to a counting device.

To obtain readings for rapid impulses a decadic reducer must be used. In this case the measurement is usually carried out within a prescribed period of time with the use of an electronic switch. Although scintillation counters can also be used for counting beta particles, they are primarily used for gamma rays. The measurement process can be arranged to provide so-called gamma-spectroscopy or spectrometric scintillation counting. In this manner it is possible to identify radioactive isotopes in unidentified mixtures that emit radioactive radiation. The cesium isotope Cs^{137}, for example, emits gamma rays with an energy of 0.661 Mev. On the other hand, radioactive cobalt Co^{60} emits two gamma ray components with energies of 1.17 Mev and 1.33 Mev.

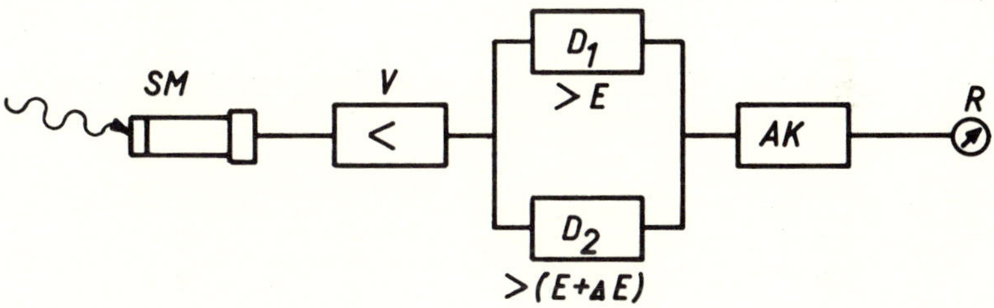

FIG. 112. Block diagram of a gamma-spectrometer.

Two typical characteristics of the radiation are: The greater its energy (usually expressed in e.v. or Mev), the greater the amplitude of the impulses in the scintillation counter. This assumes that well-stabilized supply voltages will provide a linear reading. A radiation may arise from more or less decays per second. The number of impulses per unit of time or the so-called impulse frequency (rate) is not, however, a function of the previously mentioned impulse amplitude. The gamma-spectrometer with the scintillation counter must therefore distinguish between impulse amplitudes arising from various decays. However, the rate must also be indicated. If the impulse amplitudes of various magnitudes are recorded in stages or in a chronologically continuous sequence an impulse spectrum is obtained. The instrument used for this purpose will be considered in greater detail.

Figure 112 shows a block diagram illustrating the manner of operation of a gamma-spectrometer. The scintillation multiplier SM transforms the incident quanta into current pulses corresponding quantitatively and qualitatively with the rate of incidence of the quanta and the energy level. These current pulses, after passing through a linear amplifier V, are fed to the two parallel connected discriminators D_1 and D_2. By means of a bias voltage applied to a diode, D_1 only lets through impulses whose amplitudes exceed an initially established minimum value E. The discriminator D_2 lets through only the impulses whose amplitudes are greater than E + ΔE. These signals arrive at the output. We will now consider three different impulse amplitudes (Fig. 113). Let us assume, for example, that they arise from three different energy sources. The first radiation has an energy below the threshold level E. The impulse associated with this radiation will not pass through either D_1 or D_2, since the amplitude of the impulse is too low for both discriminators. The second

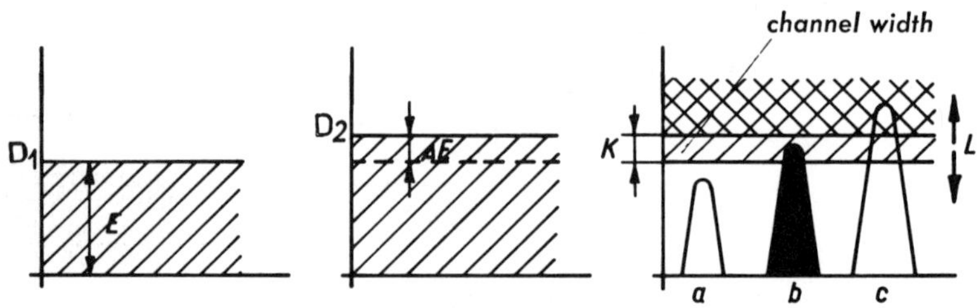

FIG. 113. Anticoincidence filter of the gamma-spectrometer. Case *a*: The impulse amplitude is too low for both discriminators. Case *b*: The impulse amplitude is higher than the threshold level for the first discriminator but lower than that of the second; there is thus anticoincidence and a measurement is possible. Case *c*: Both discriminators are conductive; there is coincidence and no signal at the output.

radiation has an energy greater than E and smaller than E + ΔE. The impulse consequently has an energy somewhat higher than the threshold E; it therefore passes through D_1. It does not, however, pass through D_2, which does not become conductive until an impulse amplitude greater than E + ΔE is attained. The third radiation possesses sufficient energy to generate a signal the amplitude of which is greater than E + ΔE. The signal therefore passes through both discriminators D_1 and D_2. The outputs of both discriminators D_1 and D_2 are connected to an anticoincidence stage (nonsimultaneity stage), which, as its name suggests, has the property of permitting a signal to pass only if the impulse comes from either D_1 or D_2. When neither D_1 nor D_2 transmits an impulse, i.e., when the impulse amplitude is not higher than the threshold E, the anticoincidence stage does not transmit a signal, and there is no signal at the output. When signals are simultaneously transmitted by D_1 and D_2, no signal is transmitted by the anticoincidence stage and once again no signal is obtained at the output. This is the case for all radiation the energy of which is greater than $E + \Delta E$. Consequently, signals are transmitted by the anticoincidence stage only when the original energy values lie in the range between E and $E + \Delta E$. No other values, i.e., below E or above $E + \Delta E$, are indicated by the rate meter R. The rate meter R therefore indicates radiation energy over the channel ΔE. The channel width determines the degree of spectral resolution, which can be made very fine by choosing a very narrow interval ΔE between the two different bias voltages. In order to obtain a spectrogram of all the energy values, the channel must be slowly shifted over the whole range of energy values, i.e., from the lowest to the highest

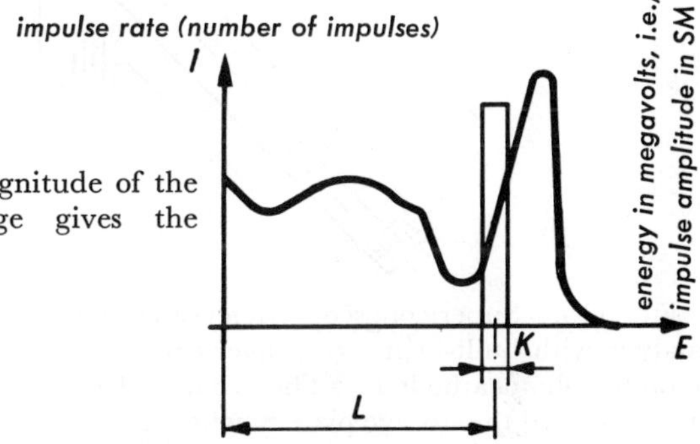

FIG. 114. The magnitude of the discrimination voltage gives the channel length.

energy impulses. To obtain such a spectrogram the measurements must be conducted within a definite measurement period. As the discriminator bias voltages of D_1 and D_2 are simultaneously raised and separated by a greater interval in accordance with the band width desired, scintillations of greater intensity, or impulses of greater amplitude, are required for a signal to get through the discriminators. At the same time the impulses are scanned for amplitude, the energy rate, i.e., the number of impulses in a given channel is transmitted to a recording device and registered on the paper. In this way the desired spectrogram is obtained; it provides both the impulse amplitude information and the impulse rate of the radiation. Each isotope or decay element has a characteristic energy or impulse amplitude. The rate is then used to determine the quantity of the element for which the radiation is characteristic. Figure 114 shows a spectrogram, i.e., the record of the impulse rates for the various impulse amplitudes. The gamma-spectrometer is used in the analysis of radioactive materials.

The error of such measurements is reduced in proportion as the number of impulses per unit of time generated by the radiation source is increased. (This relationship holds true of all statistical events.) The error is also reduced in proportion as the measurement time per channel is increased.

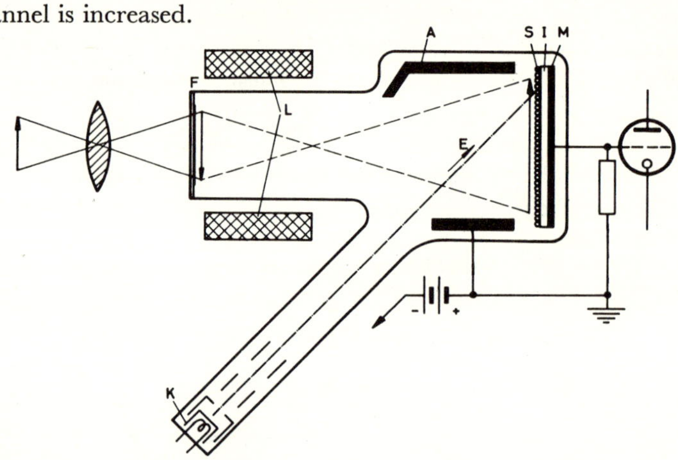

FIG. 115. Supericonoscope (television picture camera tube for image analysis with built-in image converter). The optical scenic image falls upon the photocathode F. The primary electrons emitted here are accelerated toward the storage plate S, where they dislodge secondary electrons in proportion to the intensity of the original optical image. These secondary electrons are drawn off by the anode A, with the result that an inverse electric charge pattern remains on the storage plate. The storage plate is scanned by a line-controlled cathode-ray tube in such a manner that the electric charges on the storage plate are completely neutralized. The charge differences are fed as picture signals to an amplifier tube.

Electronics includes an extremely large selection of high vacuum tubes so that it is impossible to do more than touch briefly upon the special tubes. The television camera represents a combination of a photocell, picture converter and cathode-ray tube. The cathode-ray breaks up the image of the scene on the sending side into a definitely ordered sequence of picture points or current pulses. In the receiver the process is reversed and the current pulses are retransformed in the same definite order into brightness values. The apparatus illustrated in Fig. 115 is called a supericonoscope, the basic principle of which originated with Zworykin as far back as 1928.

The optical image of the scene to be transmitted passes through a 35 mm movie camera objective and through the ground front glass plate of the evacuated tube onto the photocathode F, where electron emission occurs in proportion to the local brightness. As in the picture converter the electrons are focussed on a storage plate S located at the other side of the tube. The coil L acts as a magnetic lens and effects electron-optical magnification. The storage system consists of hundreds of thousands of microscopically small metal particles with high secondary emission coefficients. These metal particles are embedded in an insulating metal oxide and adhere to a mica layer I, which serves as a dielectric between the metal particles and a metal coating applied to the other side of the mica layer. The metal layer acts as the counterelectrode to the individual raster elements. The conductive islands, which together form a mosaic surface, act as individual storage capacitances with the metal plate M. The electrons from the photocathode F travel at a high velocity toward the metal islands, from which they dislodge secondary electrons. The number of secondary electrons is greater than the number of primary electrons that dislodge them. These secondary electrons are drawn off by a positive suction field from anode A, with the result that there is an electron deficit at the metal surfaces. This leaves a latent electrostatic image on the storage plate, because the primary electrons had initially been emitted in accordance with an optical image. The more electrons dislodged, the greater the electron deficit of the plate; and an electron deficit on a capacitor plate is equivalent to a positive potential with respect to the other plate. Since each mosaic cell has, as it were, a chance to adjust itself under the influence of the light in the course of a complete image period, the efficiency is quite high as a consequence of the accumulation (or integration).

A scanning beam E, 0.1 mm thick with an intensity of 0.1 μA of the type generated in a cathode-ray tube, is passed obliquely in line-scanning fashion[1] over the storage plates forming the latent image.

[1] For some time the interlaced scanning method has been used in television to reduce flickering. In this method the whole mosaic is not scanned in one consecutive scan in 1/25 of a second; instead one half the raster is scanned each 1/50 second. During the first half-picture the beam writes only lines 1, 3, 5, etc., and the remaining lines 2, 4, 6, etc., after the vertical flyback. The line and picture synchronizing signals ensure synchronous operation of the beam in the camera tube and in the receiver.

Because of the high velocity of the beam the number of electrons dislodged from the mosaic is greater than the number of incident electrons. The secondary electrons are drawn off by the anode of the suction field. The intense secondary emission induced at the points of contact raises the potential of each of these points to $+3$ v. When the beam leaves a given position there is a slow and continuous rain of electrons from the neighboring space charge cloud (because not all of the liberated electrons are drawn off) onto the mosaic. The highest possible limiting voltage, at which no more electrons can be absorbed, is about -1 volt (see the starting zone for diodes or triodes). The maximum potential difference that can have accumulated by the arrival of the scanning beam is, therefore, at most 4 volts for a dark picture spot. For continued illumination it is lower. The signal, which is a function of the picture content, is created by the difference in secondary electrons.

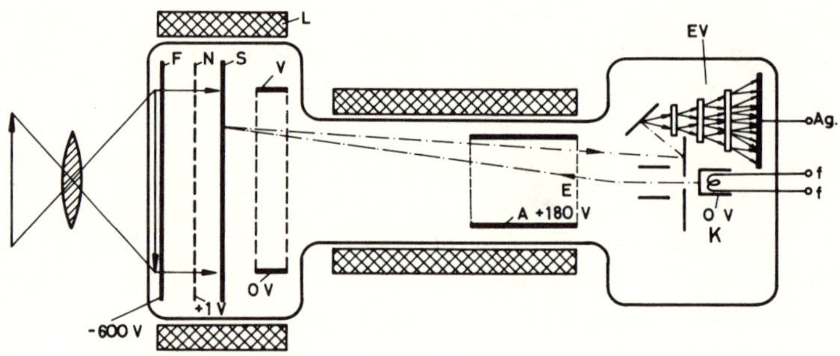

FIG. 116. Image-orthicon with a circular electron multiplier (EV) (shown here in a simplified drawing). The dash-dot line represents an instantaneous photograph of the scanning beam and the beam returning to the electron multiplier.

The charge differences are always erased by the scanning beam. The charging current varies from a single original value in accordance with the local surface voltage. The varying charge surges created by the electron pattern, which are the potential differences between the secondary electrons on the mosaic element, on the one side, and the suction anode, on the other, provide the signal voltage at the resistance of the common counter-electrode or signal plate. This voltage is known as the picture or video signal, and after further amplification it is passed on to the transmitter. Thus, the brightness values in television are transmitted in a chronologically successive manner.

The image-orthicon (Iams and Rose, 1939) or superorthicon (superorthiconoscope), as it is also known, is several hundred times as sensitive as the supericonoscope. Figure 116 illustrates the arrangement of this electronic system. The photocathode layer F with a surface of about 10 cm^2, which is suitable for normal objectives and is situated at the left side of the glass envelope, is very thin and transparent. It is connected to the negative pole of an anode current source (600 volts). By means of a very fine mesh (40 mesh/mm) wire-gauze electrode N that will not interfere with the picture and by means of the coil L of the electron lens system, the electron pattern emitted from the cathode surface is projected upon storage plate S, which possesses a zero potential. The photoelectrons are subjected to a 600 volt acceleration and consequently liberate secondary electrons from the storage plate. The storage plate becomes positively charged and its positive charge increases as the primary emission increases or as the brightness of the picture spot increases. Since the suction field N possesses a potential of +1 volt, the maximum possible potential of the storage capacitance is 1 volt (for a white spot). The suction field, which is about 5/100 mm away from the storage plate is able to trap, as it were, all the secondary electrons, because of its proximity to the storage plate. Overriding or "blinding" caused by too great a light intensity (e.g., from photographic flashes) causes an excess of secondary electrons to accumulate at the neighboring element, where they create a negative charge which becomes noticeable in practice as a black frame around bright spots. The storage plate consists of an extra-thin glass plate (about 2/1000 mm) about 4 cm in diameter; the plate possesses a certain conductivity, which makes it possible to equalize the charges within 1/25 of a second. The right side of the glass plate is scanned in television fashion by an electron beam E coming from the electron gun and passing through two mutually perpendicular magnetic fields. The narrow electron beam, which is accelerated by anode A with a potential of about 180 volts, is retarded to zero electron volts in the field of the annular deceleration electrode V and lands with almost "no velocity" on the right side of the storage plate S. In order to keep the "slow" electrons, which are the characteristic feature of the superorthicon, focussed in a sharply defined beam, it is necessary, since these electrons are sensitive to external fields, to make use of a magnetic field extending axially along the whole length of the glass container; in this manner the electrons are compelled as a result of the Lorentz force to complete several spiral rotations of very small diameter. From the scanning beam that passes over the storage plate a number of electrons, depending upon the local positive charge (which arises from the other side), remain adhering to the plate. Because of the low velocity at which the electrons are shot in, they do not release secondary electrons. The absorption of electrons at a given site ends as soon as the positive charge is neutralized and the potential reaches zero or becomes a retarding field. Further excess electrons are repelled, reverse their direction right in front of the plate and under the influence of the focussing and magnetic fields acting upon them return to the cathode. The returning beam is modulated on the storage plate by the picture brightness between 0 and +1 volt. The actual picture current is the difference between the initial beam and the returning beam. The returning electrons strike a circular target electrode, which is close to the electron gun and is coated with a layer of some material particularly well suited for secondary emission. The secondary emission multiplier EV consists of a number of oblique target electrodes; it has a naturally low noise factor and results in about a 500-fold amplification. The contrast (signal/noise interval) is extremely high in the superorthicon. A white picture source corresponds to a low output current. This tube contains a number of electronic elements including a photocell, image converter, cathode-ray tube and electron multiplier.

Plate I shows a cathode-ray storage tube (so-called scototron) developed by Standard-Elektrik Lorenz AG. The crystal layer storage surfaces—which can be seen in the middle of the tube—are written on by two electron beams coming from sources that can be seen in the foreground and by a reading electron beam from the back. During the writing or during the storage period of the information the secondary electrons dislodged within the crystal by the writing beam remain in so-called sticking sites and return very slowly to their fundamental state. If a plate composed of alkali halide crystals is bombarded by an electron beam, the crystal, which is ordinarily colorless, turns blue at the contact site. Cathode-ray tubes with screens of this sort can record an easily visible blue curve that will be maintained for a long time. The display can be extinguished by heating. Recording oscillographs of this type are called sciatrons. This system has also been proposed for large-picture television projection; here the crystal layer the transparency of which can be

PLATE I. Cathode-ray storage tube (scototron).

modulated must lie in the projection plane of an otherwise usual diapositive image. But the lag in the operation of this tube has made its practical realization impossible up to now. If a blue writing tube of this sort is combined with a scanning system we obtain the scototron illustrated in Plate I. The blue writing screen in this case also is completely transparent before recording; it is discolored by the electron beam, i.e., by the rearrangement of its electrons, and becomes opaque during recording. The blue writing screen covers a second layer composed of the luminescent screen material (ZnO) of a conventional cathode-ray tube. A scanning beam is deflected over this layer by points or lines in television fashion and the picture field is thus illuminated. The reflected light of the illuminated spot, which is a function of the blue writing discoloration, is picked up by a nearby photocell placed to one side of the tube. The photocell current contains the actual reading signal. The three different electron beams—a typical example of electronic technology—can be magnetically deflected completely independently of each other. The reading system on the back side generates the scanning light on the luminescent screen. The two writing systems in the front control the record made on the blue writing screen. The choice of two such electron guns allows the two processes to be recorded simultaneously and independently of each other. Tubes of this sort are used for transposing images from one scanning norm to another, particularly for radar images; however, they can also be used as memory and transposing devices in certain computers. The storage time, which is a function of the intensity of the electron current of the scanning beam, varies from several seconds to several hours.

Another type of storage tube possesses a large number of minute capacitances in the form of small islands of gold on a thin glass plate, the back of which is coated with metal to form a common signal plate. These capacitors are charged by the writing beam. If, for example, a radar image is recorded several times in succession, the charges of the information signal are added together, whereas the noise signals randomly distributed over the image tend to even out. A second, separate electron beam scans the amplified information signal. Both the writing and reading beams are focussed electrostatically and deflected electromagnetically. Tubes of this type are used principally for information storing as well as for the improvement of radar images.

Figure 117 shows a so-called visual storage tube. This type of tube contains two electron guns. The system on the left traces the oscillogram on the storage electrode. These electrons are intensely accelerated, i.e., they are subjected to a high acceleration voltage. The second system, on the right, passes a diffusely distributed electron beam of low velocity over the whole surface of the storage electrode. The purpose of this beam is to ensure transmission to the fluorescent

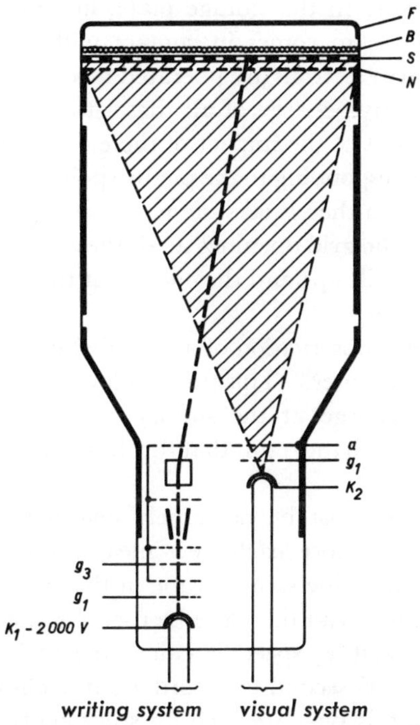

FIG. 117. Principle of a modern storage tube. F is a fluorescent screen, with an aluminum layer within the tube, which is connected with the secondary acceleration voltage of 10 kilovolts. B is a backing electrode (mesh grid). S is a storage grid. K_1 and K_2 are cathodes.

screen of the oscillogram recorded by the first system. The storage grid[1] must be charged negatively before recording. The negative charge is applied by positive pulses that are fed to the so-called backing-electrode. The negative charge repels all the electrons

[1] This consists of a mosaic composed of an enormous number of insulating islands as in a television camera tube.

coming from the system on the right. The fluorescent screen, consequently, remains dark after the pulse signal. The storage grid acts like the control grid of a triode for the fluorescent screen.

Let us now consider the whole process as it unfolds. A signal is supplied. The system on the left traces an initially invisible "record," as a result of which the rapidly accelerated primary electrons liberate secondary electrons from the storage plate, i.e., at all the points at which the writing beam comes in contact with the storage grid a number of secondary electrons are liberated by each primary electron. The originally negative charge of all these storage units is not only neutralized as a result of the impact of the electron beam, but local, positive charges corresponding to the picture points are built up. The electrons coming from the visual system on the right are then able to pass freely through the grid openings at all these points where positive charges have been built up; and they arrive at the screen, which they excite to fluorescence.

The length of time observation of a recorded curve can be delayed is between 1 and 2 minutes. This time limit is set by the fact that the negative blocking charge at the storage electrode is progressively neutralized by ions, with the result that the fluorescent screen gradually becomes illuminated.

The storage plate must be negatively charged for the required darkening of screen before each new measurement. This storage tube is superior to the blue writing tube in that it possesses a greater recording sensitivity; it can therefore be used also for measuring high frequencies. The writing speed is calculated at about 10 km/sec. The tube requires .005 second to record a curve climb of 0–5 cm.

Storage tubes of this type are used for studying transient processes (such as discharges) when the process must be studied some time after the event and one does not want to bother with photographic reproduction. The diversity of the examples of storage tubes we have just given illustrates the great versatility generally offered by electronics today.

V. VELOCITY-MODULATED TUBES

The charge carriers responsible for the current flow in a vacuum tube are all electrons. The emission current can be calculated from the formula:

$$I = \rho \cdot F \cdot v$$

where ρ is the charge density, F the emissive surface and v the velocity of the electrons.

It makes no difference which of the three magnitudes is used for control. The tubes discussed up to now were all of the type in which the charge density ρ was controlled by regulation of an adjustable negative grid bias. Tubes of this type are therefore properly known as intensity-modulated or density-modulated tubes. The negatively biased grid offers the advantage of lossless control but possesses the disadvantage that the electrons are slowed down in their path from the cathode to the anode in the area of the grid. This increase in transit time, which is essentially a consequence of the inertia of the electrons, sets an upper limit to the frequency range that can be transmitted. The limitation of these classical tubes lies in the fact that the control voltage between the cathode and the grid reverses its direction before the electrons reach the grid. It has been found that the input resistance decreases with the square of the wavelength, while the requisite control power increases with it. In cases where the amplification tends toward zero the system is no longer effective. The physical reason for this is that electrons situated between the cathode and the grid induce charges on the control grid, and these charges give rise to a displacement current I_v which is exactly in phase with the control voltage U_g. The power consumed in control is: $N = I_v \cdot U_g$. The grid-cathode stretch must be considered from a precise point of view as a parallel circuit consisting of a capacitance and admittance. In brief it has been found that lossless control of the anode current by means of an alternating grid voltage is possible only as long as the electron transit time is short in relation to the oscillation period of the control voltage. A short electron transit time can be obtained with the highest possible plate voltages by employing small cathode-grid distances. This fact has led to the development of a special sort of precision tension-grid design. In this tension-grid technique the grids, instead of being fashioned of parallel wires, are designed as mesh grids. To keep the thermal emission of the grid low, the heat conduction must be good, and the

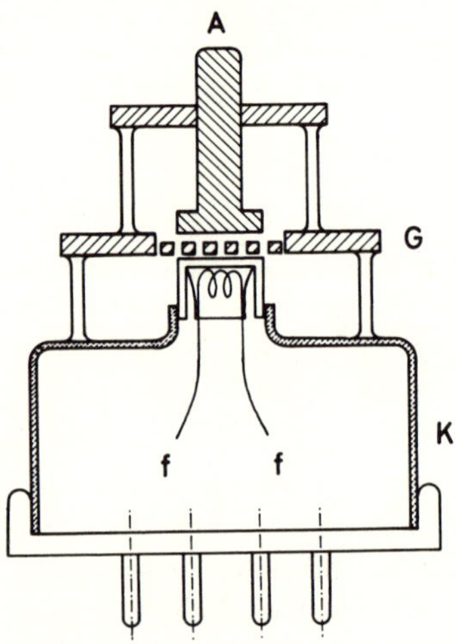

FIG. 118. Flat disk-triode. The disk-shaped electrodes (K = cathode, G = grid, A = anode) lead directly to the large low-loss metal disks. A ceramic material fused in an air-tight manner into the metal is used as the insulator here instead of glass.

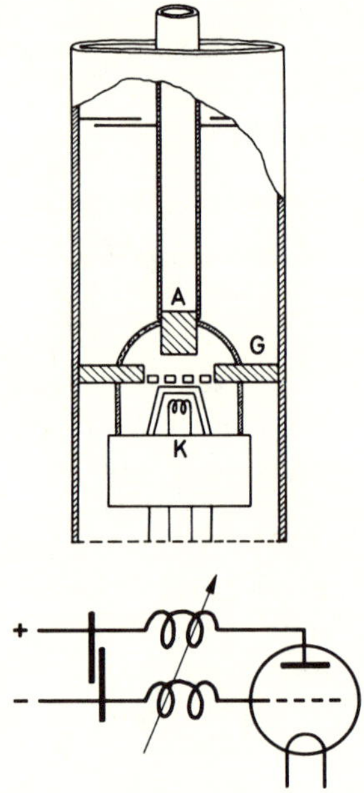

FIG. 119. Triode oscillatory circuit in the decimeter wave range.

material selected (e.g., gold) must possess a high work function even on absorption of the vaporized cathode material (barium). With this method it is possible to obtain easily reproducible cathode-grid electrode distances of about 0.01–0.02 mm. In this chapter we are also concerned with wavelengths below 25 cm. Here the inductance of the leads are of importance. At a wavelength of 10 cm an inductance of, for example, 10^{-8} henry will result in a reactance of 200 ohms. Thus, even short lengths of wire lead to high inductive reactances.

This gave rise to the development of the disk-triode, in which the electrodes are connected to the external circuit by means of large rings sealed into the glass (Fig. 118). At the time when the disk-triode was being developed, which goes back to 1935, the welding of glass and metal posed great technical problems. The great cross section of the disk or cylindrical shaped ring terminals is equivalent to the parallel connection of a very large number of lead wires and results in a low inductance. Thus it is possible to obtain an almost lossless coupling to an external oscillatory circuit arising perhaps from a wave guide system (Fig. 119). At these high frequencies the external oscillatory circuit also assumes a special form. The classical coil shrinks to an arrangement with only one winding. To keep the inductance very low the conducting coil becomes a tube. This is also the case for any anode and grid circuits. The capacitance between two such tubular conductors serves as a condenser. The arrangement is illustrated in Fig. 119. For better uncoupling between the control

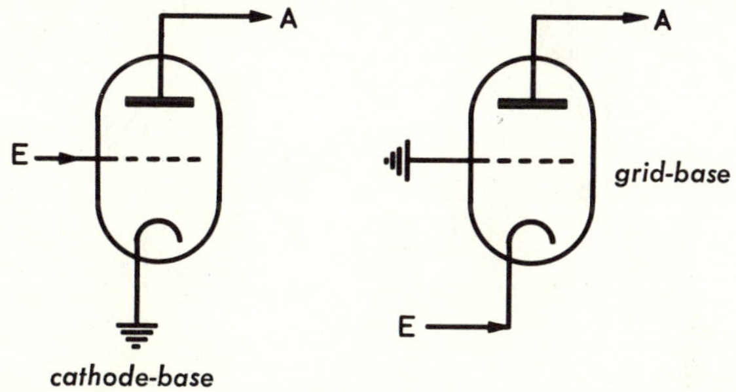

FIG. 120. When the control signal is fed to the grid of a tube with a grounded cathode the arrangement is known as a cathode-base circuit; when the grid is grounded and the control signal is fed to the cathode, the arrangement is known as a grid-base circuit.

circuit and the output circuit, the triodes are arranged in the so-called grid-base circuit. This type of circuit is highly favored for very short wavelengths. In this system the grid is grounded instead of the cathode and consequently forms a protective shield between the cathode and the anode output (Fig. 120). Nowadays modern disk-triodes, which require the greatest precision in manufacture, can be used to obtain or amplify frequencies of up to about 7000 Mc/s[1] or wavelengths of about 4 cm. The output is about 1 watt. At

[1] Mc/s is equal to 1,000,000 cps.

wavelengths below this the triode oscillator is completely ineffective, since the electron transit time is of the order of magnitude of the oscillation period.

This inadequacy of the triode oscillator led to the development of the velocity-modulated or so-called transit-time tubes. The principle upon which these tubes are based lies in the use of the velocity, the third factor of the initially mentioned fundamental formula, rather than the first factor, the density ρ, to control the operation of the tube. A control field is used to obtain slight variations in the velocity of the initially uniform electron current. This causes the electrons to bunch into groups in the course of their path. The initially constant current is replaced by a pulse current of the frequency of the control field.

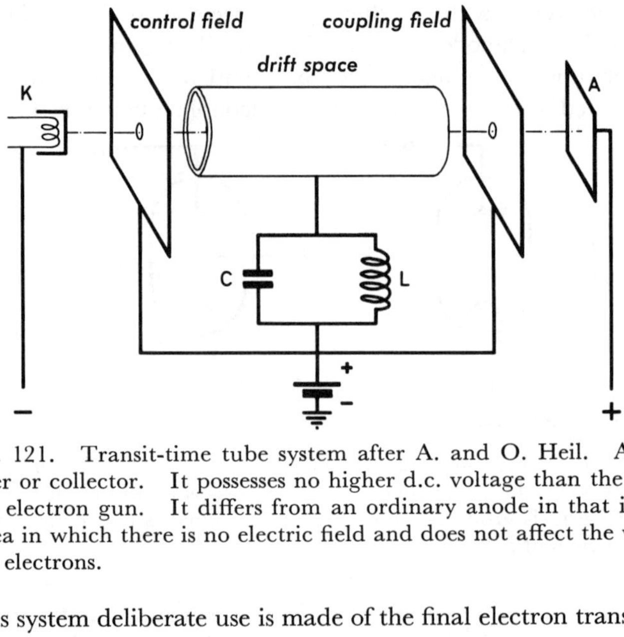

FIG. 121. Transit-time tube system after A. and O. Heil. A is the catcher or collector. It possesses no higher d.c. voltage than the output of the electron gun. It differs from an ordinary anode in that it forms an area in which there is no electric field and does not affect the velocity of the electrons.

In this system deliberate use is made of the final electron transit time and the inertia of the electrons. Tubes of this type are consequently known as transit-time tubes. Figure 121 illustrates the basic idea of such an arrangement developed in 1935 by the husband and wife A. and O. Heil. The electrons emitted by the cathode are focussed into a narrow beam along the axis of the tube and travel through the control slit into the drift space. They move through the drift space without being affected by any external fields, or variations resulting therefrom, through the exit slit to the collector A. The drift space is also known as an electron lift because in the course of passage it is possible to change the electron phase by a half-wave, i.e., by 180°, by means of the external oscillatory circuit from L and C; the electrons consequently act as if they had passed to another "level." Electrons are known to influence electric fields, and such fields are placed between the entrance slit and the entrance to the drift space and between the drift space exit and the exit slit. The electrons that arrive when the lift entrance is negative are slowed down, and those that arrive a bit later, when it is positive, are accelerated. Consequently, the electrons enter the drift space at different velocities. The control signal required for this between the entrance slit and the drift space is delivered by the external oscillatory circuit. The change in velocity of the electrons results in the development of maximal and minimal electron densities within the drift space. The faster electrons overtake slower ones that were emitted earlier, or the slower electrons that were emitted earlier drop back to faster electrons that were emitted later, which is, in effect, the same thing. This gives rise to density groups. The velocity modulation can be ended at the lift exit with a space charge concentration. The basic idea is that, with proper phase alignment, energy can be supplied to the oscillatory circuit by means of the pulse current created by the method just described. This requires tuning of the length of the drift space, the frequency of the oscillatory circuit, and the velocity of the electrons. If the "lift length" were too great, the faster electrons would again overtake the slower electrons after the concentration site, and the situation would change once again from density to velocity modulation. This would result in a periodic repetition of the same situation.

The concentrated electron packets that leave the drift space are either capable of giving up or absorbing energy, depending upon the polarity of the lift. This is equivalent to retarding or accelerating the electrons in a high frequency field. As already mentioned, the extent of the excitation produced depends upon the relation of the length of the drift space to the frequency of the oscillatory circuit, as well as upon the average electron velocity (kinetic energy) imparted by the anode voltage to the acceleration electrodes. Because of this relationship the wavelength is expressed in angular measure. For the circuit arrangement shown in Fig. 121 oscillation is generated at the lengths

$$\frac{\pi}{2},\ 2\pi + \frac{\pi}{2},\ 4\pi + \frac{\pi}{2},\ \cdots$$

and damping of the oscillatory circuit occurs at

$$\frac{3}{2}\pi, \frac{7}{2}\pi, \frac{11}{2}\pi, \cdots$$

The purpose of microwave tubes is to convert the kinetic energy of electrons into electromagnetic energy. For this purpose care must be taken to ensure that at the appropriate anode potential the electrons have a tendency to run very slightly ahead of the oscillation period

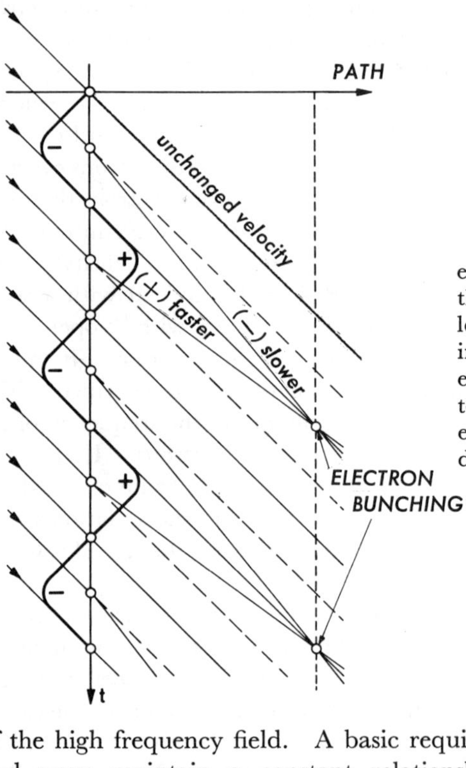

FIG. 122. Electron itinerary. The differences in the velocity of the electrons lead to different directions in the "itinerary" of the electrons and consequently to the concentration of electrons at the site of the dotted vertical line.

of the high frequency field. A basic requirement is that the beam and wave maintain a constant relationship. The electrons are consequently retarded and the resulting energy excess is used for high frequency amplification in the oscillatory circuit. Electrical engineers will find an analogy with the parallel connection of synchro-generators in electrical power stations.

The velocity-modulated beam structure, in which the mutual repulsion of the electrons also play a complicated role, is known in electronics as plasma waves. The bunching of the individual electrons brought about by the velocity control decelerates their movement away from each other and, as a result of their mutual repulsion, brings them into a state of rest with respect to each other. The amplitude of the alternating velocity of the electrons therefore grows progressively smaller down the length of the beam and reaches a null point at the concentration sites. This is precisely what we required: that the kinetic energy be transformed into a space charge with potential energy. If the drift space is very long, the potential energy is reconverted into kinetic energy as the electrons move away from each other and assume different velocities. As a result, what takes place is a periodic bunching of the electrons or a continual alternation between the two types of energy. From the standpoint of a stationary observer such a phenomenon is characterized as a standing wave within an electron current. This process can be interpreted as a wave phenomenon. That is the more significant reason for the coining of the term space charge waves or plasma waves. The term plasma wave is merely a concise expression for the periodic alternation between the potential energy of the space charge concentrations and the kinetic energy of the electrons. This alternation is illustrated graphically in Fig. 122. The graph is equivalent to a kind of itinerary of the velocity-modulated electrons. Electrons in the control field during the null phase remain unaffected. The other electrons, which enter during the various phases of the control field, are affected differently, in accordance with the phase of the field. These differences in their itineraries are represented by different acceleration slopes, which lead to the bunching of the electrons.

Thus, we have already outlined here the basic idea of the modern transit-time tubes of the modern klystron. The oscillatory circuit, which was previously separate, is incorporated into the tube as an individual circuit element in order to dispense with leads. Figure 123 shows how the normal oscillatory circuit, which initially consisted of combinations of elements such as the capacitor C and coil L of radio degenerated into the cylindrical box of present microwave technique, the so-called tank circuit. The top and bottom of this oscillatory structure represent the capacitor while the side walls represent a parallel circuit of vertical "connecting wires." The low inductance thus obtained is absolutely necessary for microwaves. In this type of oscillatory circuit or cavity resonator, as it is called, the circuit current flows vertically through the external boundaries, and the electric field, or displacement current, is at a maximum at the axis of the box between the top and bottom. The voltage curve drops from a maximum in the middle by a half-wavelength on each side. In contrast to the simple resonant circuit with capacitor and coil, the cavity resonator exhibits no absolute separation of magnetic and electric fields. The electromagnetic field of the cavity resonator becomes a carrier of both energy forms. Since such an oscillatory circuit is

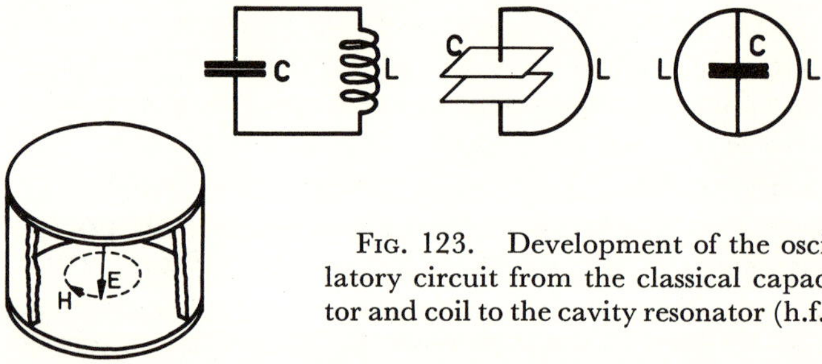

FIG. 123. Development of the oscillatory circuit from the classical capacitor and coil to the cavity resonator (h.f.).

externally closed, it emits no radiation and consequently has almost no losses. Its figure of merit Q is extremely high as compared to the classical oscillatory circuit of lower frequencies. Moreover, its mechanical stability, i.e., the inductance and capacitance values are quite constant. There is a variation of the cavity resonator in which the cavity is altered to form a rotational resonator or rhumbatron in

FIG. 124. The rhumbatron, a variation of the resonator depicted above.

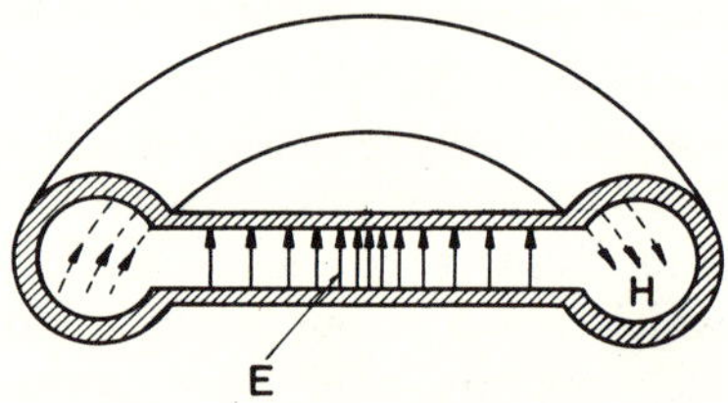

accordance with Fig. 124. For certain purposes it is possible to connect in series several such tank resonators, each of which can be grounded so that the usual insulation problems do not crop up. The Heil generator still operated with a separate, external resonant circuit. The so-called two-circuit klystron possesses two rhumbatron resonators inside the tube. One is placed in front and the other in back of the drift space. Figure 125 illustrates the principle of this type of microwave drift-tube. To allow the electrons to pass through the cavity resonator, the two rhumbatron walls are perforated in the center or designed in the form of some sort of grid electrodes. The whole system, in which the resonators are fused into glass walls, must naturally be evacuated. This is the mode of operation: The a.c. control voltage between the two grids of the first rhumbatron causes a modulation in the velocity of the electrons passing through. This is possible because the high frequency electric fields between the walls of the resonator constantly change polarity in accordance with the resonant frequency. The beam, which passes through this oscillatory circuit and is thereby velocity modulated, is concentrated into density waves as it travels through a drift space of appropriate length and

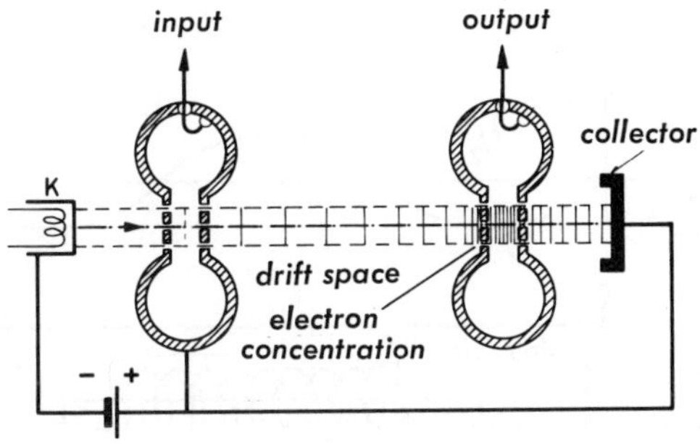

FIG. 125. Two-circuit klystron. An electron beam is projected from the electron gun through the drift space. One cavity resonator is placed before and the other after the drift space. These electrodes are quite different from the previously discussed grids.

passes through the grid system of the second cavity resonator. This resonator receives energy from the kinetic energy of the electrons. The high frequency oscillation is taken from this second resonator, and the output power is a multiple of the input power. The energy for the power increase is supplied by the source used to accelerate the electrons. The two-circuit klystron is consequently used as a microwave amplifier.

If the second cavity resonator is connected to the first, a regenerative circuit is obtained and the two resonators form a regenerative oscillator or, as it is also known, a generator. The output power is approximately inversely proportional to the frequency. At a wavelength of 30 cm, approximately 100 watts of high frequency energy can be produced for constant operation and about 1000 times that figure for pulse operation. Such klystrons are used in smaller radar installations such as those used for velocity control of vehicles. Klystron microwave tubes are also built in cascade design, with several cavity resonators placed in the trajectory of the electrons. In this way it is possible to build combination oscillator and frequency multipliers. These so-called multicircuit klystrons (mostly three-circuit) are designed for continuous-wave powers of 20–50 kilowatts at 45 db amplification. They are used in communications engineering, i.e., in direction transmission, radar, and television. They are designed for pulse powers of up to 30 megawatts at a wavelength of 10 cm. The constant current acceleration voltage during the pulse is up to 325 kilovolts and the beam current is about 185 amp. The efficiency is given as over 40%.

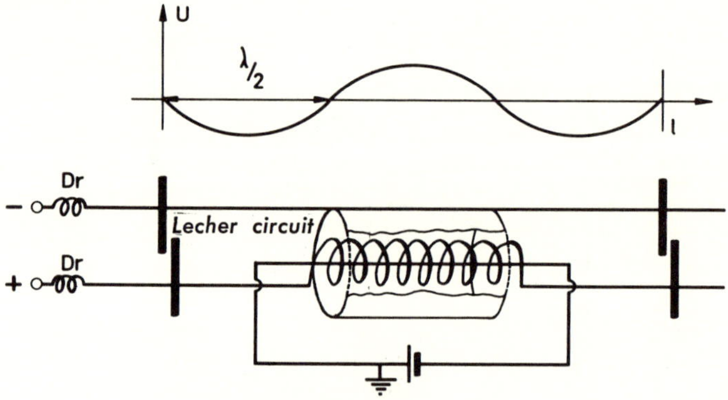

FIG. 126. Oscillatory circuit with triode arrangement; the "anode" is given a negative potential and the "grid" a positive potential. The two chokes are used to block the high frequency as a so-called low-pass filter.

The so-called reflex klystron is a variation and simplification of the two-circuit klystron and was invented, as it were, independently of this theory and almost by accident by Barkhausen and Kurz in 1920. In this arrangement, shown in Fig. 126, a classical triode is used; however, in this case, in contrast to normal usage, the grid is given a positive constant voltage and the anode a negative. Under the influence of the accelerating grid-cathode potential, the electrons at first travel from the cathode toward the grid and then pass through the grid mesh into the electric retarding field between the anode and the grid. Here they are turned back and travel in the direction from which they came. As a result of the two fields the electrons are able to oscillate back and forth several times between the grid meshes, and a parallel wire system (Lecher system) between the grid and the anode acts simultaneously as resonator and electron sorter. These tubes, which have only a historical interest, naturally had a very low efficiency. Here the wavelength is determined by the Lecher wire system. The Lecher system is an exposed oscillatory circuit that loses relatively little energy through radiation and consists of two parallel wires acting as a divided inductance and capacitance. The frequency can be varied by the wire lengths, i.e., by the slide adjusters.

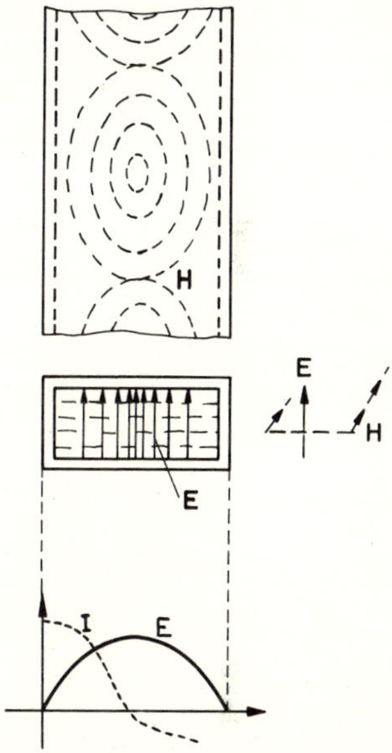

FIG. 127. Rectangular-section hollow wave guide. The lines of the electric field are drawn as solid lines. They run from one part of the inner surface to the other. The lines of the magnetic field are drawn as broken. ABOVE: Part of a top view. The transverse field varies sinusoidally along the tube. CENTER: cross section. BOTTOM: high frequency potential distribution over the cross section.

Parallel wires are also used to carry off the waves. Other developments led to the use of tubular wave guides, coaxial cables and most recently to rectangular-section wave guides. Coaxial conductors were already externally shielded. At higher frequencies, however, relatively high losses occurred as a result of heating of the inner conductor, on the one hand, and in the dielectric of the rod insulators, on the other. The rod insulators, moreover, represent separate and undesirable capacitances, which cause reflections at high frequencies.

The metallic hollow conductor wave guide dispenses with this internal conductor and with the rod insulators. The minimum dimension of the hollow conductor cross section must be of the order of one-half the wavelength ($\lambda/2$). Hollow wave guides are therefore too large and unwieldy at low frequencies but are very well suited for high frequencies from 3000 Mc/s (3 Gigacycles/s) and up, i.e., for wavelengths lower than 10 cm. Hollow wave guides are used as antenna leads in such systems as radar installations.

At a wavelength of 10 cm the transmission losses with coaxial cable are in the vicinity of 0.6 db per meter and with rectangular-section hollow conductors about 0.04 db per meter. The losses in hollow wave guides, resulting from heating in the walls, are determined by the skin effect resistance. A cross section of this modern type of

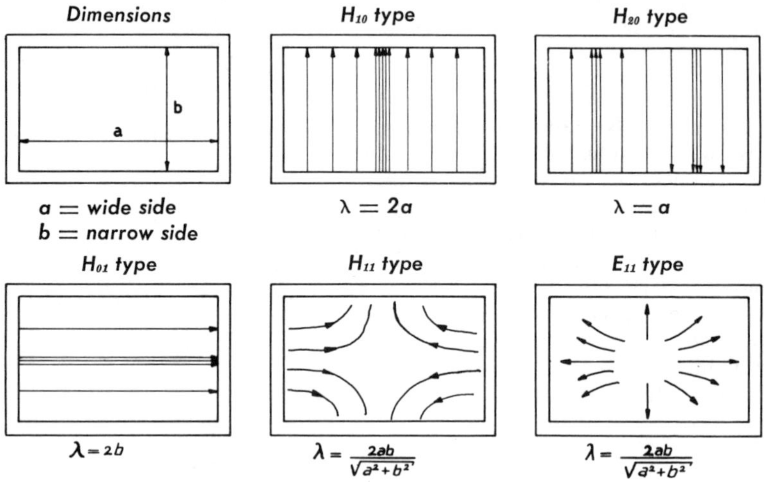

FIG. 128. The transverse electric field of several types of fields in the rectangular-section hollow wave guide.

conductor system, which is no longer reminiscent of classical electrical engineering, is shown in Fig. 127. This drawing shows the simplest relationships of the electric field E and the magnetic field H. Just as vibrating strings are in effect capable of a whole series of resonance states, here also a large number of wave types can be obtained in accordance with the type of high frequency supplied. The cross section of the wave guide is connected with the form of the electromagnetic waves. Two types of waves are distinguished, depending upon whether the magnetic component H or the electrical component E is axially oriented and thus in the direction of propagation; these are known as H-waves (or, since the electric field is perpendicular to them, transverse-E waves, abbreviated to TE-waves) and E-waves or

TM-waves, respectively. The wave is further described by an attached index number: the first numeral gives the quantitative value of the maximum of the electric field along the wide side of the wave guide, while the second gives that of the narrow side. Rectangular-section wave guides are particularly well suited for H_{10} (or TE_{10}) fundamental oscillations (Fig. 128), whereas round-section wave guides are preferable for H_{01} and H_{11} waves (Fig. 129). It can be seen from Fig. 127 that there is a phase difference of 90° between the electric field and the displacement current. As a result of the rectangular cross section the waves are clearly polarized along the wide side of the tube and will follow any twist or kink of the guide. In beam transmission (e.g., a dipole arrangement) this must be taken into account accordingly. The dimensions depend upon the frequency.

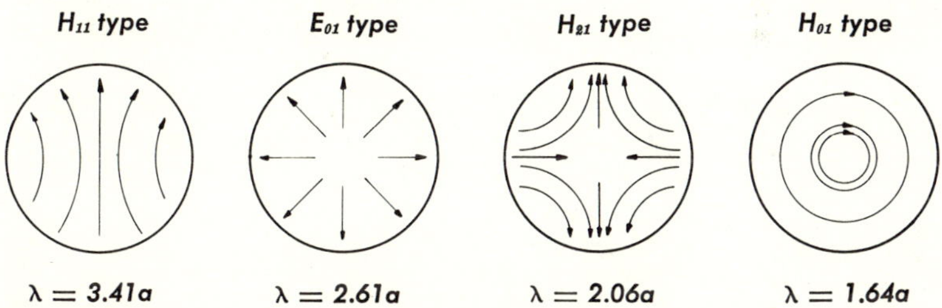

FIG. 129. There are also hollow wave guides of circular cross section. The position of the electric field images depends on the polarization of the current supplied.

The larger dimension of a rectangular-section wave guide must theoretically (calculated for a vacuum) be equal to one-half the wavelength (in practice it must be shorter). The shorter side is not so important and is about half as long. For microwaves these hollow wave guides exhibit the lowest losses, but they require the greatest precision because the slightest mechanical variations result in new electrical properties. If, for example, the cross section is narrowed by the use of two metal sheets, one at the top and one at the bottom (Fig. 130), this acts as a capacitance. If these same sheets are arranged laterally, however, an inductance is obtained in parallel circuit with the conductor. A combination of the two arrangements results in a resonant circuit or so-called resonance blind. These rectangular-section wave guides are used, for example, as connections between a microwave tube and the antenna or the focal point of such an antenna, consisting perhaps of a parabolic mirror or some similar structure.

The magic T or hybrid T, as it is also known, is a symmetrical hollow wave guide combination consisting of an E-arm (E = electric field) and an H-arm (H = magnetic field) and two lateral arms.

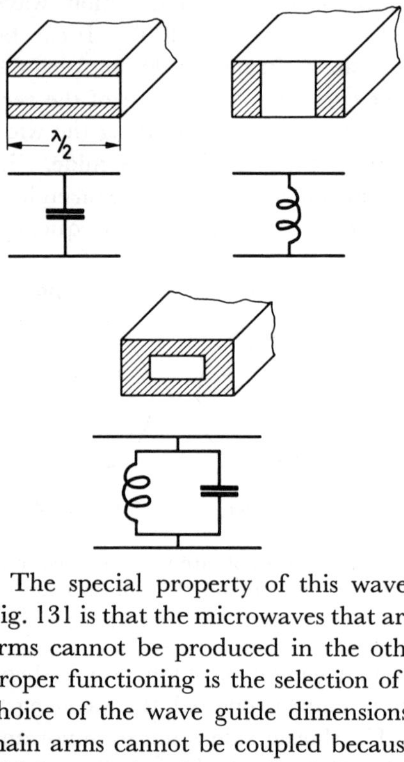

FIG. 130. Parallel connection of the capacitance and inductance in a hollow wave guide.

The special property of this wave guide combination shown in Fig. 131 is that the microwaves that are propagated in one of the main arms cannot be produced in the other. A necessary condition for proper functioning is the selection of the proper wavelength, by the choice of the wave guide dimensions, and proper matching. The main arms cannot be coupled because the H-wave (magnetic field), which enters in a longitudinal direction, cannot be propagated in the propagation direction of the E-arm (which is dimensioned for an electrostatic field component). This is easily seen if one remembers that the h.f. voltage can be divided only between two lateral surfaces of $\lambda/2$ dimension. The shorter lateral walls consequently have the same electric potential. Theoretically they should have the same zero voltage. Consequently, because of the symmetry, as in a Wheatstone bridge, no voltage can be transmitted to the large surfaces of the other main arm. Since the large surfaces of $\lambda/2$ width possess a zero potential both above and below, no field is created.

This naturally also holds good for the magnetic field, which is perpendicular to the E-field. Consequently, the E-arm cannot be coupled to the H-arm, and similarly the H-arm cannot be coupled to the E-arm.

The "magic" involved, for those who are not versed in electronics, consists in that there is no mixing of the two waves fed in at the H-arm and E-arm. Furthermore, with ideal electrical matching (frequency and compensated arm length) the symmetrical construction precludes a coupling between the two lateral arms because of the central branch

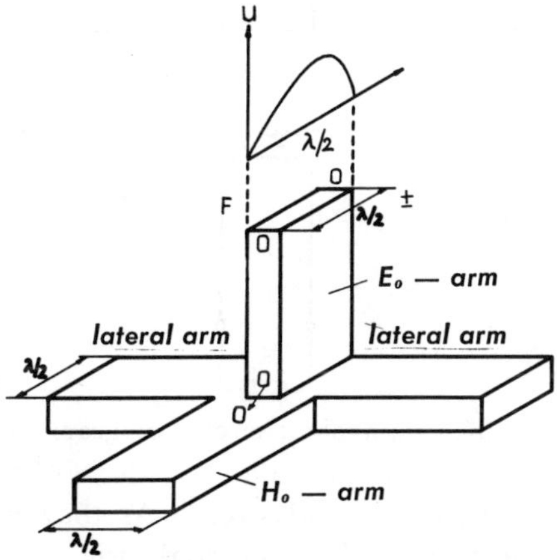

Fig. 131. Magic T fitted together from exactly prepared rectangular-section wave guides.

exchange or because of reflection. It follows, therefore, that energy transmission between opposite arms is completely impossible. However, wave energy entering one of the four arms is shared equally by the two arms that do not lie opposite the entrance arm. Energy received by the E-arm is equally divided by the lateral arms I and II. The magic T can therefore be used as a type of bridge circuit for testing purposes, which is a reason why this unit is highly favored in microwave technology. Figure 132 shows an equivalent circuit diagram for the magic T.

Now that we have had a brief look at the subject of wave conduction we can return to the reflex klystron, to which we were introduced in the historical form of the Barkhausen-Kurz tube. The modern reflex klystron is, as shown in Fig. 133, essentially half of a two-circuit klystron. Only one cavity resonator is necessary. An electron beam passes through the cavity resonator toward an electron-repelling, negative reflector electrode, the so-called repeller. The retarding field thereby created is situated at about the half-way mark of the

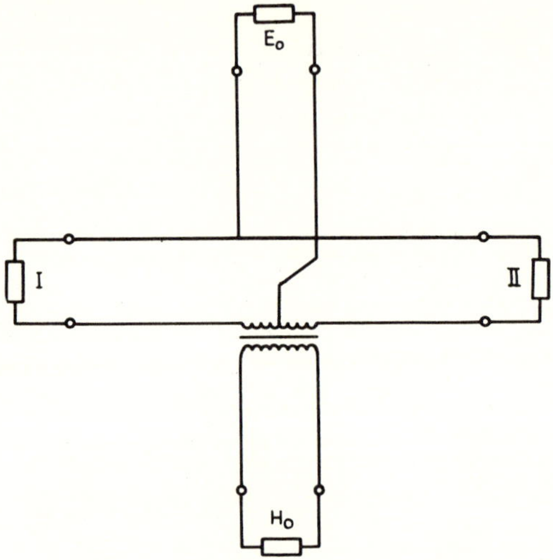

FIG. 132. Simplified equivalent circuit diagram for the magic T.

drift-space of the two-circuit klystron. As they pass through the resonator, the electrons, which entered originally at a constant velocity, are again velocity modulated so that they possess diverse velocities as they emerge. As a result of this velocity modulation, the

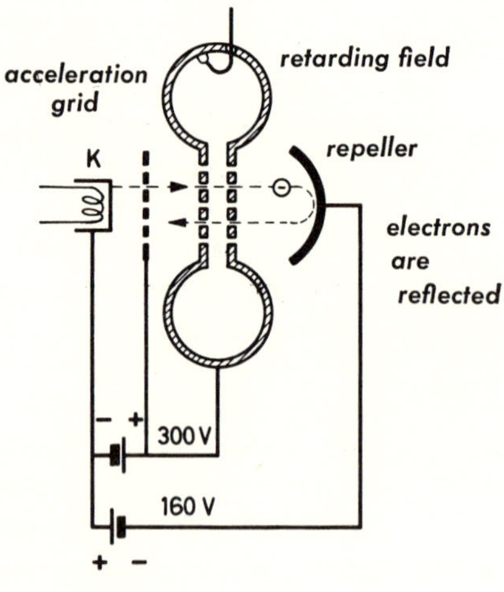

FIG. 133. Reflex klystron. The repeller acts as a reflector (it can be compared to a mirror).

slower electrons are overtaken by the faster electrons following them, which leads to bunching of the electrons or to phase focussing in accordance with a definite path length behind the resonator. If the repeller has a definite negative voltage, they are forced to reverse their path at the appropriate time, so that on the return path the concentration points will arrive in proper phase at the control cavity resonator. The electrons are again retarded and consequently able to supply potential energy to the h.f. oscillatory circuit. Since only a single circuit is required for both the input and output signal, the reflex klystron functions exclusively as a self-exciting oscillator.

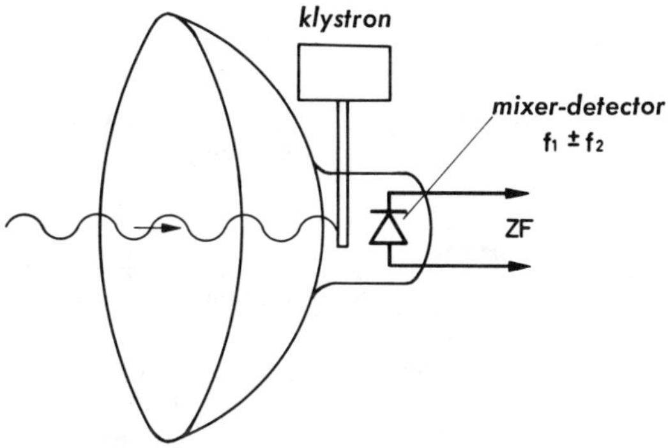

FIG. 134. Parabolic antenna with receiver input stage in which a reflex klystron acts as the mixer oscillator and in which there is an additional crystal diode mixer. The intermediate frequency (ZF) is $f_z = f_E \pm f_0$, where f_E is the receiver frequency and f_0 is the klystron frequency.

This microwave tube is especially well suited for low power uses, for example, as a mixing oscillator in microwave receivers. In Fig. 134 the principle of the mixer stage of a radar receiver is illustrated. The reflex klystron is also favored as a laboratory oscillator in such applications as microwave spectroscopy. It has the additional advantage of simplicity of design, since the single oscillatory circuit requires only one circuit. A crude adjustment of the frequency can be obtained by turning a set-screw or by mechanical change of the dimensions of the barometer-design cavity resonator, as shown in Fig. 135. However, it is also possible to vary the frequency by changing the reflector voltage by about $\pm 1\%$. This simple electronic detuning method can be used for frequency modulation. The method is therefore excellent for the transmission of television from the camera truck to the transmitter, for directive radio beacons that frequently connect several television networks (Eurovision), for multichannel telephony, as well as for frequency tuning, etc. The modulation voltage that must be superimposed upon the repeller voltage (−160 volts) is several volts per Mc/s frequency variation. Today reflex klystrons are designed for microwaves of up to about 2.5 mm and a power of about 40 mW down. Usually their efficiency is below 5% and drops as the frequencies get higher. A klystron multiplier with an initial circuit tuned to the second harmonic of the input wavelength of 1 cm, will give a power of 0.1 mW at 5 mm wavelength if supplied with 200 mW power.

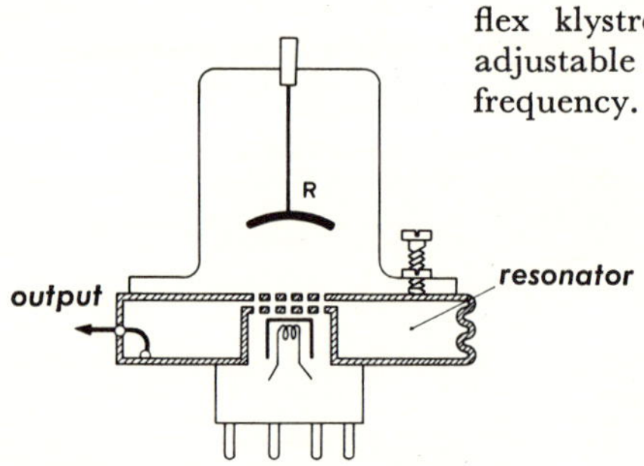

FIG. 135. Section through a reflex klystron with a mechanically adjustable resonator for varying the frequency.

The traveling-wave tube was invented in 1942 in the Clarendon Laboratory at Oxford by R. Kompfner. This type of tube operates in the following manner. An electron beam is produced by an electron gun as in the cathode-ray tube. In this case the electron gun consists of a Wehnelt control electrode and one or more acceleration anodes, in addition to the heated cathode.

The resulting electron beam traverses the cylindrical axis of an evacuated glass tube about 30 cm long and about 25 mm wide (Figs. 136*a*, *b*). This beam, which travels through the center of the tube, is enclosed by a spiral wire coil (the helix). This helical line is fixed by means of three ceramic bars that pass through it and thus forms a long, single-layer coil with separated windings. The electron beam, which consists of an enormous number of individual electrons, consequently also traverses the axis of this single-layer coil and travels toward a catcher, the so-called collector, where there is an average anode current of 20 mA. The coil and the collector are both given a positive bias.

In order to keep the beam focussed, i.e., optically speaking, to avoid dispersion, the tube is surrounded by an axial magnetic field of about 400–500 g. This focussing field, which is produced by an external magnetic coil, is also known as the beam-directing field. Such electrons as move radially outward are thus subjected by the magnetic field to restraining forces that compel them back into a focussed beam (electron-optics, compare with the convex lens). The high frequency signal that is to be amplified is fed in at the beginning of the helix and

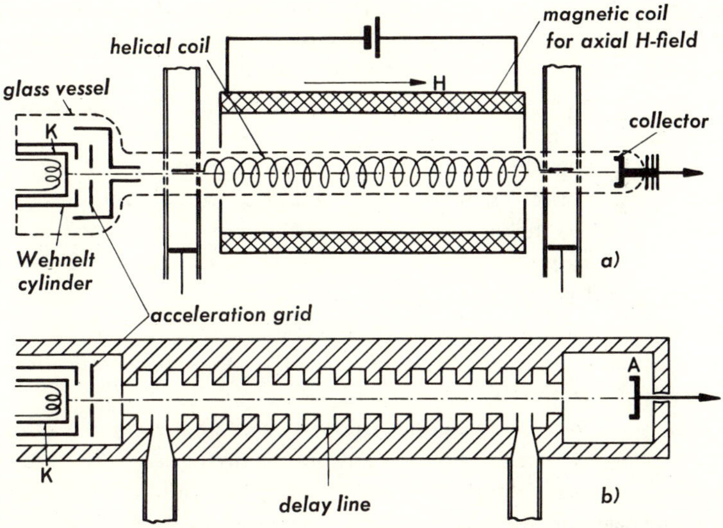

FIG. 136. (*a*) Traveling-wave tube for centimeter waves with adjustable input and output. The adjustable wave guide for the input signal can be seen on the left and the one for the amplified output signal can be seen on the right. (*b*) Traveling-wave tube for millimeter waves.

the signal, which is increasingly amplified in its passage along the helix, is uncoupled at the end.

To clarify the manner of operation we will first consider the task performed by the helix.

As is commonly known, an electromagnetic wave is propagated in empty space with the velocity of light *c*, i.e., 300,000 km/sec. When, however, such an electromagnetic wave is fed into the wire of the helical coil, the signal wave is propagated along the coil. Here, however, for each wind of the coil the wave advances in the axial direction (the direction of the cathode ray) only by the advance or pitch of the coil, and the axial propagation velocity v_a is consequently reduced to

$$v_a = \frac{S}{\pi \cdot D} \cdot c,$$

where S is the pitch (as in a screw thread) and D is the diameter of the coil. Geometrically $D \cdot \pi$ represents the perimeter or path from one winding to the next along the helix.

The axial transit time T of the wave is computed from the mechanics equation $T = S/v_a$, where S represents the axial path per winding. The axial velocity can be reduced as desired by the choice of an appropriate pitch and diameter. The helical coil is consequently an ideal delay line for the axial propagation of the wave. In practice a propagation velocity of the order of 10% of the speed of light is desirable. This is the usual velocity with which the electrons are propagated along their axial path. According to our earlier description, the electron velocity is a function of the magnitude of the anode voltage, by means of which the electons are accelerated. The electron velocity v_e in m/sec is computed from the balance between the kinetic and potential energy:

$$v_e = 6 \cdot 10^5 \sqrt{U}$$

At the customary voltage of about 2000 volts of these tubes one obtains a velocity of exactly 10% of the speed of light.

When such values are chosen the velocities of the traveling high frequency field and the electrons are practically equal. The amplification in the traveling-wave tube is then a result of the energy exchange between the electron beam and the h.f. signal wave that travels along the coil. The longer the beam and the wave travel synchronously along parallel paths, the greater their mutual inductive effect and the greater the energy exchange. The amplification process is illustrated in Figs. 137*a*, *b*, *c*, and *d*. In each part of the figure the electron beam and the signal wave travel from left to right. The signal wave is assumed to be a sinusoidal high frequency oscillation propagated with v_a, i.e., the so-called phase velocity along the axis of the helix.

If the electron beam possesses exactly the same velocity as the wave, the wave remains at rest relative to the electrons. Such a case can be described from the standpoint of the electron as a standing wave, as is shown in Fig. 137*a* (compare with two parallel wave trains traveling at the same speed). In Figs. 137*b*, 137*c*, and 137*d* a section through the helix is depicted and the force lines of the electric field strength E are represented as at relative rest for an instant. The letters *A–H*

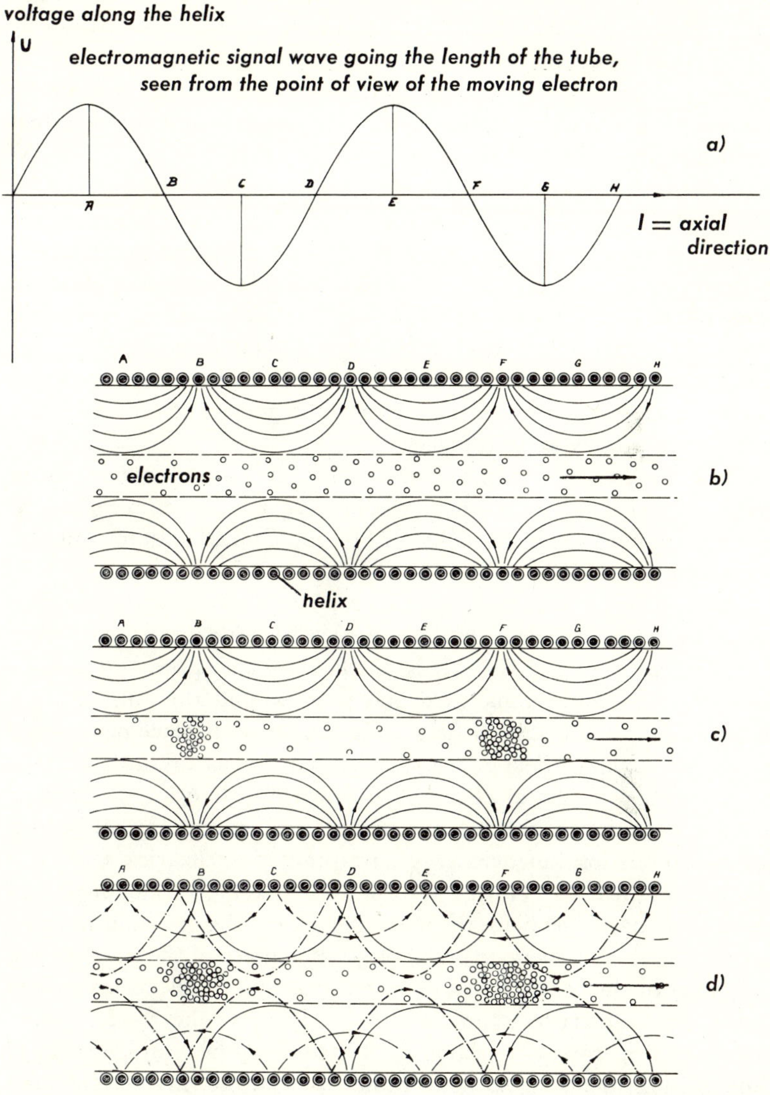

FIG. 137. Amplification process in the traveling-wave tube. See text for explanation.

of Fig. 137*a* correspond to the letters of 137*b*, *c*, and *d*. As in the klystron the electric field of the wave (see Fig. 137*c*) gives rise to progressive bunching of the electrons. In accordance with local polarity of the helix a given electron is accelerated when the field is positive, and retarded when the field is negative. As the arrows in the figure indicate, acceleration always occurs when the direction of the electric field is from left to right, and retardation always occurs when the field runs from right to left. Thus, in region A the electrons are accelerated and move into space B. At B there is no field, however; the electrons are not accelerated any further and therefore remain momentarily at relative rest. The electrons in region C are situated in a retarding field and their velocity is progressively reduced, so that they move relatively in the direction of space B, i.e., from a practical standpoint in such a manner that they are subsequently overtaken by the faster electrons. This continuous and progressive interaction leads to the bunching of electrons. Thus, at points B, F, etc., more and more electrons are bunched together, i.e., their density increases markedly at such places.

These bunched electrons form electric fields and consequently induce electric fields on the helix as shown by the dashes in Fig. 137*d*. These fields induced on the helix are 90° out of phase with the initially present field (continuous line).

The two fields are added together in accordance with common vector methods to form a resultant field, which is represented by the dash-dot line in Fig. 137*d*. It is this field which appears greater in comparison to the original field and thus exhibits the amplification effect. This new resultant field with the phase displacement $-\pi/2$ exercises a braking effect on the electron bunches as a result of its new phase position. The electrons slow down and lose kinetic energy. Since in accordance with the laws of mechanics energy cannot be lost, the apparently lost kinetic energy is transmitted as electrical energy to the field on the coil. However, for it to be possible for kinetic energy to be given up to the field, the mean velocity of the electrons must be adjusted in such a manner, by appropriate choice of the acceleration voltage at the helix and the other electrodes, that the electrons tend to travel somewhat faster than the phase velocity of the waves. Only under such operating conditions will the energy be transferred from the electrons to the signal field, rather than in the reverse direction. An analogous situation is the parallel connection of two generators in an electrical plant. As a result of the tendency of the velocity of a generator to increase as a consequence of the opening of the water feed, this generator gives up energy. When the velocity is reduced it absorbs energy and the generator becomes a motor. In the traveling-wave tube, under the correct operating conditions, the total energy of the electrons consequently decreases along the beam, and the total energy of the high frequency signal increases along the helix, which in gross physical terms means there is amplification of the signal. The amplified output energy, from a fundamental viewpoint, is provided by the anode voltage source. To summarize, the synchronization of the electrons and the wave results in a bunching effect. If the electron velocity of the electron groups is somewhat greater, the high frequency field absorbs energy from the electrons. The electrons are then situated, in spite of the periodic acceleration phase of the bunching, in an average braking field. If the electron velocity is reduced—for instance, by reduction of the anode voltage—the energy is transferred from the high frequency signal field to the electrons, and the electrons are thereby accelerated once more to synchronization. (The analogy is to a motor.)

This reverse operation also finds application in physics. The nuclear physicist makes use of so-called linear accelerators in which charged particles or electrons are passed through high frequency fields and thereby accelerated to high final velocities. Instead of delay lines free high frequency fields are employed to accelerate the particles to final velocities approaching the speed of light, which can of course never quite be reached.

A traveling-wave tube constructed for frequencies between 8000 and 10,000 Mc/s possesses an amplification of about 30 db. This corresponds to an amplification factor of 32. Traveling-wave tubes for other frequencies, for example, lower frequencies of 3600–4200 Mc/s are also available; these tubes yield even greater amplifications. Values in excess of 40 db are reported.

Advantages of this type of tube are the great band width and the low noise factor, which is given at about 4–7 db. This makes it possible to amplify weak signals that would otherwise be drowned in noise. In general the amplitude of the wave increases uniformly along the helix, since the energy exchange between electrons and the high frequency field of the helix has been going on longer.

At the beginning of the helix, however, this assumption does not apply, since the free electron beam is still a purely continuous current as it enters the delay coil; the alternating current component, i.e., the sinusoidal oscillation, is first formed with progressive bunching of the electrons by the high frequency field. On the other side a limit is set to the amplification by the possibility of feedback. To avoid self-excitation a constant resistance is incorporated into the helix from the middle to the end with resultant damping. Traveling-wave tubes with a power of several watts are favored for use in radio-range chains and television relay stations. Similarly, they are used in such installations as low-noise preamplifier stages. The band width of a traveling-wave tube amplifier is not limited by the physics of this type of tube, but by the coupling lines for the input and output, for which rectangular cross section wave guides are most commonly used.

Related to the traveling-wave tube is the so-called carcinotron or "crab tube" (Fig. 138). The wave propagated along the helix, as

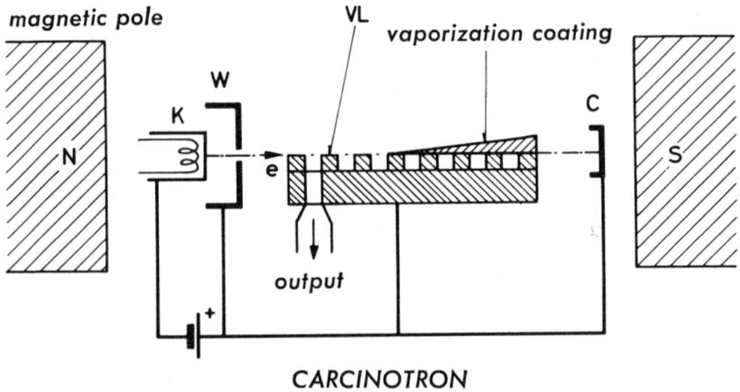

FIG. 138. Carcinotron in a magnetic focussing field. (There are several other types of backward-wave tubes besides the carcinotron.)

we have seen, represents a standing wave for the similarly advancing electrons, which possess the same velocity. Physically, a standing wave can be obtained and explained as two waves traveling in opposite directions. It is therefore conceivable to use a second wave traveling in the opposite direction to accompany the signal wave.

This backward wave results in internal back coupling, as in the reflex klystron. If the damping is dispensed with in a traveling-wave tube, oscillation is obtained. In the typical carcinotron the oscillation energy is uncoupled not at the output side but at the input side of the tube, and the frequency is set by the choice of the electron velocity, and thus the level of the voltage. The carcinotron is therefore a transmitter tube that can be modulated easily by variation of the acceleration voltage.

Both these tubes play an important role in modern electronics.

The traveling-wave tubes that we have just discussed with helix or filter lines have in common the use of a straight, thin electron beam that is focussed by means of a magnetic field that is parallel to the axis of the beam. If a magnetic field transverse to the direction of the beam is employed in accordance with an idea of Hull (1921), we obtain the magnetron, a favorite diode oscillator in radar engineering. This so-called magnetic field tube under the right operating conditions has a high efficiency and is capable of delivering up to many thousands of kilowatts of power for pulse operation. As we will see, in this case also there is interaction between the wave and the electron beam over a long path, which in this case is circular. The treatment of the

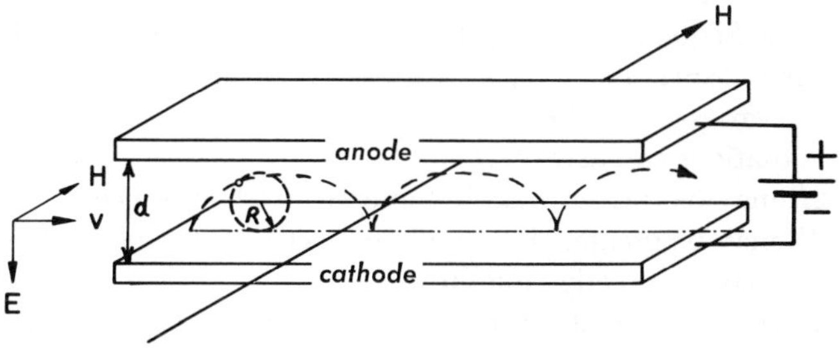

FIG. 139. The behavior of an electron between two plates, the lower acting as cathode, the upper as anode. There is a magnetic field perpendicular to the electric field created by the plates.

plasma waves is theoretically similar to the situation in the klystron and the traveling-wave tube. This is the reason why a correct theoretical explanation of the magnetron was impossible until after the development of the traveling-wave tube. The basic principle is illustrated in Fig. 139. The bottom plate is connected to the negative pole and acts as the cathode, while the upper plate, which is connected to the positive pole, acts as anode. The flat electron beam coming from the cathode, or, for descriptive purposes, one of the electrons of this beam, is deflected by the magnetic field H in the space between the anode and the cathode. Under the influence of the mutually perpendicular, constant electric and magnetic fields E and H, i.e., under the influence of the coulombic and Lorentz forces resulting from the fields, the beam or the electron describes a cycloidal path and acts as a traveling wave. In the following formula the symbol B is used in the conventional sense to represent the magnetic flux density. It is directly related to the magnetic field strength H and differs from it in accordance with the formula $B = \mu\mu_0 H$, in which μ_0 represents the absolute permeability and μ the magnetic permeability of ferromagnetic materials. The constant μ_0 no longer occurs when B is introduced into the following formula. The radius R of the circle of rotation described by the electron is determined from the formula

$$R = \frac{m}{e} \cdot \frac{E}{B^2}$$

and the frequency of the cycloidal rotation ω_0 is given by

$$\omega_0 = \frac{e}{m} \cdot B$$

in which $\omega = 2\pi f$.

Thus, the frequency of the cycloidal rotation is shown to be independent of the electric field. The diameter R of the circle of rotation, however, varies with the magnitudes E and B. Three different typical situations can arise here:

(1) $2R$ may be smaller than the interelectrode distance d.

(2) $2R$ may be greater than the interelectrode distance d.

(3) $2R$ may be exactly equal to d.

In the last case the electrons at the crest of the cycloid touch the surface of the anode and a cyclotron oscillation can be profitably obtained. The critical anode potential can be calculated from the formula

$$U_{A_K} = \frac{1}{2} \cdot \frac{e}{m} \cdot d^2 \cdot B^2$$

The cycloidal orbit traversed by the electrons results from the superimposition of the rotational motion upon the path of transit along the cathode. Since the electrons are emitted and start at all times, in-phase and out-of-phase rotation orbits that act to neutralize each other can be created. Though the out-of-phase, energy-absorbing electrons are sorted out near the cathode under the influence of the initial traveling wave and the system itself, it has been found advantageous here also to coordinate the primary, uncoordinated cyclotron oscillations by means of an appropriate oscillatory circuit. For this purpose the anode is designed as shown in Fig. 140; grooves and holes are cut in the anode so that it is divided into segments that act as small resonance circuits connected in series (as a delay line). The grooves act primarily as capacitances and the perimeter of the holes as inductances. Consequently, the sections are small cavity resonators that are coupled together by the high frequency field during operation.

On each occasion that an electron traveling along a cycloidal path passes an anode opening it encounters a transverse field component (+ −) of the high frequency alternating current field. If the cycloidal movements are synchronized with the high frequency field that is propagated from one delay element to the next, energy is supplied to the field. This process is repeated at each groove until the electron reaches the anode. The condition that must be satisfied is that the crest of each cycloid must coincide with a resonance opening and with the maximum amplitude of the high frequency field. Again there is an interaction, as the high frequency alternating voltage reacts upon the electron current, modulating its velocity and density. Theoretical calculations indicate that in this case also there are other possible oscillation states, which we cannot go into here.

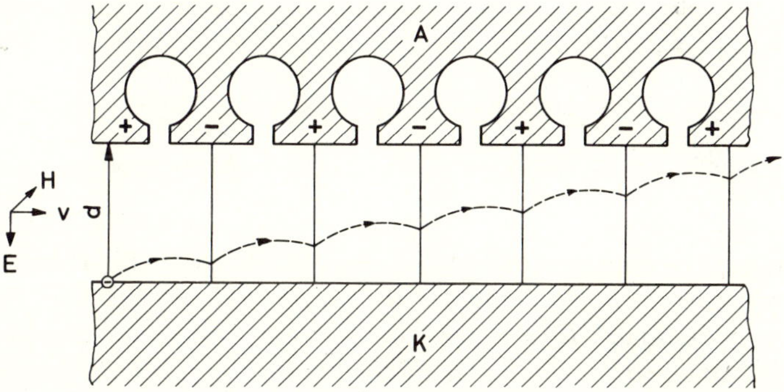

FIG. 140. The same as in Fig. 133, but with the resonators cut into the anode block.

The arrangement with two flat electrodes is inconvenient even when the anode is provided with the so-called pockets, because the electrons pass out of the system at the transit velocity. As shown in Fig. 141, it is advantageous to bend the two electrodes into two axially symmetrical cylinders forming a circular system in which the input and output coincide. The cathode is placed inside the anode, and the magnetic field runs transversely through the anode axis.

At a positive anode potential the electrons would travel radially in a straight line to the anode if there were no magnetic field. The axial magnetic field deflects the electrons into circular paths.

Figure 142 gives schematic illustrations of the electron paths for three different magnetic fields. In the absence of a magnetic field, $H = 0$, the electrons would travel from the cathode to the anode by the shortest possible path, i.e., along a radius. Under the influence of a magnetic field the electrons are deflected from their radial paths, and if the critical magnetic field intensity H_k is surpassed they will not reach the anode. As curve III shows, they return to the cathode

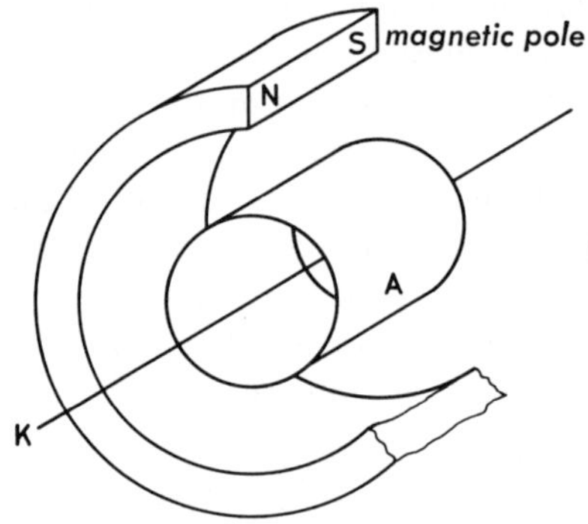

FIG. 141. The same as Fig. 133. Instead of flat electrodes a rotationally symmetrical arrangement is employed.

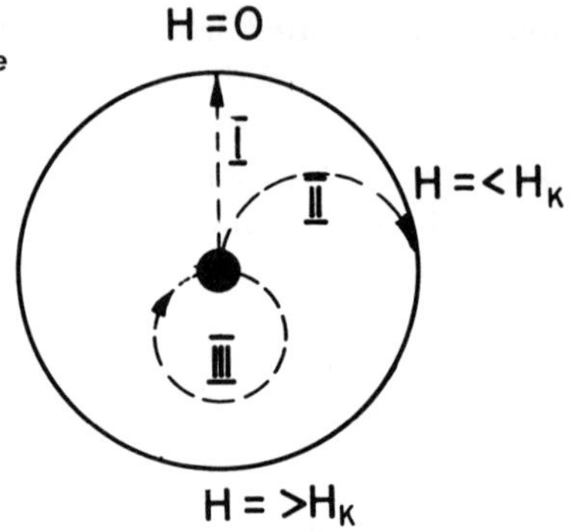

FIG. 142. Electron orbits between the cathode and the cylindrical anode as a function of the magnetic field.

and may complete several loops. Figure 143 shows the curve of the anode current as a function of the magnetic flux density at a fixed anode potential. If the magnetic field is too strong, no current flows from cathode to anode; and if it is too weak, the entire emission current is obtained. Between these two extremes is the critical state II. Here the curve shows a nearly abrupt leap from the zero value to the

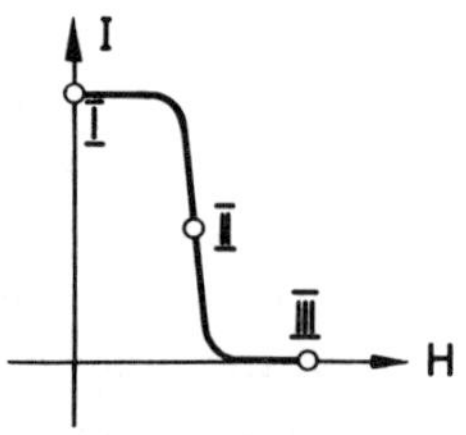

FIG. 143. The anode current as a function of the magnetic field strength.

saturation current. The sharpness of the transition is only slightly softened by the space charges of the electrons. Figures 144 and 145 illustrate the conceptual development of the anode system in terms of equivalent circuits from the original and simplest version, the solid, cylindrical anode, to the multi-slot cavity resonator in which during oscillation there is an alternating field between the individual segments, that can be divided into two inversely rotating fields. Since the delay line is self enclosed, i.e., the output is directly in contact with the input, the instrument acts as a regenerative traveling-wave oscillator. The rotating "electron spokes" produced by the formation of groups

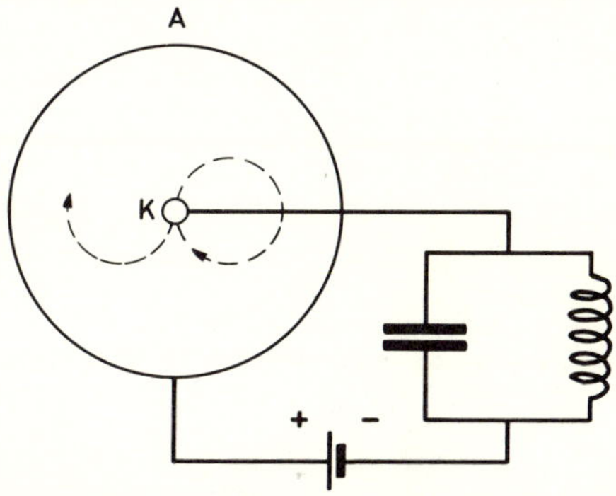

FIG. 144. Coordination of electron motion by means of a resonant circuit.

deliver their power by way of the pole of the resonant circuit to an inductively coupled loop. The operating zone of such so-called cavity resonator magnetrons is in the centimeter range; its efficiency is very high, being about 80%. The continuous power output is of

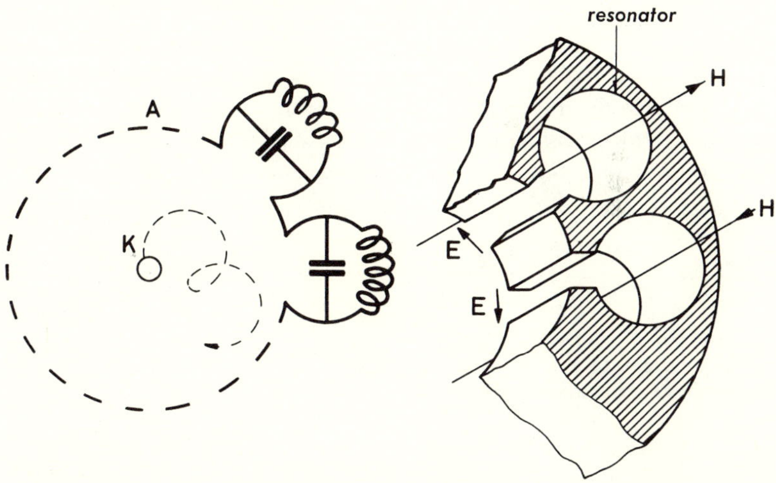

FIG. 145. Cut-away section of the copper anode block containing the machined grooves that act as cavity resonators.

the order of several hundred watts. Under shock loads, during pulse operation, it is capable of delivering many thousands of kilowatts of h.f. power. Figure 146 shows a top and side view of a magnetron in which may be seen the cylindrical multi-slot anode with its eight resonator cavities that form a self-enclosed delay line of the filter type. In the center is the coaxially situated cathode filament *f*. The north and south poles of the magnetic field are aligned with this same axis. In practice the positive anode, which is made of copper and

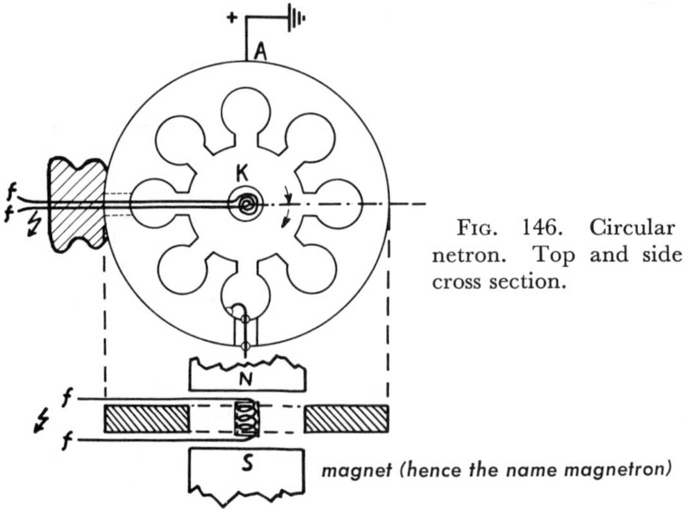

FIG. 146. Circular magnetron. Top and side view, cross section.

provided with cooling fins, is grounded. The cathode is consequently insulated against high negative voltage from the ground. The anode voltage is extremely high and is at least 30,000 volts. The principal method of supplying the anode voltage is shown in Fig. 147; this is

FIG. 147. Pulse operation. The anode voltage is fed into the cathode as a negative pulse.

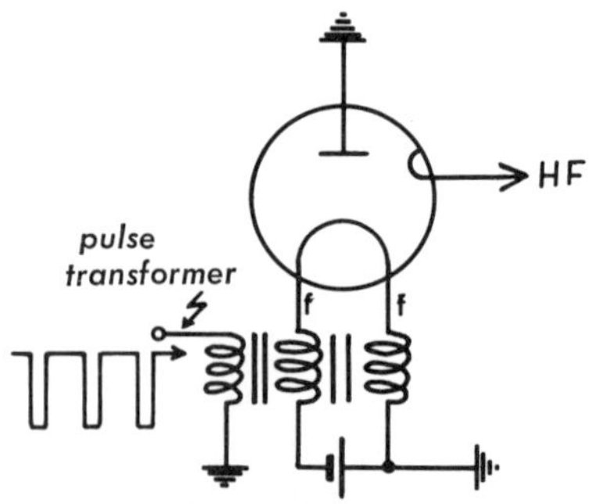

the usual method in radar installations. The anode voltage, which in other cases is usually positive, is here supplied to the cathode as negative potential by a so-called pulse transformer. Thus, for example, 2000 pulses per second of 30,000 volts potential are supplied for a constant output length or pulse length of about 1 microsecond in the circuit.

At present there are millimeter wave magnetrons for wavelengths of 2.5 mm that deliver peak pulse powers of from 1–2 kilowatts and possess an efficiency of 1%.

For still shorter wavelengths in the millimeter wave range, passive frequency multipliers are used. Semiconductor diodes are favored for this purpose. Up to 18 harmonics can be obtained as a result of the non-linearity. The limit for such tubes is approximately 1/100 mm, where the damping rises to 60 db.

Attempts are always being made to invent new tubes. Figure 148 illustrates the basic principle of a new type of reflection transit-time

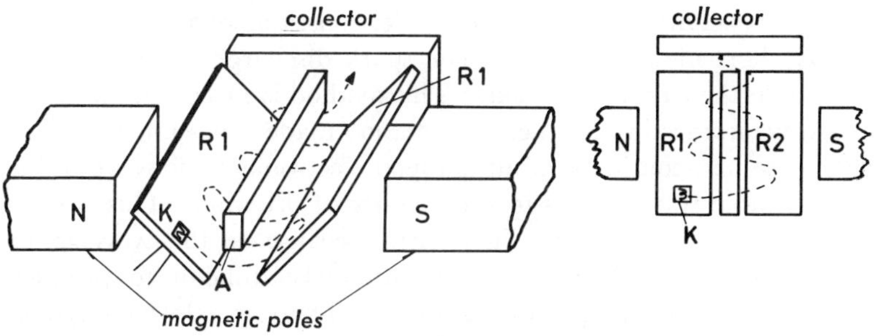

FIG. 148. Strophotron, a special reflection transit-time tube (perspective view and top view).

tube, the strophotron. A parabolic electric field is formed between the anode A and the two reflectors R_1 and R_2. Electrons are emitted from the cathode K and as a result of the transverse magnetic field travel underneath the anode instead of directly to it.

The phase-focussed electrons oscillate between the two reflector electrodes. The frequency of a mechanical pendulum is independent of the amplitude of oscillation, and analogously here the phase-focussed electrons oscillate at a fixed frequency. The electrons reach the collector electrode only after their kinetic energy has been transformed into high frequency energy. The frequency can be varied over a wide range and is, in this case also, a function of the magnitude of the electric field. Tubes of this type are capable of delivering about 1 to 2 watts high frequency power over the range of 1000 to 2000 Mc/s at a respectable efficiency of 20–30%. Still another type of system operates with the so-called electron-wave tube. The delay line is replaced by a second electron beam. Only at the input is there a short delay line, the purpose of which is to modulate the previously mentioned electron beam by an h.f. signal. Electron groups are formed and create a field similar to that of an h.f. wave traveling along a filter line. An electron beam with somewhat higher velocity, that is focussed independently of this field, is fired into the field. The usual interaction results in amplification. The h.f. energy is uncoupled at the output by means of a short uncoupling delay line. If frequencies with wavelengths below 1 mm are desired, other systems are used, some of which, however, have not yet been perfected. The simplest and oldest method (1924), the so-called mass radiator of Madame Glagelova-Arkadyeva, uses a Hertz spark gap for high frequency excitation of miniature dipoles consisting of a large number of very short metal particles. The spark discharge causes these to oscillate in their natural period. This method is successful up to the long-wave infrared range, i.e., this method made it possible to obtain with these non-continuous, but damped electromagnetic wave trains demonstrable wavelengths in the vicinity of 0.1 mm. This is considered a great scientific achievement because as the wavelength becomes shorter it becomes more and more difficult to produce demonstrable high frequency radiation. There is a lower physical limit to the wavelengths that can be produced by traditional electrical means. To be specific, in the light-wave range nature generates quantic rather than coherent radiation. The waves are of a double nature and exhibit the characteristics of both wave and particle. Electrodynamics merges into quantum theory. Under these conditions it is understood that there is no longer a continuous interaction between the electron beam and the h.f. field; instead the energy exchanges proceed in discontinuous quantic leaps. Nevertheless, it was necessary to devise new methods for obtaining the shortest possible wavelengths by generating undamped waves. Undamped waves are most effectively produced electronically with tubes. A magnetron with very small slots in the anode block seemed to be effective for this purpose. If one of the upper harmonics of the approximately 5 mm wave obtained in this manner is filtered out, it is possible to obtain waves in the range of 1 mm and a little below.

Still another method is based upon nonuniform acceleration of electrons in accordance with Fig. 149. This leads, on energetic grounds, to the emission of electromagnetic radiation. One such tube employing this principle is known as an undulator tube. The electron beam from an electron gun is phase-focussed by means of a filter type delay line and passed with an extremely high acceleration voltage through a row of magnets of alternating polarity. The periodic magnetic field results in a harmonic motion of the electron beam that is perpendicular to the translational motion. This results in nonuniform acceleration and forces the electrons to emit electromagnetic waves whose frequency is increased as a result of the Doppler effect, in the direction of the translational motion and thus

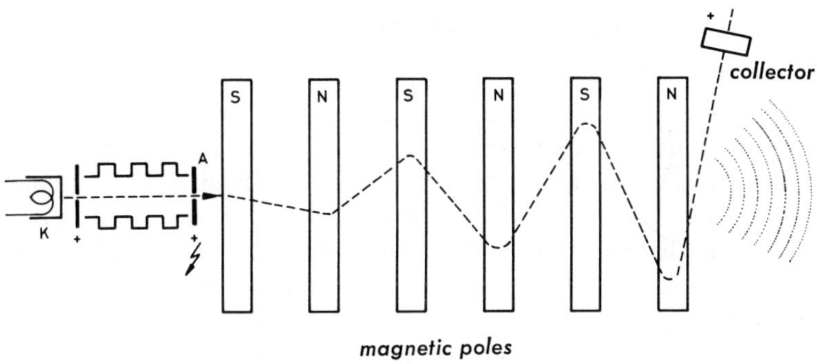

FIG. 149. Undulator tube. The generation of 0.1 mm waves by means of the relativistic Doppler effect.

for an observer stationed to one side. To generate extremely high frequencies, for example, of the order of 300,000 Mc/s, an acceleration energy of 5 Mev must be expended. According to the latest data, the pulse output of such a system is nearly one watt. The undulator tube is part of the attempt being made by microwave engineers to fill up the gap in the electromagnetic wave spectrum between the radio range (about 0.1 mm wavelength) and the range of thermal and light radiation emitted by atoms and molecules (about .001 mm). This should give some indication of the direction now being taken in the study of microwaves.

VI. SEMICONDUCTOR ELEMENTS (TRANSISTORS)

1. Germanium and semiconductor diodes

Semiconductors now appear capable of replacing electron tubes. Here the electronic phenomena take place in the crystal lattice rather than in a vacuum. When new elements such as these offer competition to existing instruments they must offer certain advantages. Without at this point going into all the details, it may be said of all semiconductors that the absence of heating offers distinct advantages, in that heat losses are consequently eliminated and the element is ready for immediate operation. Their small dimensions, light weight, and relatively great physical strength, since there are no glass parts, offer further advantages. It has also been found that the operating life is considerably higher than that of the hot-cathode tube. Semiconductor components are represented by two electrode elements (the semiconductor diodes) or three electrode elements (triodes or transistors). Transistors are also made with more electrodes.

The germanium rectifier, the successor to the galena detector of Professor Braun (1906), familiar for many years as a demodulator in radio receivers, is still physically simple enough to be understood easily.

The basic material is germanium (Ge), which is located between the conductors and nonconductors in the periodic system of the elements (Chapter I, Fig. 3); it possesses 32 electrons, 4 of which are in the valence shell. The existence of germanium was predicted in 1871 by Mendeleyev, the creator of the periodic system, fourteen years before its actual discovery in Saxony as mineral germanium oxide mixed with silver. It was named germanium in honor of the country in which it was discovered. The color of germanium lies between that of silver and zinc. It has a melting point of about 950°C and a specific gravity or density of 5.34 g/cm^3. Pure metallic germanium is obtained by a complicated purification process. The oxide is first reduced with hydrogen, and a zone melt process is then employed to remove almost all of the remaining impurities. In the course of the zone melt process most of the foreign atoms (in this case undesirable impurities) pass from the solid-liquid interface into the liquid phase. The process is conducted in the following manner: a germanium ingot is heated in sections in a collier; it is introduced into a conveyor-equipped high frequency furnace provided with separate heating zones, where it is drawn together with the collier over a separator pipe. The concentrated impurities are taken off in the liquid phase,

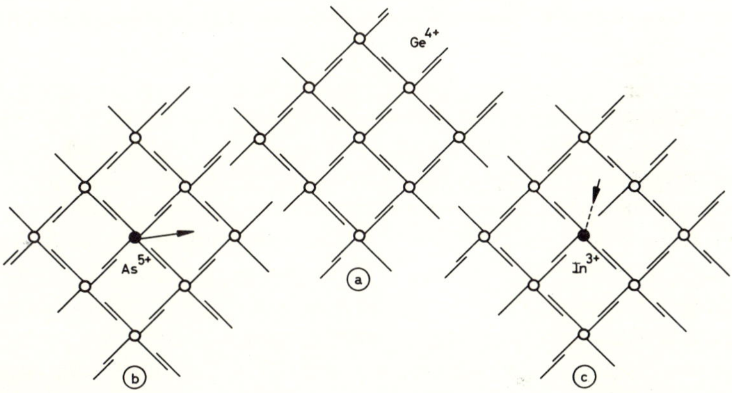

FIG. 150. Schematic representation of the atomic structure created by the bonds formed by pairs of electrons in the valence band of a germanium crystal (plane projection): (*a*) pure Ge crystal; (*b*) a foreign pentavalent atom (arsenic) has slipped in as a defect in the structure; (*c*) a foreign trivalent atom (indium) has been incorporated as a defect.

and the solidified material consists of a more purified germanium. The impurity concentration, i.e., the ratio of foreign atoms to atoms of pure germanium can be reduced by this method to $1:10^9$. In the course of hardening the germanium crystallizes into a cubic lattice system. It has the same structure as a diamond, which is a typical crystal of this sort. A diamond is an allotropic form of pure carbon, which is also tetravalent. Generally when the same types of atoms are joined in a crystal the structure is based upon the valence. Silicon is also one of the tetravalent elements. However, we will take up germanium first. Each atom or ion is surrounded by four neighbors and connected to an ionic lattice by means of the four outer valence electrons. As Fig. 18 in Chapter I shows, the four bonding electrons that together form an energy band have almost the same energy. They differ from each other by no more than approximately 10^{-22} e.v. The manner in which the atoms are "held together" is shown in Fig. 150. In the drawing the three-dimensional structure is simplified and represented by its two-dimensional projection on the picture plane. The circles represent the ionized atoms, and the valence electrons are represented by lines, because these illustrate more graphically the nature of a bond in a force field. The atoms are held together in a semiconductor crystal of this type by the forces resulting from the tendency of each electron in the valence band of an atom to form a "pair" with an electron of a neighboring atom. Germanium has four valence band electrons that give rise to these bonds. To break one such bond requires an energy of 0.75 e.v. At absolute zero germanium is an insulator, whereas metals are usually superconductors at absolute zero. But even at room temperature, when thermal motion dislodges electrons from their positions, germanium still retains extraordinarily high resistance. The specific resistance of germanium is approximately 50 ohm cms. (The pure germanium that exhibits this specific resistance is known as i-Ge, which is an abbreviation of intrinsic-Ge.) The number of freely mobile electrons and holes increases exponentially with the temperature and declines progressively as the width of the forbidden zone increases, as shown in Fig. 17. For practical purposes this thermal semiconductivity is insufficient. But if several pentavalent foreign atoms are incorporated into the crystal lattice either during the hardening of the crystal or through diffusion, each foreign atom possesses one quasi-free electron that is available for electrical conduction. Among the pentavalent atoms that may be used are antimony (Sb), arsenic (As), etc. Each of these elements possesses five electrons in the valence shell, and consequently the positive nuclear charge is also one unit higher. As shown in Fig. 150*b*, it is this extra, fifth electron which does not fit into the lattice structure and wanders about as a mobile, negative charge carrier. It might be mentioned in the interest of completeness that the changes in conductivity at absolute zero are also in accord with the experiment and that when arsenic atoms are introduced into the crystal the bond energy per free electron is only 0.01 e.v. This fifth electron requires little energy—the energy supplied by the heat at room temperature is sufficient—to be raised to the conduction band, where it acts as a negative charge carrier. The remaining four electrons necessarily serve as bonds to the crystal lattice. In this way the original insulator is transformed into a defect-site semiconductor. A ratio of 1 such foreign atom to 10^7 germanium atoms is sufficient to reduce the extremely high, and initially purely temperature-dependent specific resistance of about 0.5 M ohm to the order of approximately 100 ohm or 100 mΩ/mm^2.

As is evident from the foregoing, the conductivity or charge carrier density of a semiconductor can be determined by the amount (concentration) of foreign substances that have been introduced during production.

The starting materials must be of almost inconceivable purity before the impurities are introduced. There must be no more than 1 atom of impurity to 10^8–10^9 germanium atoms. Some sort of picture of this can be obtained if one tries, as suggested by Professor Strutt, to imagine the presence of 20 criminals among the entire population of the earth. Germanium that has been deliberately doped with a controlled quantity of an impurity such as arsenic is known as an *n*-type semiconductor because it possesses an excess of negative charge carriers. The pentavalent impurity is known as a donor. If the pure germanium is doped with a trivalent foreign material, a so-called acceptor, instead of a pentavalent impurity, the trivalent material takes the place of a germanium atom in the lattice. This means that there is a deficit of one electron in the germanium crystal. Thus the situation is reversed (Fig. 150*c*). Indium (In) is an example of such a trivalent substance; others are easily found in the periodic system. The four germanium neighbors will always have four valence electrons available for bonding into the ionic lattice. But the foreign atoms have only three valence electrons. Consequently each of these foreign atoms is short of one bonding electron. These deficit sites or so-called holes of the foreign atoms are under certain circumstances capable of trapping neighboring electrons and making up the deficit. This, however, necessarily results in a new hole in the neighboring atom, which in turn strives to capture an electron from one of its neighbors. The hole shifts its position as the empty places are successively filled with electrons from other sites, and a chain reaction is created. This is known as hole migration. Since from an electrical viewpoint a hole is the absence of a negative charge, it has the opposite properties of an electron and is identical with a positive charge. Thus, the germanium, which in practice is doped with only a very small quantity of this trivalent impurity, acquires free, positive charge carriers, and this germanium is known as the *p*-type. This hole conduction, which is a new conduction mechanism for us, travels in accordance with its sign in the direction opposite to that of electron conduction. The hole current consequently flows from the positive to the negative electrode, or in the opposite direction from the electron flow in a radio tube. Conduction of electricity by positive charges was demonstrated as long ago as 1889 by Professor Rowland. Using static electricity, he created negative charges on a round disk of insulating material; he then caused the disk to rotate at a high velocity. The moving train of negative charges formed a circular path and also a circular current. This current gave rise to a magnetic field in the same way that such a field is created by the flow of electrons through a coil. The orientation of the field was determined by means of a compass. In the second part of the experiment the negative charges were replaced by positive charges. On rotation of the disk, it was observed that the orientation of the magnetic field was reversed. To obtain the same orientation of the magnetic field as in the first instance, the direction of rotation had to be reversed. The obvious conclusion to be drawn from this experiment is that a positive charge travels in the opposite direction from a negative charge. In defect-site conduction in semiconductors electron conduction is contrasted to hole conduction, which travels in the opposite direction, and this will have to be borne in mind in further considerations of the subject. In semiconductors, then, there are two types of current carriers: electrons and holes. Ionic conductors are not included among the semiconductors. The following table summarizes the classifications that recur throughout this chapter.

Impurity spots	*Examples of foreign atoms*	*Designation*	*Current carriers*	*Type crystal*
pentavalent	antimony	donors	electrons	*n*-Ge
trivalent	indium	acceptors	holes	*p*-Ge

A current-direction-dependent resistance, a diode, is obtained when an *n*-type and *p*-type germanium are intimately joined to form a single unit. The contact surface forms a so-called junction or barrier that exhibits special physical properties.

From a technological viewpoint these diodes require the production of *n–p* transitions. If a melted drop of indium is heated on a foil of crystalline germanium that has already been doped with trace quantities of a pentavalent element to form *n*-germanium, some of the liquefied germanium penetrates the liquid indium. After congealing, the germanium recrystallizes in the transition zone as *p*-germanium, because some of the indium remains dissolved. By means of this alloying process an abrupt *p–n* transition is obtained.

The germanium diode shown in Fig. 151 consists of *n*-germanium on the left of the junction and *p*-germanium on the right. The charge carriers (electrons and holes) are indicated by plus and minus signs. The free electrons of the *n*-germanium diffuse across the boundary

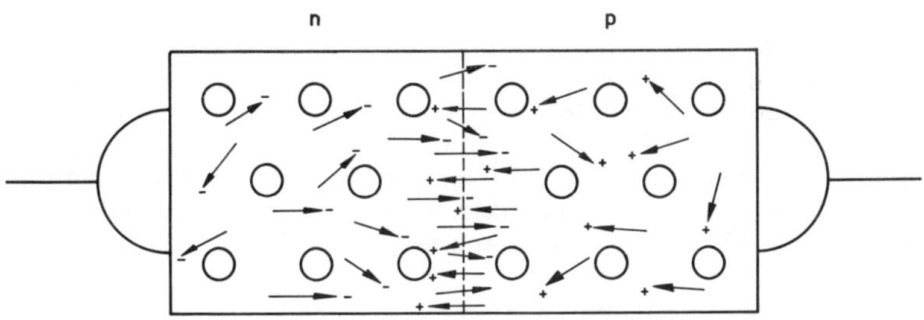

FIG. 151. The junction of an *n*- and *p*-germanium forms a semiconductor diode.

layer even into the right side of the *p*-germanium. Conversely, the positive charge carriers, which are also mobile, seep across the boundary layer into the *n*-side. Since the atoms, from an electrical viewpoint, were originally neutral and the number of electrons was originally exactly equal to the nuclear charges of the protons, the diffusion of charge carriers leads to a disturbance of the electrical equilibrium. The *p*-side of the junction receives more negative charges, and the *n*-side receives more positive charges. The diffusion of the electrons toward the right side and the diffusion of the holes toward the left create electric fields in the area of the boundary that oppose the equalization processes, giving rise to an energy equilibrium. The present illustration simplifies the processes at work, and later we will have to introduce refinements. If the circuit is coupled across an external potential source, as shown in Figs. 152*a* and 153*a*, so that the negative pole is connected to the *n*-side, the positive charge carriers, the holes, are drawn to the left across the junction and into the current circuit. Meanwhile, the positive pole on the right side of the *p*-germanium draws electrons out, so that these charge carriers are also enabled to flow freely through the junction, in this case from the left to the right. The supply of charge carriers is maintained at a constant level by the voltage source. A germanium diode connected in this

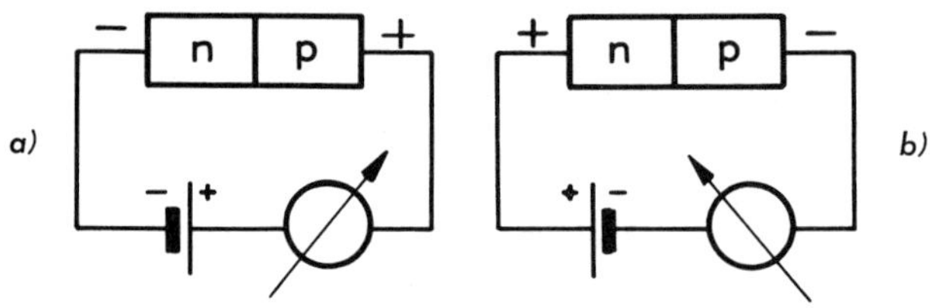

FIG. 152. (*a*) A diode operating in the forward direction. (*b*) A diode operating in the reverse direction. If an external voltage of such polarity is applied to the free sides of the two semiconductors that the charge carriers of both sides are drawn into the junction, a current flow occurs (forward direction). On the other hand, if the voltage is of such polarity that the two types of charge carriers are drawn away from the junction, impoverishment occurs and the current is reversed.

fashion operates in the forward direction. The resistance in practice is very low. The situation is quite different when the poles are reversed, as in Fig. 152*b* and 153*b*. The positive pole of the voltage source is now connected to the *n*-side and the negative pole to the *p*-side. The excess electrons that have migrated into the *p*-side are

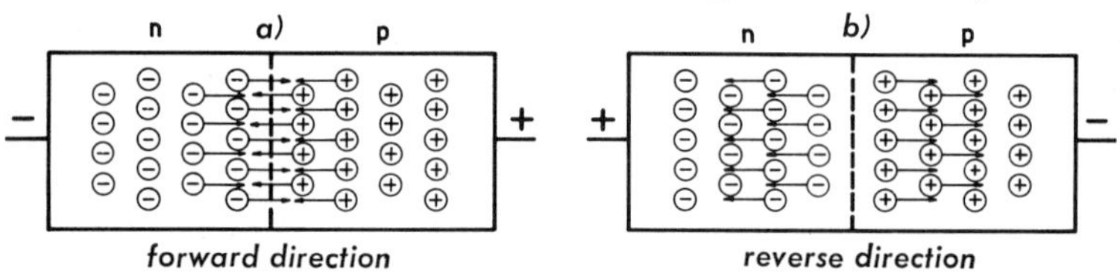

FIG. 153. Same situation as in Fig. 152, showing movement of electrons.

sucked back and deplete the zone of the junction. The same holds true for the right side, where the holes are drawn back toward the negative pole. The junction is thereby emptied of charge carriers. Since there are almost no charge carriers present, the current flow becomes extremely low. This so-called reverse current direction is called the blocking direction, and the layer in which the phenomena take place is known as the blocking layer. The influence of other factors keeps the current from being blocked completely but in practice the resistance is very high. The characteristic of this type diode with different current and voltage scales is given in Fig. 154. A diode of the type just described is known as a surface rectifier. If the diode is supplied with an alternating current, it will rectify the current. Only the half-waves in the forward direction will pass unobstructed, whereas the reverse motion will be suppressed.

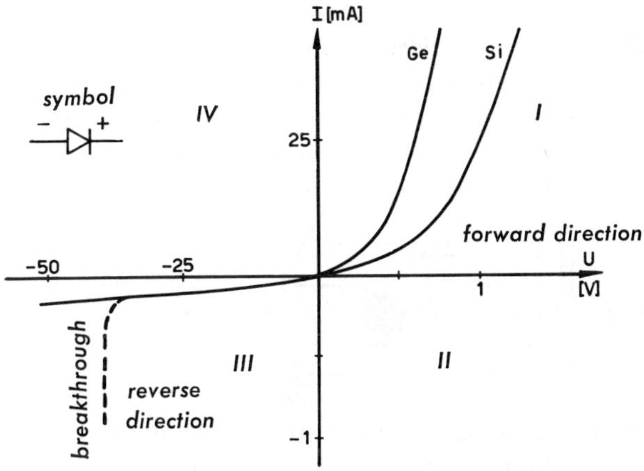

FIG. 154. Current-voltage characteristic for forward and reverse operation of a diode. Notice the difference in scale of the quadrants I, II and III, IV.

In explanation of the rectifier mechanism it can also be said that when the current is traveling in the forward direction the internal electric field is broken down, with the result that the diffusion current prevails. The electrons diffuse increasingly into the bordering *p*-zone, and the holes flow into the *n*-section. When the current travels in the blocking direction the situation is reversed; the diffusion-suppressing electric field is strengthened, as a result of which both types of current carriers are forced back to their positions.

In the point contact rectifier or point contact diode (Fig. 155) a thin tungsten or platinum wire spring is pressed against an *n*-germanium crystal. In the production process this wire is fused to the germanium crystal at the point of contact by means of a brief current pulse. In the small area of the weld, i.e., around the metal point, the germanium is transformed into *p*-germanium. The contact surface between the barrier layer formed around the point and the basic germanium crystal is extremely small. This results in an extremely low capacitance, which makes this circuit element very well suited for high frequency use. As in a crystal detector the weld is not absolutely necessary. It is of course necessary for the impurity in the basic germanium crystal to consist of a pentavalent element such as antimony for an electron current to flow only in the direction crystal-to-point. With *p*-germanium the current direction must be reversed and the point is made negative.

While point contact electrodes are used for high frequencies, contact surfaces similar to those in barrier junction selenium rectifiers are used for high power output at lower frequencies. The rectifiers of English railroad locomotives are equipped with such germanium diodes and deliver more than 750 kilowatts of power.

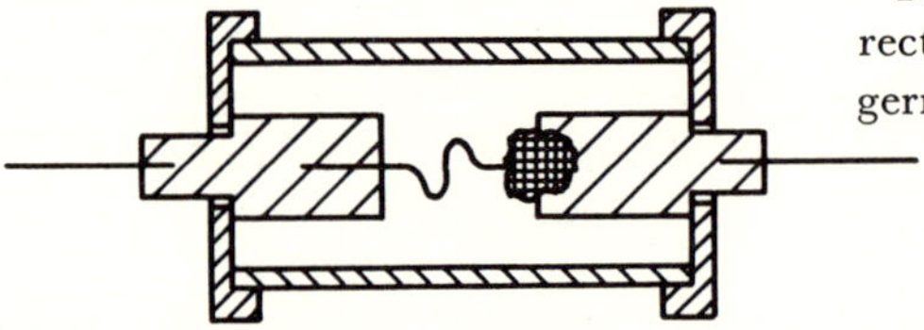

FIG. 155. Point contact rectifier (detector) with a germanium crystal.

Since the operating temperature of germanium must not exceed 75°C, germanium is replaced as a raw material by silicon, which is also tetravalent, when especially high power is required. Silicon, which belongs to the same column of the periodic system as germanium, has the same structure but exhibits greater thermal stability and can be used at operating temperatures up to 150°C. The reason for this is that silicon possesses a larger forbidden energy gap between the valence and conduction bands than germanium. It is therefore more stable and its valence electrons do not become conduction electrons when the temperature is raised. The melting point of silicon is approximately 1430°C and the boiling point 2300°C. Silicon, at 28.06, has a lower atomic weight than germanium. The nucleus, with an atomic number of 14, is surrounded by three successive shells containing respectively, 2, 8 and 4 electrons, or a total of 14 electrons.

Our description of its conduction mechanism may have given the erroneous impression that a germanium diode can in fact serve as a commercial current source. This, however, is not the case. To be sure, at room temperature a diffusion voltage of about 0.3 volts is created at the barrier zone or so-called space charge zone, since the charge carriers of the one zone diffuse across the barrier zone into the other zone. The remaining ionic lattice, i.e., the atom fragments, create a field that retards further exchanges. The voltage is connected to the individual particles and follows their concentration in the *p–n* transition zone; consequently it drops with the densities of *p* and *n*. Equations for the correct picture of the physical phenomena involved can be obtained from quantum mechanical considerations, but also from the characteristic of Fig. 154. If the basic requirement is satisfied, that in the voltageless state ($U = 0$) the current I is also zero, we arrive at the mathematical formulation of the so-called rectifier equation:

$$I = I_0 \exp\left(\frac{e \cdot U}{kT}\right) - I_0$$

This follows from the basic principle of the thermodynamic equilibrium. I represents the current flowing through the rectifier; I_0 the saturation current in the reverse direction; e the electron charge; k the Boltzmann constant $= 1.38 \cdot 10^{-23}$ wattsec/deg K; T the absolute temperature; U the voltage across the transition. Experiments show good agreement with the formula. An understanding of the conduction mechanism in the area of the junction between two different types of semiconductors involves greater difficulties than are met with in homogeneous semiconductors or crystals. When charge carriers are present in it, a crystal behaves in a manner similar to a metal. In Chapter I the total number of conduction electrons was compared to an electron gas because in the metal they move in all directions with various velocities. As with gas molecules, the average velocity of the electrons increases as the temperature is increased. For many phenomena and calculations the statistical behavior is quite sufficient. In these cases the conventional principles of gas kinetics can be applied. In other cases the problem is more effectively considered from the standpoint of an individual electron. In addition to the charge carriers made available by the foreign atoms, one must at the same time take into account the holes and electrons that are liberated or that disappear as a result of the temperature level. The process is quite complicated. We will, therefore, attempt to clarify first several generally valid individual phenomena, beginning with the phenomena of the quasi-free electrons.

Once it is in motion the free electron travels at a uniform velocity until it collides with one of the lattice atoms of the crystal and rebounds elastically in accordance with the laws of reflection. It then travels in a new direction, once again following a straight path, until it collides again with a lattice atom obstructing its path. In this way it travels back and forth inside the crystal structure. By means of the experimental and calculated data for gases it is possible to compute the mean free path length between any two successive collisions in a germanium crystal. It has been found that the average path length traversed by electrons in a germanium crystal at room temperature is about several thousand times as great as the distance between individual atoms. According to Chapter I, interatomic distances are of the order of about 10^{-8} cm. This should give some idea of how far apart the atoms are from each other and how small the electrons are. The kinetic energy $E = (m \cdot v^2)/2$ of a freely mobile electron, which is proportional to the temperature, in view of its origin, can also be computed. The relation $E = k \cdot T$ gives an electron velocity of approximately $v = 100$ km/sec at a room temperature of 20°C. The average time between two collisions can be calculated from the velocity and the mean path length. Statistically, the electron collisions, which occur in all directions, neutralize each other externally. As mentioned in Chapter II, these collisions manifest themselves externally only as statistical fluctuations in potential or as noise. The situation changes when an electric field is applied. Under these conditions forces are brought to bear upon the electrons. The charge carriers are no longer accelerated uniformly in all directions; they are now accelerated in one direction and retarded in another, and this corresponds to electrical conduction. The so-called drift velocity of the electrons in an electric field is proportional to the field strength. In practice this directed velocity is much lower than the velocity of the random thermal motions. As we remember from Chapter I, the electron drift velocity in metals is extremely small and is of the order of magnitude of several cm/sec.

It must be remembered that ideal, pure semiconductors contain an equal number of quasi-free electrons and holes.

External excitation such as that caused by light quanta $(h \cdot f)$ or thermal energy $(k \cdot T)$ may activate electron-hole pairs and give rise to spontaneous current in pure germanium. By analogy with the photon (the light quantum), the magnitude $k \cdot T$, which represents the thermal influence, is considered as a type of thermal energy quantum and is known as the so-called phonon. It is to be understood in this way: that no electron can be liberated in the absence of a describable cause, such as a phonon. The effect of the temperature T, that will liberate electrons from the atomic bond is given by the following formula: $n^2 = 10^{32} \cdot T^3 \cdot e^{-8700/T}$. In this formula n represents the number of electrons per cm^3. The Boltzmann constant k is taken into account in the exponent of the exponential function, as in all thermodynamic considerations. The conductivity σ (in ohm^{-1}) is a function of the diffusion mobility as well as of the number of electrons and holes that are liberated. The relationship is expressed by the formula: $\sigma = e(\mu_\ominus \cdot n + \mu_\oplus \cdot p)$, where n represents the number of electrons and p the number of holes per cm^3, while μ represents the mobility, i.e., the velocity at a field strength of 1 volt/cm. The flowing current I is calculated from Ohm's law: $I = \sigma \cdot U$. Earlier measurements with a so-called filament transistor of nearly pure germanium gave the following approximate values for the diffusion mobility:

Electron mobility $\mu_\ominus \cong 3600$ cm^2/voltsec

Hole mobility $\mu_\oplus \cong 1750$ cm^2/voltsec

For silicon, the corresponding values are:

$$\mu_\ominus \cong 1400 \text{ cm}^2\text{/voltsec}$$

$$\mu_\oplus \cong 440 \text{ cm}^2\text{/voltsec}$$

Nowadays the Hall effect is utilized (see Chapter I) to measure the electron and hole mobilities. The sign of the Hall potential differs accordingly for electron or hole conduction, and the Hall constant also differs.

The formula derived in Chapter I for the Hall effect

$$U_{\text{Hall}} = \frac{\mu_0}{e \cdot n} \cdot \frac{I \cdot H}{d}$$

can be solved for

$$n = \quad \text{or} \quad p = \frac{\mu_0 \cdot I \cdot H}{e \cdot d \cdot U_{\text{Hall}}}$$

where n or p is the concentration we are attempting to determine, i.e., the number of electrons or holes per cm^3. Since there are $4.5 \cdot 10^{22}$ atoms of germanium in each cubic centimeter of pure germanium and each free electron or hole belongs to a foreign atom, the concentration can be precisely determined for each alloy.

The drift velocity v with which the current carriers are propagated as a result of the electric field E (in volt/cm) is obtained from the formula $v_{\text{cm/sec}} = \mu \cdot E$.

We will hear more of the drift velocity and its significance further on. The mobility of the charge carriers is one of the factors that sets the upper frequency limit. If the mobility is doubled the upper frequency limit is also doubled. In this connection germanium is superior to silicon. For this reason new semiconductors or mixtures of various elements that exhibit especially high charge carrier mobility are being sought after. Another important property is the size of the gap between the valence and conduction bands (see Fig. 17), i.e., the forbidden zone. The size of the forbidden zone determines the amount of energy required to move an electron from a bound condition in the valence band to the conduction band, where it is available to carry current. The greater the distance between the two bands, the higher the permissible operating temperature. For an electron to cross a larger gap more thermal motion is required to supply the necessary energy. The operating properties of the semiconductor are limited by the amount of undesirable spontaneous current (dark current) which results from a kind of spontaneous ionization. On the other hand, semiconductors with a small band gap possess an advantage in that little energy is required to stimulate a current flow. In internal photoeffects (see p. 234), where light quanta are absorbed, elements with small band gaps are desirable. Indium-antimony exhibits a high mobility of 23,000 cm^2/voltsec, with a band gap of only 0.17 e.v. Also in Hall generators, i.e., in elements capable of being influenced magnetically, this property is advantageous.

Now let us consider the junction of a germanium diode, as shown in Fig. 151. As a result of thermal motion there are free bonding electrons and holes present in both *n*- and *p*-germanium at normal room temperature. As we have already mentioned, the left side of the junction in the zone of the *n*-germanium acquires more of the positive charges, which are firmly bound to the crystal, and the right side, more of the negative charges. Since the germanium atoms, and the foreign atoms as well, are by nature externally electrically neutral, it is logical for an excess of electrons to accumulate on the left side. They are consequently called the major carriers. If a diode is operated in the forward direction the charge carriers diffuse into the zone of opposite conduction type. The holes migrate into the *n*-zone and the electrons into the *p*-zone. Since there are numerically fewer of these elementary charges than of the current carriers arising from the foreign atoms, the former are called minor carriers. On the right side the situation is reversed. The holes are the major carriers and the electrons are the minor carriers. Since the charge carriers are in motion, electrons, as probability calculations show, will continually drop into the deficit slots and recombine. From an electrical point of view they are then neutralized. On the basis of the probability of recombination it is possible to determine with considerable accuracy the average life ($\approx 10^{-4}$ sec–10^{-5} sec) and the average diffusion length ($\approx$0.1–1 mm) for the mobile charge carriers. Thermodynamic equilibrium requires that the number of charge carriers that recombine is always equal to the number of new charge carriers created. Thus it follows as a basic principle that the more often an electron finds a hole and the more of the two types of carriers available, the higher the number of recombinations. If a *p–n* diode is operated in the reverse direction, the minority carriers that are within one diffusion length of the junction tend to migrate across the junction. If the "reverse potential" applied is high enough, it is possible in this way to generate the saturation current I_0. It can be seen from Fig. 154 how in the absence of other influences the curve becomes a saturation line in the third quadrant. The saturation current is attained, as we have said, when all the minor carriers collect at the junction. There is a barrier junction when there are no more charge carriers available. The low current carried through the barrier junction by the minority carriers is limited by the diffusion. If the reverse potential is increased, the electric field across the barrier junction is strengthened. A breakthrough occurs after a certain field strength is attained. In Fig. 154 this effect is marked by a dotted line. A few years ago C. Zener discovered that a high external potential accelerates the electron-hole pairs and free electrons so greatly that they collide with other atoms of the crystal lattice, creating new charge carriers and thereby multiplying themselves. This effect occurs at a critical field strength. The multiplication of the two types of current carriers on both sides of the junction results in an abrupt rise in the current. It has also been found that this great field strength also liberates electrons from solid bonds. This effect is observed even in homogeneously doped, single-type semiconductors. Such germanium resistances which become conductive as a result of contact ionization of the impurity spots above a certain breakthrough potential and at low temperatures (4.2°K) are called cryosars. The type of contact ionization we are dealing with here is comparable to that of a glow discharge in a gas-discharge tube. We must be very careful here to distinguish between two different effects, the field emission under discussion here and the so-called tunnel effect, which is encountered in semiconductors with highly doped *p–n* transitions and requires a small potential hill to function. The action mechanism of these semiconductors will be described with the Esaki diode.

The field strength at the *p–n* transition required for the breakthrough in the case under discussion (in reverse operation) is of the magnitude of several 10^8 volts/cm. This phenomenon, which is manifested as an almost instantaneous onset of current resulting from so-called field emission, is called the zener effect. The breakthrough can be compared to the processes occurring in a gas discharge, where after a certain definite limit is exceeded there is a sudden avalanche-like onset of contact ionization. Fortunately, the breakthrough is reversible. If the blocking voltage is lowered again, the zener current

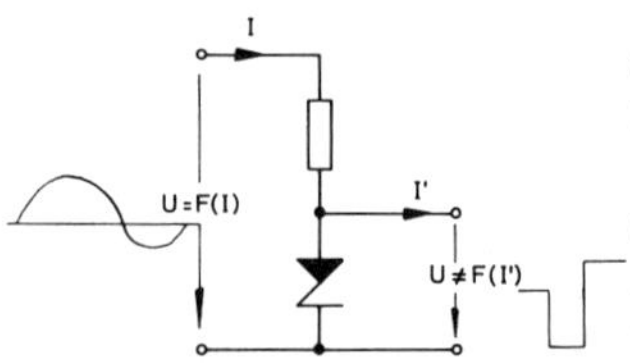

FIG. 156. Symbol of a zener diode and the circuit arrangement of a voltage stabilizer with a zener diode. Since there is a sudden onset of zener current at the barrier junction, the zener diode is suited for cutting off voltage peaks above the zener voltage; it can therefore be used to obtain square-wave voltage from a sinusoidal voltage curve.

returns to the level from which it had risen. The breakthrough voltage depends upon the thickness of the barrier layer but also upon the incorporation of appropriate intermediate layers. For a given construction it is extremely constant and can be used as a voltage standard. Zener diodes are used for stabilizing voltage fluctuations, particularly for voltages between 6 and 30 volts, depending upon the design. Figure 156 shows the circuit of a voltage stabilizer using a zener diode. The basic material used for the zener diode is silicon, with its lower atomic weight, because it possesses a more favorable characteristic than germanium. This is because in silicon the energy difference between valence band and conduction band is greater than in germanium. Figure 157 gives the characteristics of silicon zener power diodes. The internal resistance, the so-called differential resistance in the zener zone, is extremely low as in a stabilizer cell and is of the order of 1–20 ohms, depending upon the type. The maximum zener current at the present time is from 0.2–0.8 amp, depending upon whether or not the diode is operated with heat dissipation.

If overvoltages of alternating currents and, consequently, in both directions, are to be prevented by the use of such semiconductor diodes, the two diodes are connected in the reverse direction in series but with reversed poles. The avalanche-like breakthrough occurring in the range of permissible operation should therefore not be confused with breakdown or other destructive processes in a circuit element. The relationships at the junctions of semiconductors can be better observed by comparing the relationships of the energy level diagrams of the various materials. It can be seen in Fig. 16 how the electron orbits (the energy levels of which are represented by horizontal lines), which are still sharply defined in the individual atoms, spread to form wide energy bands as a result of energetic interaction in the atomic

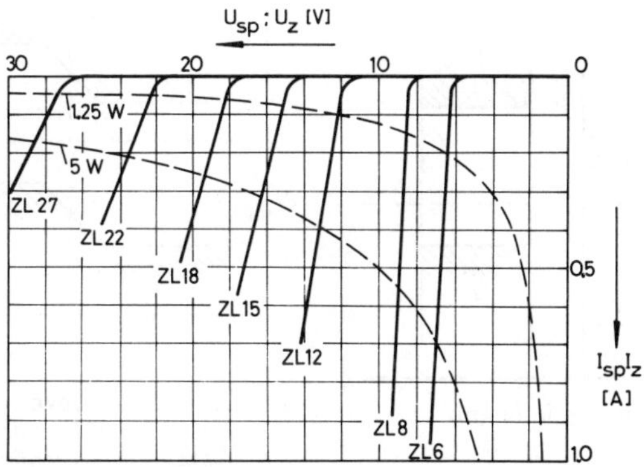

FIG. 157. Characteristic of a zener diode. The reverse current is a function of the reverse voltage.

aggregate. Of special interest here is the gap between the valence and conduction bands with regard to the impurities.

The Fermi level in semiconductors with current carrier concentrations (doped semiconductors) lies within the gap. It sets the energy limit at which at a temperature of absolute zero all the places are occupied by electrons. With increasing temperature more and more electrons move into higher zones at the expense of the lower zones (spontaneous current). If a semiconductor is doped with donors, the Fermi limit is raised from its initial position in the middle of the gap and brought closer and closer to the conduction band (Fig. 158*a*). At a definite concentration of charge carriers the Fermi level is actually displaced into the conduction band, and the semiconductor then exhibits the electrical properties of a metal. In a semiconductor that has been doped with acceptors, i.e., a *p*-type semiconductor, the Fermi level drops, as Fig. 158*b* shows, below the middle of the gap. At a certain concentration the Fermi level is displaced to such an extent that it enters the valence band. If *n*- and *p*-semiconductors are brought together, the two Fermi levels are equalized as a result of the thermodynamic equilibrium, as long as no external voltage is applied. A potential wall arises, as shown in Fig. 159. When an external voltage is applied, the Fermi levels are displaced in relation to each other, and the potential wall is also displaced. Figure 160 shows band models

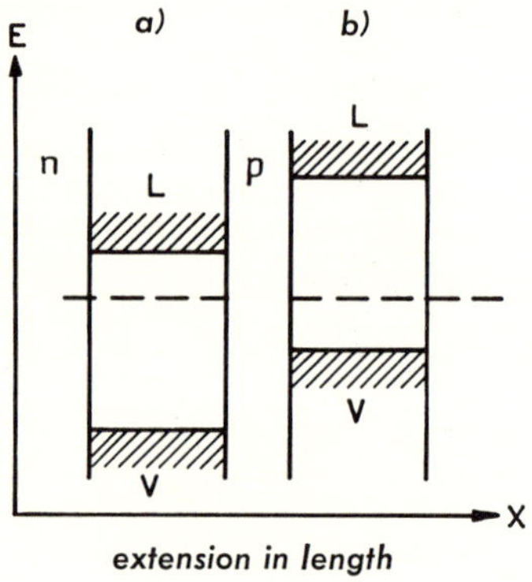

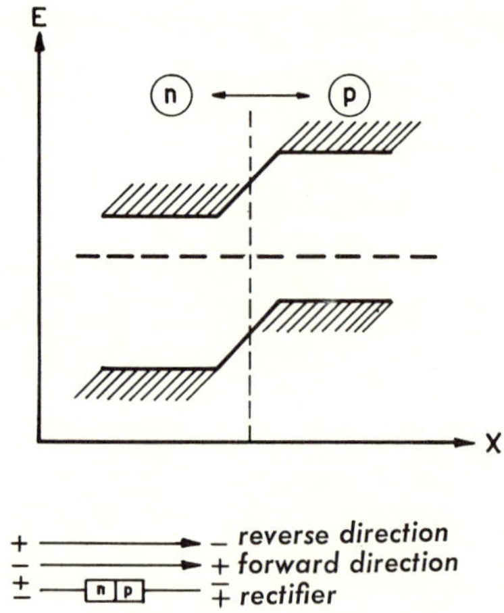

FIG. 158. Two separate semiconductors brought together.

FIG. 159. *n–p* transition in the band system. Two semiconductors joined without preliminary voltage.

of four typical cases in which the Fermi level is not deformed. Figure 160*b* shows how the potential wall is lowered when the poles of the current source are so arranged that current passes through the diode. In Fig. 160*c* the current source is coupled in the reverse direction and the potential wall and the Fermi level on the *p*-side are raised. In Fig. 160*d* the reverse voltage is raised to such an extent that the lower edge of the conduction band in the *n*-zone is lower than the upper edge of the valence band in the *p*-zone. Though it would be reasonable to assume that a charge carrier would require a great deal of energy to scale so high a potential wall, it has been found that there is another conduction mechanism that can be called into play that will lead to a flow of current. This mechanism is the so-called tunnel effect, which is particularly intense when there are thin walls between the two boundary zones or the two potential energy systems. Quantum mechanics shows in connection with the wave conception of small particles that even if there is insufficient energy a thin potential wall will allow passage or will be tunneled through. The thinner the potential wall, the greater the tunnel current. In spite of all expectations the *p–n* transition will then no longer block the current as the reverse potential is increased; on the contrary, a current flow is obtained. This current has been studied by Zener with his tunnel diode.

Silicon and germanium are among the most important elements in semiconductor technology. Both these elements are very expensive; 1 kg of germanium (the pure crystal) costs about $758 (3250 Swiss francs) and 1 kg of silicon about $3965 (17,000 Swiss francs). Silicon

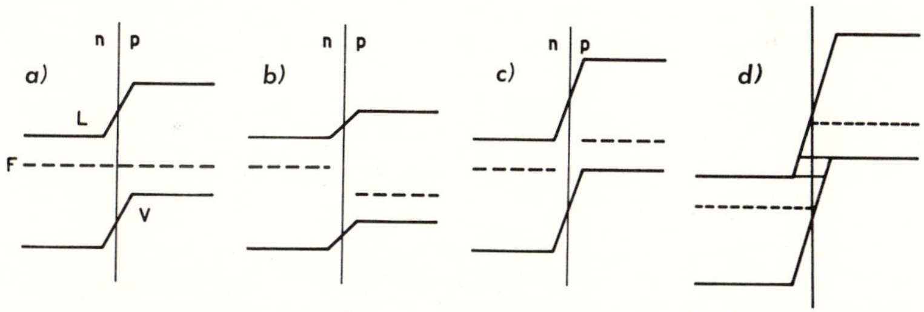

FIG. 160. Band models in which the Fermi level is not deformed. (*a*) Without external potential; Fermi level equally high. (*b*) With potential in forward direction. (*c*) With potential in reverse direction; potential wall is raised. (*d*) Lower edge of conduction band in *n*-zone is lower than upper edge of valence band in *p*-zone.

compounds are very common and account for about 40% of the earth's crust. But purification of such compounds as quartz SiO_2 to the pure crystal is very expensive.

In classical semiconductor rectifiers for power transmitters, which we will merely mention here in the interest of completeness, selenium is used. Because of the poor electrical conductivity of selenium a selenium rectifier requires a very thin (0.1 mm) selenium layer joined on each side to a large surfaced metal contact. On the low resistance side a material such as nickel or bismuth, which does not act as a barrier layer, is used. The compounds of these metals with the selenium, which form selenides, are very good hole conductors with high conductivity. Since nickel and bismuth are expensive it is customary to make use of iron plates provided with a thin coating of the contact material. The selenium layer mentioned previously is then coated onto this. For the other side of the selenium layer, where the rectification is to occur, cadmium is used as the metal electrode. The selenide of cadmium in contact with the pure selenium forms an excellent barrier layer. Cadmium, however, is also expensive, and in practice it is replaced by a low melting cadmium alloy that is applied with a spray gun. The characteristic of a selenium rectifier lies between that of the germanium and silicon rectifiers. In the rectifier characteristics the curve on the forward

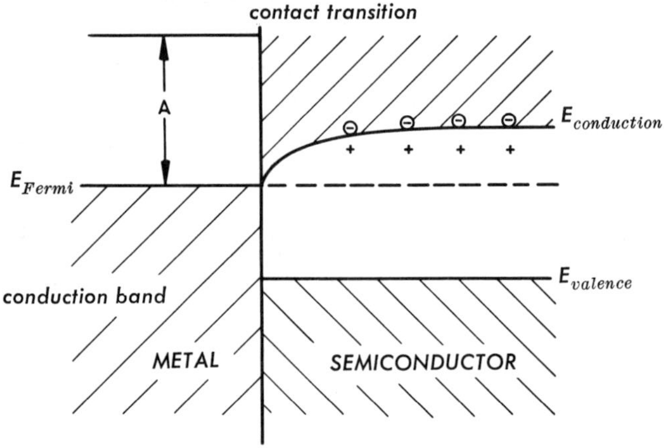

FIG. 161. Band system of a suitable metal–semiconductor contact transition.

direction side, after an initial curvature, becomes more or less a straight line. The extension of this portion to the voltage axis gives at the point of intersection with the abscissa the gap or gate voltage. It shows a gap width in accordance with Fig. 17. The barrier layer lies between a selenium layer (−) and a counter-electrode of tin alloyed with cadmium (+). Also familiar are cuprous oxide (CuO_2) rectifiers, which are used, among other applications, as rectifiers in moving-coil measuring instruments for alternating current measurements. Here the barrier layer is formed between the copper and the cuprous oxide layers. Here the threshold voltage is lower than in the selenium rectifier, i.e., the forward-direction current starts to flow at a low positive voltage.

Lead contacts to the semiconductor electrodes must not form barrier layers with them. For this reason the energy systems of the metal and the semiconductor must be suited to each other in the manner shown in Fig. 161, so that no new boundary layer is formed. Therefore the choice of a suitable ohmic contact material is severely limited. Since the operating power is limited by the temperature, rectifiers are frequently provided with cooling fins. In this way the heat is dissipated into the surrounding air. Another factor limiting the operation of these rectifiers is the maximum permissible voltage that will not result in breakthrough or breakdown. If higher potentials must be rectified, several diodes must be connected in series. Because of the ohmic scatter of the individual elements a safety factor must also be allowed for in the calculations.

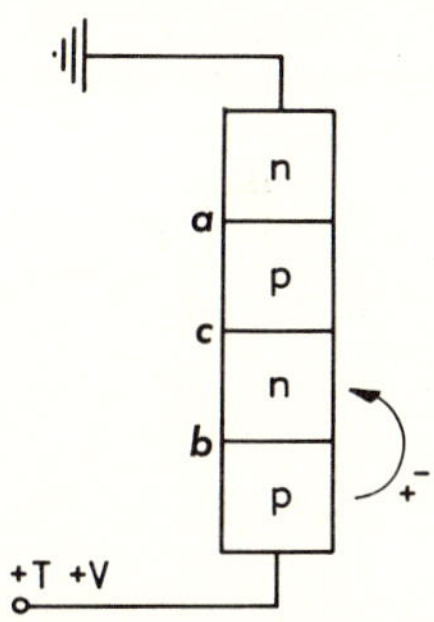

FIG. 162. The *a* and *b* transitions are connected in the forward direction, and *c* is connected in the reverse direction.

Another new modern semiconductor circuit element is the four-layer bistable diode. It operates on the barrier layer principle. Its characteristic is shown in Figs. 162 and 163. The drop in the curve between zones I and III is typical of this type of element; the resistance here is negative and values of up to approximately −300 ohms are attained. As the voltage increases the current decreases here instead of increasing. Like the dynatron this type of circuit element with its negative characteristic can be used to reduce damping in an oscillatory circuit. Ohmic resistances are eliminated and oscillations are thereby produced. With R- and C-elements saw-tooth curves are obtained. It is this property that enables the four-layer diode to replace tubes or transistors. As in the magnetron special measures are necessary to separate the input and output circuits. This circuit element is particularly effective as an electronic switch. As the voltage is increased one passes from a high impedance of the order of magnitude of several megaohms to a low impedance of the order of 10–20 ohms. The condition of the circuit is thus a function solely of the supplied operating voltage or current. The hatched section in the negative zone of the curve indicates that the rise and drop in current do not lie exactly in the same curve. The conditions here are similar to those in a piece of iron that is magnetized and demagnetized; hysteresis is involved in this case also.

The tunnel diode or Esaki diode, which also possesses a characteristic with a negative zone and is also particularly well suited for high frequency operation, is somewhat similar. Values of up to 10,000 Mc/s are obtained at operating temperatures between −273 and 100°C. The noise factor is very favorable, lying at about 5 db. Figure 163 would represent an Esaki characteristic if the positions of the ordinate U and the abscissa I were exchanged. Also, in this special diode the reverse current is greater than the forward direction current. Naturally, the diode is used only in the interesting area of

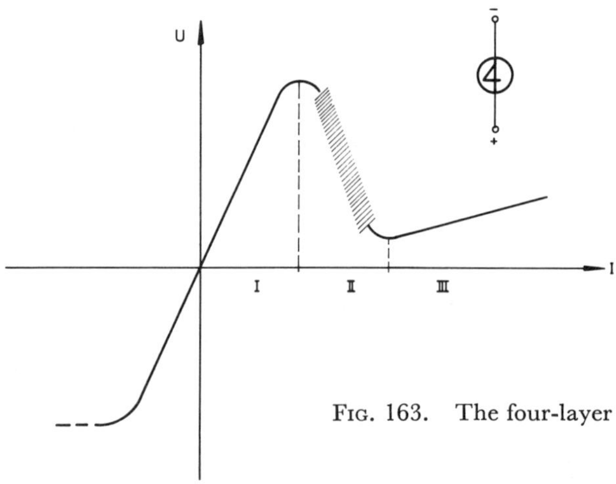

FIG. 163. The four-layer diode.

the forward direction current. The current maximum, i.e., the peak between zone I and zone II is situated at about 0.035 volt.

The most significant aspect of the work of the Japanese scientist Leo Esaki are his studies of the *p–n*-junctions of semiconductors with highly doped *n*- and *p*-zones. The incorporation of many impurity spots results in semiconductors containing many available, quasi-free current carriers.[1] A diode of this type has a very low ohmic resistance and exhibits the electrical conductivity properties of a metal.

We will now compare the potential curves shown in Fig. 164 of a

[1] A pentavalent foreign atom will supply one electron for current transport after being supplied with a small quantity of energy. A trivalent foreign atom under similar circumstances will supply one hole or a deficit of one electron.

lightly doped semiconductor diode and a heavily doped semiconductor diode.

The threshold is the measure of the minimal energy required for electrons to pass from the *n*- to the *p*-zone. Only the fastest electrons are capable of passing over the potential threshold. Even in the equilibrium condition there are always electrons that are capable of climbing the barrier, but other electrons fall back to the bottom of the slope, with the result that in the middle no current flows (noise source).

The height of the potential barrier depends upon the charge carrier density, which is to say, upon the concentration of impurities in the semiconductor. When the semiconductor is heavily doped and there are many foreign atoms present, the diffusion pressure of the charge

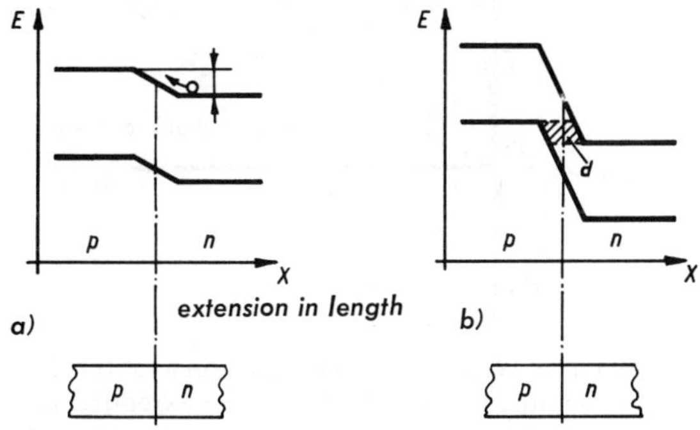

FIG. 164. Comparison of potential curves of lightly and heavily doped semiconductor diodes.

carriers is high. The diffusion and the electric field currents, which maintain the equilibrium of the boundary zone, both become high. The potential threshold, as shown in Fig. 164, is relatively high compared to that of a lightly doped diode.

If a voltage is applied to the diode in the reverse direction the energy bands are further separated and the potential barrier is raised. If the poles are reversed and the diode is operated in the forward direction, the potential wall is lowered. This is the typical behavior of semiconductor diodes; it is shown in Fig. 165 as the line O E F G (normal diode current) and describes the general conduction mechanism of conventional semiconductor diodes.

If the semiconductor is heavily doped and there is a very thin barrier layer, which is shown by a thin intermediate zone *d*, a new supplementary conduction mechanism is observed. Figure 166 shows that current carriers at this point are able to tunnel through at low energies from the *n*-layer to the *p*-layer without having to pass over the potential wall. This current path, which is known as the tunnel effect, is possible because the valence and conduction bands are very close to each other.

Wave mechanics assigns waves to electrons. The existence of these waves can be confirmed by the demonstration of interference

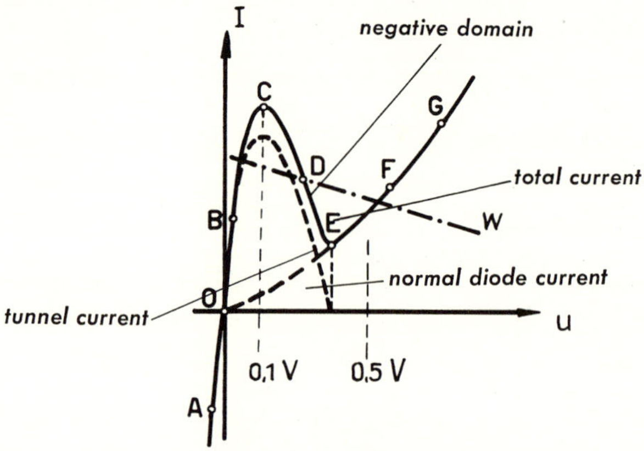

FIG. 165. The curve OEFG is the typical course of the $U - I$ characteristic for normal semiconductor diodes. The extended line shows the superposition of tunnel current and normal diode current. W = the line of resistance that arises at the corresponding work resistance. In this way a bistable circuit (flip-flop) can be constructed.

phenomena. In the light of this physical theory it can be shown that an electron moving from the right side of this energy barrier will have a high probability of reappearing on the left. The degree of probability increases enormously as the width of the barrier is decreased. In order to understand the reason why electrons pass through the transition zone, our figurative model of a *p*- and *n*-semiconductor, as illustrated in Fig. 151, must be refined somewhat further.

In the previous energy schemes no special attention was given to the energy values of the current carriers of the impurity spots as compared to those of the normal lattice occupants. We will now take this into consideration.

The pentavalent foreign atoms in *n*-type semiconductors require only a small additional energy to give up their extra electron to the conduction band. The magnitude of the total internal work function is illustrated in Fig. 167. Since little energy is required it follows that we are concerned with an energy level that is not far below the conduction band L ⧹⧹⧹. In this case just a little additional energy is enough to raise the current carrier into the conduction band.

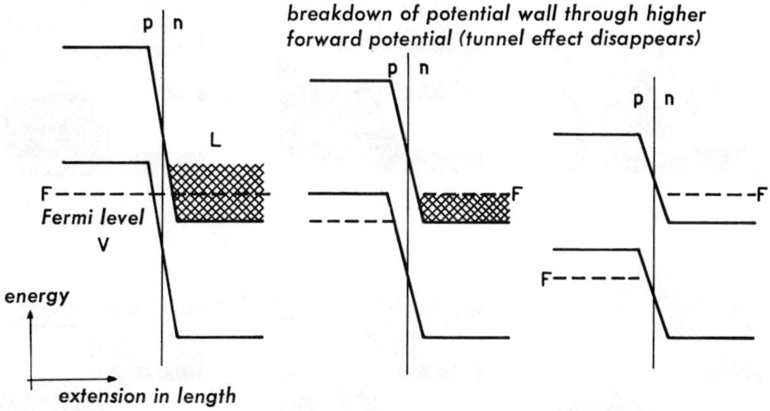

Fig. 166. Tunnel effect.

In the trivalent impurities (p-semiconductors) a little additional energy is all that is required for electrons to be captured from the normal neighboring atoms, which results in the creation of holes and gives rise to hole conduction. This may be observed, as shown in Fig. 167b, right above the valence band.

Fig. 167. Level-position of the n- and p-layers.

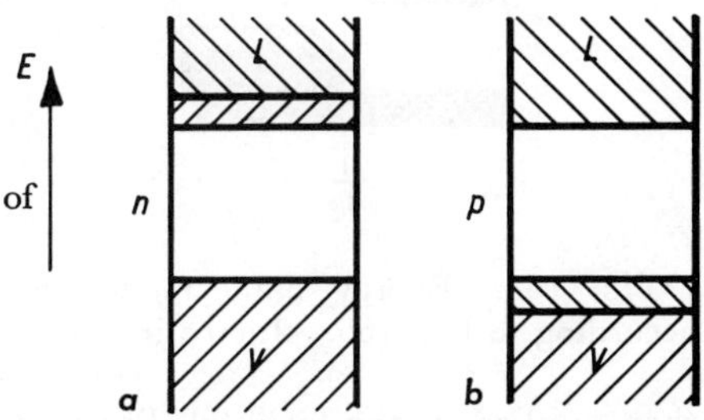

If we consider a highly doped n- and p-transition zone, in accordance with Fig. 168, we find that the energy levels of the foreign atom, which are hatched here, are very close to each other. In the absence of an external biasing voltage the Fermi levels of the n- and p-layers must be situated along a straight line (dashed lines).

However, when there is heavy doping, the Fermi level of the *p*-layer is situated in the valence band and the Fermi level of the *n*-layer in the conduction band (a so-called deformation). If a small voltage is

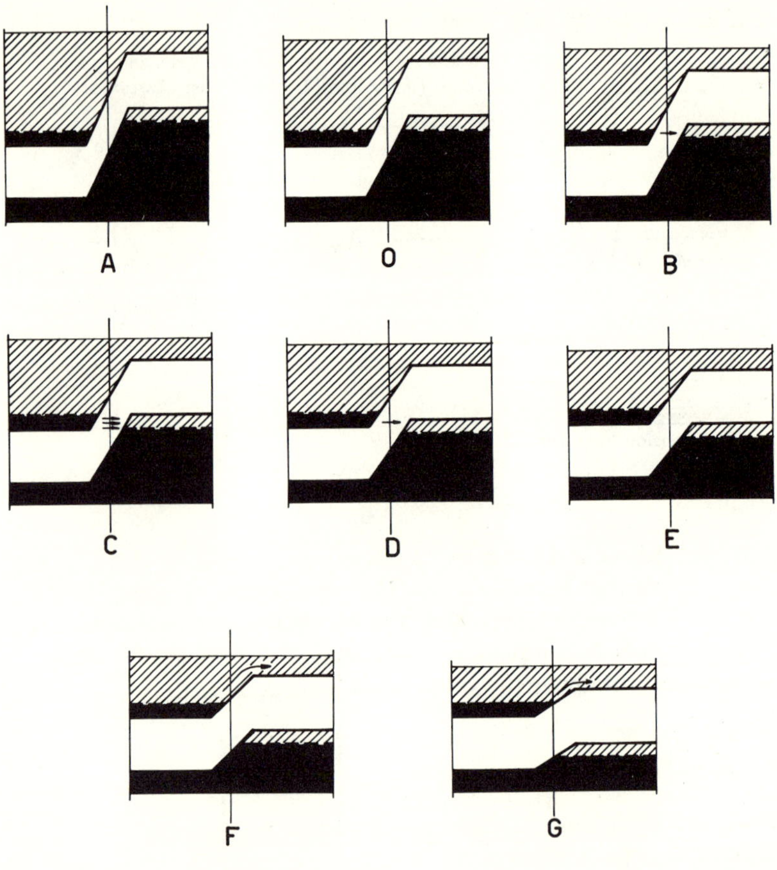

FIG. 168. Energy band models for the different working points according to Fig. 165. Increasing voltage in the forward direction.

then applied in the forward direction, the potential wall is leveled, the lower edge of the *n*-layer being raised and the upper edge of the *p*-layer being dropped, so that with an appropriate biasing voltage the transposed energy bands are brought exactly into line with each other. Here the free defect-site electrons on the *n*-side are on the same energetic level as the unoccupied holes on the *p*-side.

If the transition zone is moreover extremely thin, the probability of tunneling through is very great. A thin zone is energetically equivalent to a low tunnel-energy threshold. These conditions are satisfied by the heavily doped diodes studied by Esaki.

If the voltage in the forward direction exceeds approximately 0.1 volt, the hatched surfaces that were previously directly opposite each other move away from each other once more. The probability of tunneling through, or the tunnel current, consequently decreases again as the voltage increases above 0.1 volts; the tunnel current consequently exhibits a hill-shaped U-I characteristic (Fig. 165), the left side of which presents an ohmic course and the right a negative curve course. A negative curve course in a U-I diagram means that we are concerned with a negative resistance (increasing voltage—decreasing current). The final characteristic of an Esaki or tunnel diode is the sum of the usual diode characteristic and the previously cited tunnel effect.

Figure 168 shows the structure of the band models at various initial voltages (working points on the characteristic) of a tunnel diode. The first point taken is a reverse voltage, then zero voltage, and then the voltage is set at three different points in the forward direction.

Figure 168C illustrates with especial clarity the upper limit of the tunnel current, i.e., the point beyond which the current is not increased by raising the voltage, but falls again. From this point on, the tunnel energy bands (foreign atom hole and electron levels) are no longer directly opposite each other, but draw apart.

Figure 168D shows how an increasing potential in the forward direction results in the leveling of the potential wall, i.e., the Fermi levels are shifted toward each other. The hatched overlap, which is the determining factor for the tunnel effect, becomes smaller until, as in Fig. 168E, it disappears completely.

As soon as the conduction band boundary on the *n*-side rises to a level higher than the valence band boundary on the *p*-side, the tunnel current disappears completely. As the initial forward direction voltage is increased the normal diode current that is created by the familiar diffusion effect increases, so that the curve gradually moves upward.

As a result of the heavy doping, which had previously been used essentially only in Zener diodes, a steeply rising current curve is also encountered in the reverse direction in Zener diodes. The effect here, as in the Zener diode, is attributable to the tunnel effect, and is characterized by the massive onset of a breakthrough current at a certain potential, which depends upon the extent of the doping.

Tunnel diodes offer the advantage of faultless performance at high frequencies. This is not surprising since the conduction mechanism in this case does not involve the diffusion through of neighboring minor carriers. The electrical conductivity of the tunnel diode is based exclusively upon the activity of the major carriers or basic carriers, as a result of which the velocity as in a metal is approximately equal to 300,000 km/sec.

Tunnel diodes are very rugged, two-electrode, electronic elements. They operate without disturbance over a broad temperature range and are quite insensitive to radioactive radiation. These diodes are, however, designed only for low power uses at present and are, moreover, extremely difficult to handle because the separation of input and output in such two-electrode elements can only be accomplished effectively with well-conceived separators.

At present they raise additional difficulties: To obtain very thin potential walls in the band models, heavy doping with foreign atoms is necessary; this results in an element with a much lower ohmic resistance than any other conventional circuit element.

The maximum–minimum current ratio is approximately 1:10.

If a working point in the area of the falling characteristic is selected (by appropriate choice of the forward direction voltage, approximately 0.3 volt), a negative resistance is encountered, which results in a current that does not behave in a usual manner, but decreases as the voltage is raised. The selection of this working point makes it possible to reduce the damping of oscillatory circuits that exhibit an unavoidable ohmic resistance. A diode operated in this manner is also known as an active element. It is not only capable of producing oscillation in oscillatory circuits, but it is also capable of amplification in amplifier circuits. The small size and solidity of construction make the tunnel diode a suitable circuit element for use in rockets. The negative resistances of these tunnel diodes are already being used for improvement of circuits. The use of amplifiers of this type allows operation over frequencies ranging from 0 direct current into the microwave range. At present tunnel diodes are available commercially for extremely short switching times and for frequencies of 2 Gigacycles/sec.

Today oscillators with these circuit elements are constructed for frequencies of up to 2.5 mm, though the output power is still of the order of microwatts. The basic materials in the construction of tunnel diodes are the arsenides of silicon, germanium, and gallium. In all probability future development will lead to the use of the tunnel effect in multielectrode systems, where input and output are more easily separated, as well as in diodes.

At present an intense study is being made of mixture proportions or doping ratios with all the appropriate elements of the 3rd, 4th, and 5th groups of the periodic system, in order to determine the optimal semiconductor properties for transistors, for Esaki and normal semiconductor diodes, and for solar batteries. The most important favorable properties are a wide gap between valence and conduction bands and high charge carrier mobility. The greater the gap width, the more energy is required to free an electron from the valence band, where it is bound to its own atom, and move it into the conduction band, i.e., into the overlapping or common orbit with the neighboring atom. The required energy for germanium is 0.7 e.v. and for silicon is 1.1 e.v. The larger the gap width, the higher the maximum permissible operating temperature becomes; thus, germanium operates unexceptionably at temperatures up to about 80°C and silicon at temperatures up to about 150°C.

Recently, gallium arsenide has appeared as a suitable material; it has a gap width of 1.32 e.v. Gallium phosphide has an even larger gap width of 2.92 e.v. This material possesses a current carrier mobility four times as high as that of silicon, and consequently its upper frequency limit is also four times as high. The current carrier mobility of the most common semiconductors, as determined by the Hall effect, compares as follows:

Germanium:	Electron mobility	3600 cm^2/voltsec
	Hole mobility	1750 cm^2/voltsec
Silicon:	Electron mobility	1400 cm^2/voltsec
	Hole mobility	440 cm^2/voltsec

The electron mobility of gallium phosphide is below 1000 cm^2/voltsec, and this semiconductor is distinguished as a rapid functioning element. On the other hand, indium antimonide, which is commonly used in Hall elements, has a high electron mobility of 57,000 cm^2/voltsec. The gap width here, however, is very small, about 0.17 e.v., and this disadvantage outweighs the previously mentioned advantage in transistor use. It appears, rather unfortunately, that high mobility can be obtained only at the expense of a small gap width. However, the final word of technology has not yet been spoken.

Before we go on from two-electrode to three-electrode semiconductor elements, we would like to consider certain aspects of the concept of "electronics."

In a rough classification the domain of electronics includes all current flows resulting from electrons, positive and negative ions, and holes in an evacuated or gas-filled chamber or in certain semiconductors. The phenomena resulting from electron or ion flow in metallic conductors or liquids, and from the displacement currents in condensers, arising from the relations between the bound atomic nucleus and its electron shells, belong to the domain of electrical

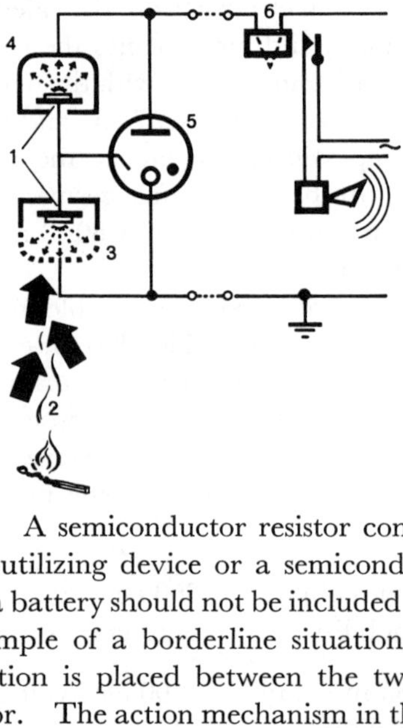

FIG. 169. Fire alarm.

science. A semiconductor resistor connected in series with another current-utilizing device or a semiconductor rectifier that is used to charge a battery should not be included among electronic instruments. An example of a borderline situation is the case where a radium preparation is placed between the two opposite plates of an open capacitor. The action mechanism in this case is somewhat different. The α rays ionize the air molecules, which split into positive ions and electrons that either travel a certain distance and recombine into neutral molecules or travel to the appropriate pole of the capacitor plates, which were initially charged with a d.c. voltage source. The capacitor collects the current carriers, and consequently the current flow can be measured in the external circuit. In principle this arrangement (see Fig. 169) could be used as a fire alarm. Combustion gives rise to particles that are large in comparison to the air molecules.

These particles (soot), which are usually invisible, move in the capacitor field much more slowly than the air particles. This leads to a longer stay and increases the probability of their recombination in the course of their path. In practice this means that the presence of such combustion particles would lead to a smaller current than that produced by the pure air molecules. This resistance change could be used to control an alarm system. A capacitor acting in its usual capacity as a passive element cannot be considered an electronic instrument, but it is an electronic instrument in this rather uncommon use because its functioning is essentially controlled by the externally influenced charge carriers. This concept of "control," taken together with the original rough definition, is an excellent aid in completing our understanding of the distinctive character of electronic instruments. In the next section we will take up semiconductor combinations that are capable of being controlled and that clearly belong to the domain of modern electronics.

2. Transistors

As we have shown, a local, variable concentration of charge carriers is formed at the boundary zone of, for example, a *p–n* transition. This concentration of charge carriers is comparable to the space charge formed in front of the cathode of a tube, and it is this requisite physical condition which enables semiconductors to be used in the place of tubes.

If three alternating layers of the different types of germanium are connected in series in a single unit as in Fig. 170, one obtains a so-called *p–n–p* surface-transistor (junction transistor). These transistors are characterized by two junctions: a *p–n* and an *n–p* junction. To function properly the distance between the two junctions, i.e., the intermediate *n*-zone, must be smaller than the average diffusion length for the charge carriers involved. The left "electrode" (as shown in Fig. 170) is known as the emitter and the right electrode as the collector. The middle layer, consisting of approximately 1/100 mm of *n*-germanium, is known as the base region. The first junction is operated in the forward direction; the emitter is supplied with a positive potential of approximately 0.1 volt with respect to the base. The second junction is operated in the reverse direction, a negative bias of 2–30 volts being applied to the collector. The available, freely mobile, charge carriers in the *p*-layer on the left are primarily positive (holes).

Since the emitter has a positive potential with respect to the base, these charge carriers are repelled from the left and attracted to the right; they are consequently accelerated toward the *n*-layer of the base. These holes are moreover injected into the *n*-layer and, as a result of diffusion, into the *p*-layer of the collector. Here the accumulation of holes is sucked away by the negative pole, or, more precisely, the holes at the boundary between the germanium layer of the collector and the metal contact are filled up with electrons from the metal and finally disappear.

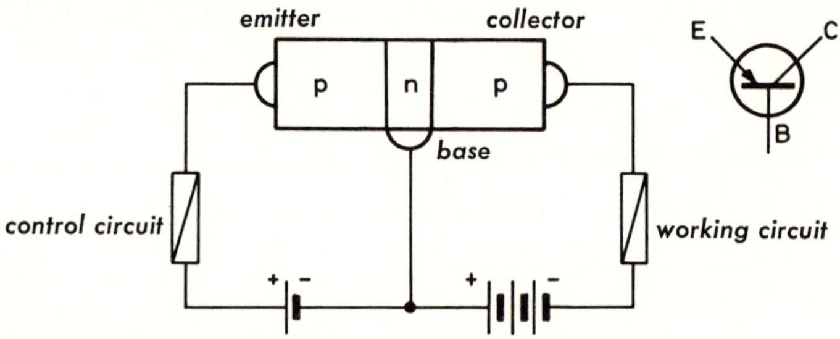

FIG. 170. Illustration of a *p–n–p* transistor with the schematic circuit symbols.

This brings us to the 3-electrode semiconductor element, which is equivalent to a triode and in which input and output are separated. The amplification effect of these elements was discovered by W. Shockley in the United States. This new circuit element was given the name transistor.

It follows from what we have said that the distance between the two junctions must be much smaller than the diffusion length. Otherwise the holes and electrons would recombine in the intermediate zone (the base) and would be lost to the collector as current carriers. In practice a certain number of these carriers are unavoidably neutralized. This is evident from the fact that a change in the emitter current results in a somewhat smaller change in the collector current. For the purpose of the following considerations it will be sufficient to assume that most of the current injected at the emitter reappears at the collector. Thus, all changes in the emitter current will be reflected in the collector current. The current amplification factor α lies in fact between 0.95 and 0.99. The voltage amplification figures are more favorable. The emitter circuit, possessing as it does a positive initial voltage and operating in the forward direction, exhibits a low ohmic resistance of about 50 ohms. The collector circuit, which is operated in the reverse direction, exhibits a high ohmic resistance. In practice the values are of the order of one megohm. Since the voltage is calculated from $U = I \cdot R$, the static voltage amplification v corresponds to the ratio of the output resistance to the input resistance, with the condition that the emitter and collector currents are equal, since

$$v = U_2/U_1 = \frac{R_1 \cdot I}{R_2 \cdot I} = R_1/R_2$$

The same factor holds for the power amplification. $v = N_1/N_2$ as a result of the relation $N_1 = I_1{}^2 \cdot R_1$ and $N_2 = I_2{}^2 \cdot R_2$. Because it is necessary to employ external resistances, the practical values are lower.

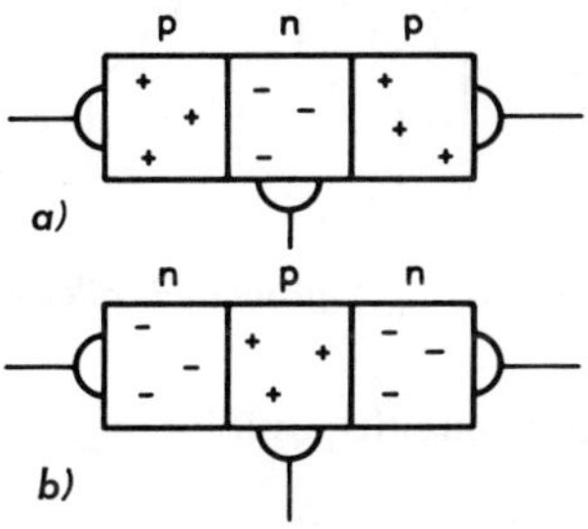

FIG. 171. (*a*) *p–n–p* transistor; (*b*) *n–p–n* transistor.

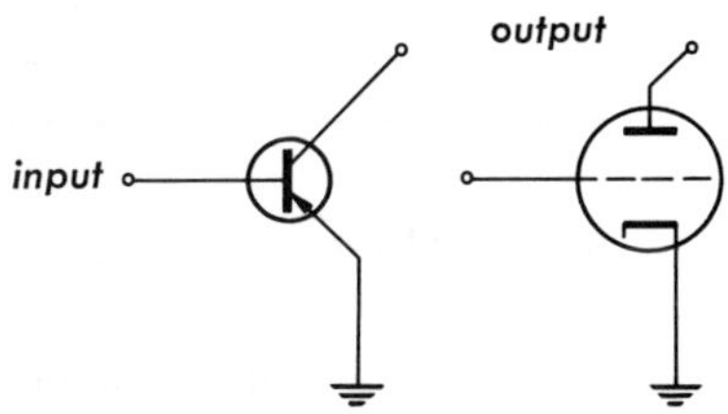

FIG. 172. Analogy between transistor and tube.

There are also *n–p–n* transistors as well as *p–n–p* transistors (Fig. 171). In these transistors negative current carriers are injected at the emitter and travel through the *p*-germanium layer in the direction of the positive collector. In this case the current is carried by electrons rather than holes. From the foregoing illustrations it can be seen that with appropriately controlled doping it is possible to cause the reverse junction current to consist almost entirely of either hole current or electron current. The diffusion velocity μ determines the frequency limit. Since for electrons $\mu_{\ominus} \cong 3600$ cm^2/voltsec and for holes $\mu_{\oplus} \cong 1700$ cm^2/voltsec, it is apparent that *n–p–n* transistors are better suited for higher frequencies. Thus, the frequency limit of *n–p–n* transistors is higher because electrons have a greater mobility than holes. Naturally, the voltage sources for these transistors must be connected in exactly the opposite polarity of the *p–n–p* transistors. There is a certain analogy between a transistor and a high vacuum triode, if it is considered that the *p–n–p* transistor is operated with reversed potential (Fig. 172). The emitter assumes the function of the cathode, and the base, that of the grid, which in this case, however, cannot be subjected to lossless control as in a tube. The collector as the output can be compared to the anode. Just as a tube can be operated with a grounded cathode, grounded grid, or grounded anode, a transistor can also be operated in a circuit with a grounded base, emitter, or collector (Fig. 173). Nevertheless, new circuit

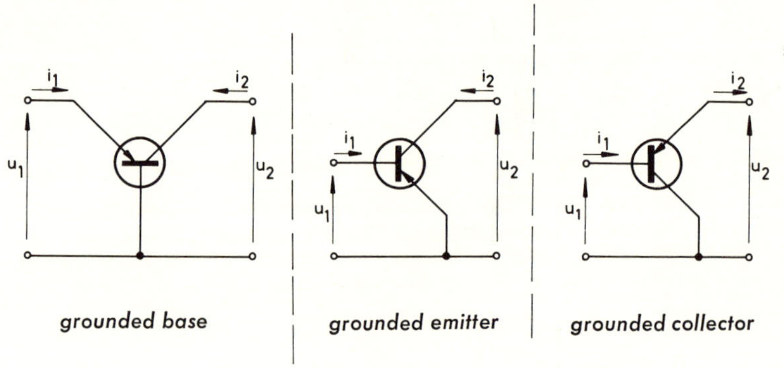

FIG. 173. Comparison of the three circuit possibilities.

arrangements are necessary. As with the control voltages in tubes, the control currents here must be linear. This condition is usually satisfied by the use of appropriate series resistances. Thus, the transistor is a current controlled element. Consequently there is a linear relationship between the input current and the output current. Since the input characteristic i_1/u_1 is not of a linear nature, a sinusoidal control current would result in a greatly distorted input signal voltage. It would therefore be pointless to make use of negative feedback in which the input voltage is stabilized in the ratio i_2/u_1. To avoid distortions the circuit must be designed so that the output current controls the input current. The various properties of a transistor with respect to the significant feedback relationship between output and input are characterized by the *h*-parameters or *h*-coefficients, which make use of the following abbreviations:

h_{11} = the input resistance with short circuited output, in ohms

$$= \frac{u_1}{i_1}, \quad (u_2 = 0)$$

h_{12} = voltage feedback with open input

$$= \frac{u_1}{u_2}, \quad (i_1 = 0)$$

h_{21} = current amplification with short circuited output

$$= \frac{\dot{i}_2}{\dot{i}_1}, \quad (u_2 = 0)$$

h_{22} = output conductance in 1/ohm or mho

$$= \frac{\dot{i}_2}{u_2}, \quad (i_1 = 0)$$

These values are used in h-matrix calculations to determine the properties of transistors on the basis of an appropriate four-pole equivalent circuit diagram. The transistor descriptions furnished by the manufacturer contain the necessary equivalent circuit diagrams and the related formulas, into which the h-values are to be introduced. With the h-values the type of wiring arrangement is required, and therefore the first letter of the reference electrode is placed after the index number; thus, for example, h_{21E} refers to the current amplification with short circuited output in a grounded emitter circuit. The h-values make it possible to determine essentially the input and output resistance values for various wiring arrangements and load conditions. The following formulas are used to calculate the current and voltage amplification:

(1) Grounded base circuit

$$\alpha = \frac{I_C}{I_E} \approx 0.9 - 0.99$$

$$v = \frac{U_2}{U_1} \approx 10^4 \quad \text{(voltage amplification)}$$

(2) Grounded emitter circuit

$$\alpha = \frac{I_C}{I_B} \approx 100$$

$$v = \frac{U_2}{U_1} \approx 10^3 \quad \text{(power amplification)}$$

(3) Grounded collector circuit

$$\alpha = \frac{I_E}{I_B} \approx 500 \quad \text{(current amplification)}$$

$$v = \frac{U_{EC}}{U_{BC}} \approx 1 \quad \text{suited for use as an impedance transformer}$$

where α = current amplification and v = voltage amplification. The family of transistor characteristics as shown in Fig. 174, right, might be compared with those of a pentode in a I_a/U_a field. The curves rise very sharply and then turn in a horizontal direction. The horizontal portion of the characteristic is quickly reached, and the useless section is therefore smaller than with tubes. In this respect *n–p–n* transistors are particularly advantageous. When the collector voltage becomes greater than approximately 0.2 volt, which is a multiple of the inherent thermal potential kT/e, the collector current

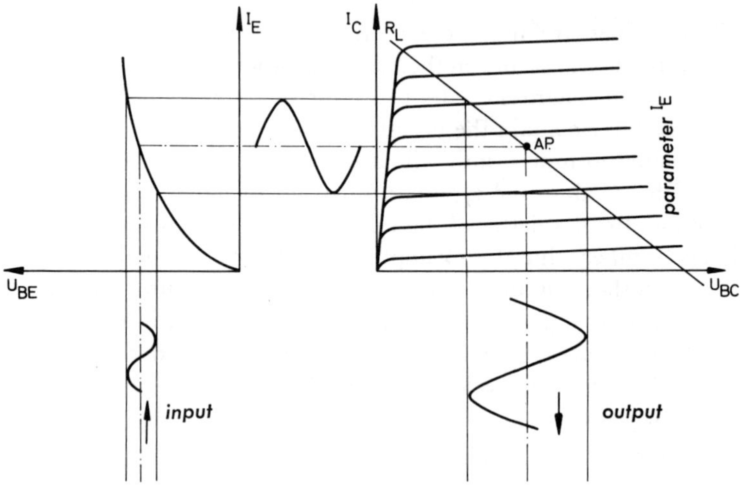

FIG. 174. Principle of the amplification process in terms of grounded base circuit transistor characteristics. U_{BE} = potential base/emitter, I_E = emitter current, I_C = collector current, U_{BC} = potential base/collector.

becomes independent of the collector voltage as a result of the saturation in the collector region. It is therefore possible to construct transistor oscillators and amplifiers with outputs of the order of several milliwatts. The efficiency of transistors, which can go as high as 49%, is greater than that of tubes, which goes up to 25%. Disadvantages of transistors are the difficult reproducibility and the temperature-dependent characteristic, which has a particularly unfavorable effect when a grounded emitter circuit is employed. The lower operating voltages of transistors make battery operation possible, as a result of which their most important use is in portable instruments. Another advantage of transistors is that there are no heating filaments that can burn out, and there is no anode battery. The microphonics of tubes is absent from transistors. Moreover, the appearance of transistors has brought about miniaturization of other components as well, such as resistors, capacitors, etc. Thus, it has become possible to make smaller and lighter instruments.

Because transistors, which have a very small volume, also generate less heat than tubes, it has been found advantageous to introduce them into electronic computers. Considering the thousands of transistors used in these instruments, it is estimated that the elimination of the cooling systems alone results in energy savings of 90% and space savings of almost 50%. The same considerations have led to the introduction of these elements into remote controlled rockets. The life span of a transistor is primarily determined by the thermal diffusion processes and the electrolytic transportation of material. Under a moderate load these processes are extremely slow. Supplementary cooling or cooling fins for dissipation of the generated heat are used to keep the temperatures within the limits set by their physics (inherent current) and technology (melting point, etc.). A life span of tens of thousands of hours can be expected.

At the present time transistors are suitable for low signal powers. Today, however, attempts are being made to design transistors for high output powers also. At present the maximum signal power obtained is approximately 50 watts. Aside from problems of frequency, the upper limit is essentially a thermal problem.

From the standpoint of thermal noise (kT) transistors are markedly superior to heated tubes, which require operation at high temperatures. Thus, it is to be expected that in the course of development semiconductor amplifiers will quite generally offer more favorable noise ratios than tubes.

In modern surface transistors small base thickness and other measures are employed in an attempt to prevent the recombination of the electrons and holes. Figure 175 illustrates the relationship between noise and frequency. Beginning at low frequencies, the noise factor decreases (at the approximate rate of $1/f$). Here we are concerned with a Schottky effect at the semiconductor surface of the type discussed in Chapter II (in connection with tubes).

The upper frequency region f_u is dominated by the shot effect. The low noise factor in the middle region, which barely exceeds 2–3 db, combined with the absence of all microphonics, and the relatively low input impedance make the transistor a suitable microphone amplifier.

In the course of development so-called bred or drawn transistors have also been developed. A monocrystalline piece is immersed in a melt and slowly withdrawn. The melt material adheres firmly to the solid piece and grows into a crystal. The *p–n* or *n–p* junctions are formed during "monocrystal drawing" of germanium. The pure germanium is doped successively in accordance with a definite program with the requisite impurities, such as indium and arsenic or antimony. This process is very simple if the melt consists of *n*-germanium. If a small quantity of indium is added to the melt during the drawing process, the trivalent impurity spots are then suddenly in the majority. A *p*-type crystal then grows onto the *n*-crystal. This process is then

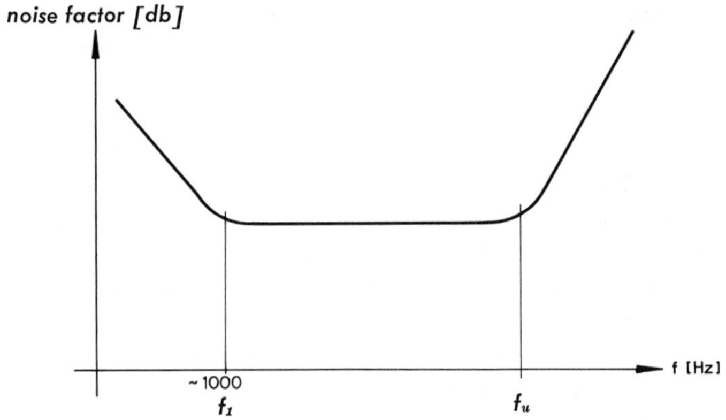

FIG. 175. The noise factor as a function of frequency. Attempts are being made by means of improved methods of production and more favorable conditions to extend the central region (of thermal white noise) into the high frequency range ($f_1 - f_u$).

followed by one in which a pentavalent metal is added. In this way another *n*-type crystal is formed. In this method it is essential for the crystal structure to be maintained throughout the entire process. When the crystal layers obtained in this manner have been cut up into bars and provided with ohmic contacts, the transistors are completely ready for use as circuit elements.

One transistor commonly used in radio sets (Fig. 176) contains essentially a small crystal slab of *n*-germanium about 4 mm^2, that serves as the base electrode. A pill of indium is alloyed onto each of the two sides, as a result of which the junction is formed inside the *n*-germanium. The effective base inside the germanium must be very thin for high frequency transistors and must not exceed 0.02 mm. The indium pills serve at the same time as the solder for the ohmic contact connection. The indium droplet on the emitter side in practice is somewhat smaller than the one on the collector side in accordance with the output power. The collector capacitance is kept very low by the selection of a very high ohmic resistance in the direction of the collector for the middle electrode, i.e., the germanium base. The life span of the electron–hole pairs before recombination should be as long as possible and at least 100 μsec. The life of the electron–hole pairs depends to a great extent upon the method of production employed. The life span is usually shorter when the resistance of the material is low. When a layer of extremely pure or intrinsic germanium is placed before the collector transition of a *p–n–p* transistor, the new transistor is known as a "*pnip* type" (the *i* standing for

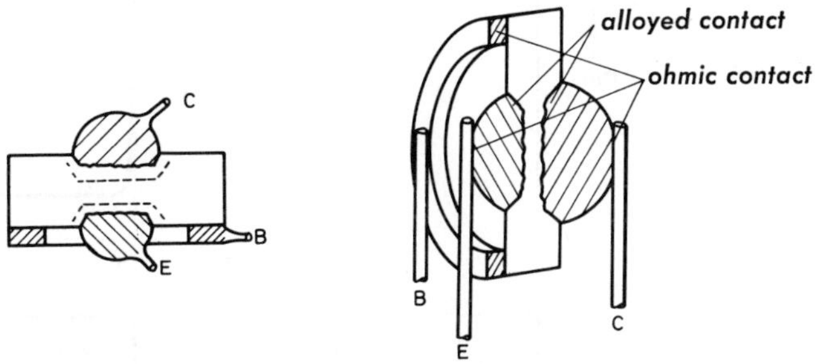

FIG. 176. Alloyed *p–n–p* transistor in cross section and in a perspective view.

intrinsic). The holes diffused from the emitter pass more easily and, therefore, with greater velocity through the *i*-layer to the boundary of the collector transition. The collector capacitance is reduced by this method from 2.5 pf to approximately 0.5 pf. The empty zone makes possible the use of a higher collector potential and consequently allows an increase in power. The physical relationships of the *p–n–i–p* transistor will be better understood if we begin by considering the previously discussed *p–n–p* transistor, where the concentration of impurity spot atoms is assumed to be constant in the base. As in a good conductor there is almost no field in the intermediate zone. The charge carriers move solely as a result of diffusion through this middle layer. Since the diffusion process is relatively slow and passage requires a high transit time, the frequency limit even of thin layers is relatively low. To raise the frequency limit, the charge carriers must be accelerated by an electric field. This can be accomplished by making use of a base in which the impurity spot concentration, instead of being uniform, decreases in the direction of the collector. According to Boltzmann's law, an electric field is thereby produced that accelerates the injected charge carriers and raises the frequency limit in the manner desired. In order to create a constant field strength, the base is doped with an exponential concentration of the foreign atoms.

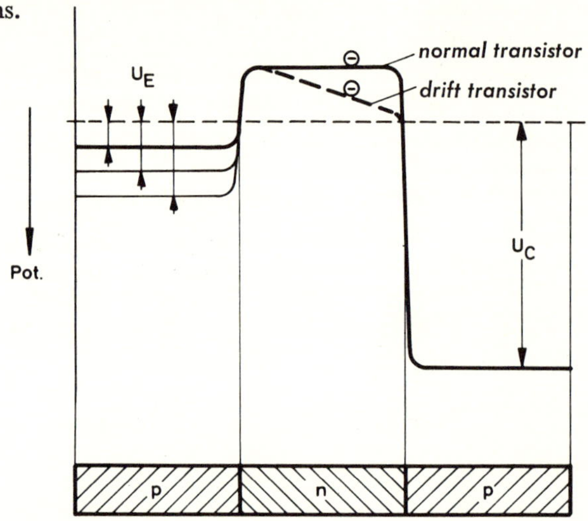

FIG. 177. Potential curve in a *p–n–p* transistor. U_E = the potential difference between the emitter and the base. U_C = the potential difference between the collector and base. BETWEEN: the diffusion or drift potential. Control is effected between the emitter and the base. If the internal field is diminished by means of the applied voltage, the potential wall is lowered and more charge carriers can pass over. On the other hand, if the internal field is strengthened by means of the applied voltage, the potential wall is raised.

To summarize, it can be said that the charge carriers move as a result of diffusion in the original transistors and as a result of an electric field in the more recent types (see Fig. 177). Transistors employing the second method are known as drift transistors. In the manufacture, however, the tendency is to increase the concentration of impurities in the direction of the emitter and to create there an effective boundary transition between the *p*- and *n*-layers. When the concentration of impurities is varied in this manner an electric drift field is created. Figure 178 gives a doping distribution or so-called impurity spot profile for the whole length of the transistor. The abscissa represents the relative length of the transistor, while the vertical axis from zero upward represents the ratio of dilution of the *p*-impurities (acceptor atoms) and from zero downward the *n*-impurities (donor atoms). The zero transitions represent the boundary surfaces. The diagram shows very clearly that the *p–n* junction between the emitter and the base is the more significant one.

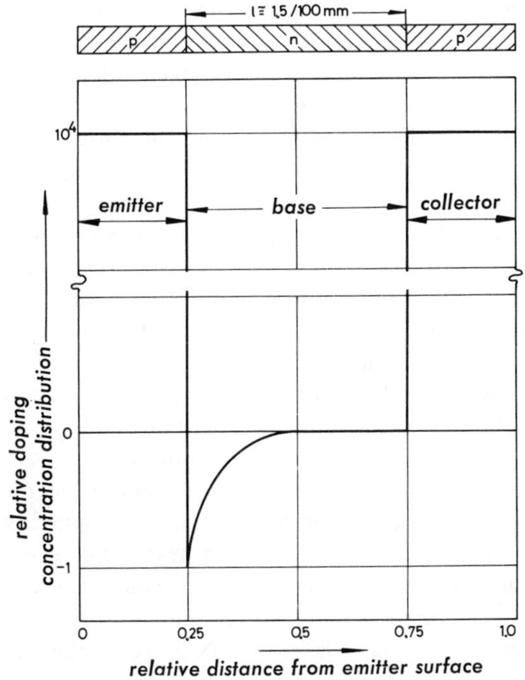

FIG. 178. Impurity spot distribution for the *p–n–p* zones of a drift transistor. In the base layer the concentration of *n*-impurity sites gradually drops toward zero. An exponential concentration curve results in a constant field strength (a drift effect).

The injected positive charge carriers are consequently more intensely accelerated. They then diffuse through the increasingly pure germanium of the base until they reach the collector. Manufactured drift transistors operate at frequencies up to approximately 100 Mc/s. This can be improved upon by such refinements as reduction of the base thickness to a microscopically thin layer by means of a precise etching process and by application of thin emitter and collector coatings of micrometric accuracy. The instantaneous frequency limit for such custom designed transistors is approximately 1000 Mc/s. The frequency limit is to be understood as meaning that frequency at which the amplification drops to 70% of the value at 1 kilocycle/s.

Transistor tetrodes, four-layer transistors, unipolar transistors, and phototransistors are manufactured as well as transistor triodes. Back in 1950 before the surface transistors had been discovered, point-contact transistors were being constructed by William Shockley, the transistor scientist. They resemble point-contact diodes and possess two thin, narrowly separated tungsten wires that rest upon the germanium crystal, to which they are welded or moulded, as is commonly said. In the course of welding the wires to the crystal a *p*-zone is formed around the points, with the result that a *p–n–p* transistor is formed. Here also the interaction of the three zones results in a triode-type control. Special effects obtaining here result in a current amplification factor of up to 3.5, which by way of analogy can be compared to secondary emission with a dynode. Transistor circuits can also be arranged in such a manner that two stable working points are obtained on the characteristic. When so designed they can be used as switches of either the trigger type or as flip-flops. The various types of circuit arrangements cannot be discussed in this book; possibly they will be taken up in a later volume.

Figure 179 illustrates the design of a modern mesa transistor (named for their resemblance in form to the flat-topped, steep-sided mountains of the southwestern United States). The mesa transistor represents a further development of the classical surface transistor schematically illustrated at the left of the figure; in the classical surface transistor the base is situated between the emitter and collector.

The mesa transistor is characterized by a collector that acts as the mechanical support for the entire electronic element. This results in a relatively large surface that is capable of dissipation of the heat, which is significantly greater at the collector output side of the transistor. If this bottom "electrode" is, moreover, placed upon a metal plate, the heat dissipation is nearly ideal. The base layer, which is usually placed in the middle, is now placed simply on top of the collector (as in Fig. 179*b*), rather than beside it, as in the classical transistor, and acts as an ohmic contact (thus, without a barrier junction). The bottom plate (edge length 1–2 mm), which consists of germanium that has been doped with trivalent aluminum, gallium, or indium to make it a *p*-type conductor, is the collector. A layer approximately 0.002 mm thick of the upper side of this small plate is converted into *n*-germanium. The layer depth is controlled by diffusing into the plate pentavalent antimony or arsenic over a definite period of time and at a prescribed heat (approximately 65°C). A gold-antimony alloy, which is a good electrical conductor, is applied to the side of the thin base layer (see Fig. 161) by means of a masking piece and acts as the ohmic base contact. At the same type the trivalent element aluminum is alloyed onto the *n*-base layer (on the right in the figure), where it combines with the germanium to form the *p*-emitter zone. The site of the aluminum acts as the ohmic contact for the emitter. These contacts can only be fashioned under a microscope.

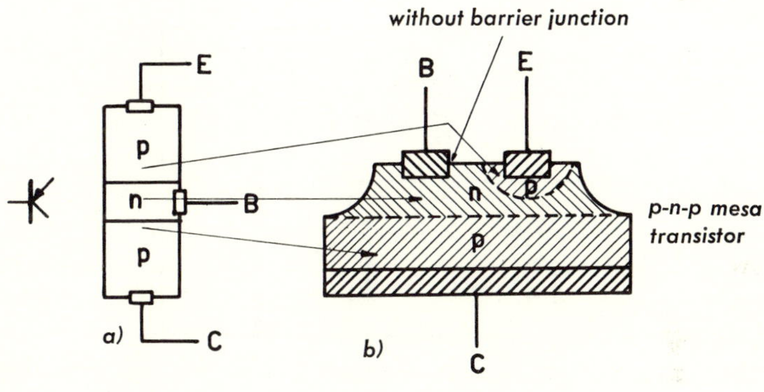

FIG. 179. *p–n–p* mesa transistor.

After this process is completed all the excess material is etched away, and the typical mesa form is obtained.

Mesa transistors are suited for use as oscillators for frequencies of up to 500 Mc/s, and are also suited for use as amplifiers for television tuners, telescopic sight image amplifiers, television and audio amplifiers.

The mesa transistor is constructed for input stages and the lowest power stages. The noise is very low and amounts to about 3.5 ($kT_0 = 1$) where k = constant and T_0 = absolute temperature.

Figure 180 shows the circuit diagram of an oscillator for 500 Mc/s, using a mesa transistor.

Most recently mesa transistors have tended to be replaced increasingly by the so-called disk or planar transistors. Because of the greater thermal stability and other technological advantages these transistors are now based solely on silicon.

In addition to the attention paid to the production of smooth junctions and *n*- and *p*-zones of reproducible thickness, an increasing emphasis is now being placed upon protection of the sensitive junction surfaces from the effects of air and moisture. It is known from chemistry that silicon on absorption of oxygen forms a stable, i.e.,

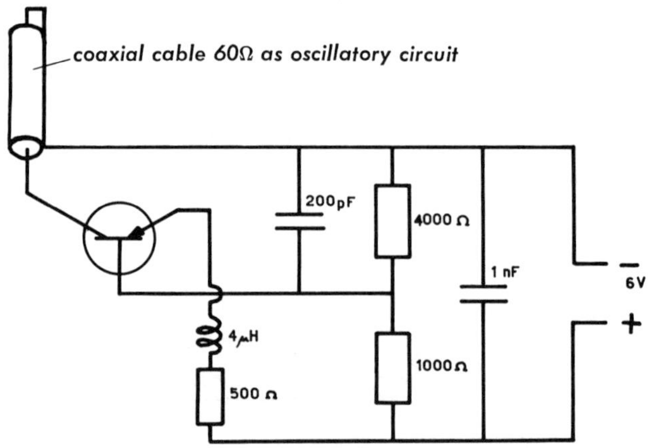

FIG. 180. Circuit diagram of an oscillator for 500 Mc/s, using a mesa transistor.

chemically inactive, dioxide (SiO_2). (The behavior of silicon in this respect can be compared to that of carbon, which on combustion oxidizes to CO_2, carbon dioxide.) The boundary layer surfaces are consequently protected with layers of SiO_2, which are, so to speak, produced at the same time as the boundary layers. The diagrams in Fig. 181 illustrate the production phases.

Nowadays surface transistors are used exclusively, and the power-limited point-contact transistors, that had long been used for higher frequencies, are now things of the past. Today transistors are the greatest users of semiconductor materials. However, these newly discovered elements also find other electronic applications, for example, as photocells. The germanium phototransistor, for example, possesses an extremely high efficiency in comparison to other systems. Under certain conditions the quantum yield approaches 1, which means that one electron–hole pair is liberated by each quantum.

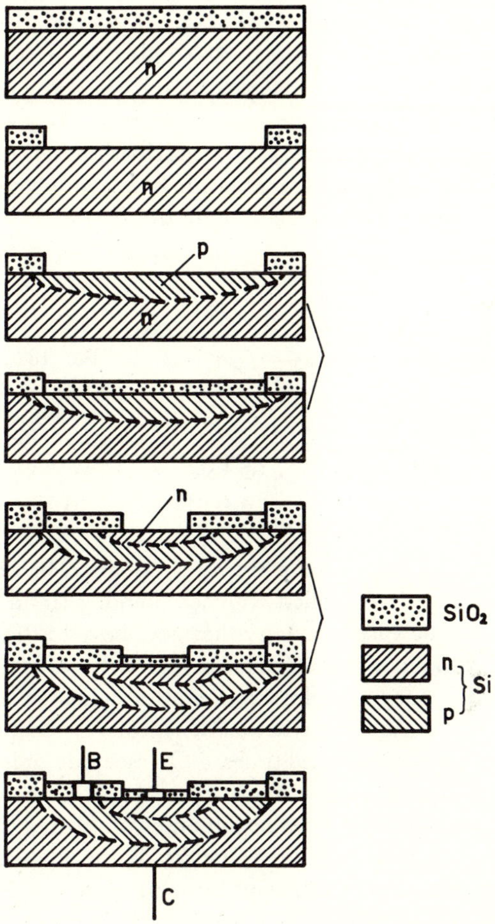

FIG. 181. The various steps in the production of a disk transistor.

Many types of semiconductors besides germanium are used. The cesium antimonide crystal (Cs_3Sb) with its cubically centered ionic lattice is very characteristic of the semiconductor photoeffect. The light quanta here are no longer absorbed at the surface as in the regular photocell but in the interior of the crystal. If a bonding electron is freed from its atom on the absorption of a light quantum, assuming that the energy transmitted to the electron is sufficient to dislodge it from the crystal lattice into the outer space and that it is not trapped anywhere, a hole is left behind. It is possible for this hole to capture a neighboring electron and thus create a neighboring hole, resulting in hole conduction. This hole current is proportional to the incident light. Figure 182 provides a schematic illustration of the energy bands of a cesium antimonide crystal. It is assumed here that there are no structural defects or impurities present in the crystal, and we will also assume that the experiment is conducted at absolute zero. In the absence of incident light the crystal is an insulator.

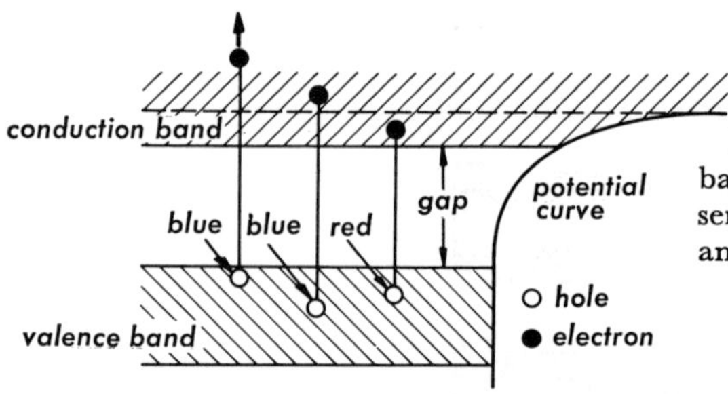

FIG. 182. The energy bands in photoelectric semiconductors (cesium antimonide crystal).

Above may be seen the initially empty conduction band and at the bottom the energetically filled valence band. If an electron is supplied with an increase in energy that goes beyond the dotted line, i.e., the work function, the electron is emitted. If it possesses the requisite kinetic energy, it does not matter whether it is emitted directly from the surface or whether it is able to escape only after bouncing back and forth several times. On escaping, it leaves behind an empty place or hole or, one might also say, an unoccupied energy level. Electrons that have been less intensely activated, for example, by incident light whose quantic value $h \cdot f$ is too low, after a short time drop back into the same hole or into some other hole and recombine in the usual manner. Electrons located between the emission threshold (on top) and the lower limit of the conduction band serve, like the holes, as current carriers. The current flow consists of a double motion. The electron current flows between the crystal atoms (in the conduction band), and the holes result in a displacement of the vacant sites among the valence electrons. The ohmic resistance of the crystal is therefore reduced. Such semiconductors are consequently used as light-dependent, so-called photoresistors.

When electrons are freed from their atoms by light and are then capable of moving as quasi-free charge carriers within the material, thus making the crystal conductive, we speak of an internal photoeffect. The photocells discussed in Chapter II, on the other hand, made use of an external photoeffect. The internal photoeffect is put to use in such applications as the resistron, the industrial television camera tube. The semiconductor used is selenium, the conductivity of which is altered by the incidence of light. Figure 183 illustrates the basic design of a resistron. The picture to be transmitted is projected onto the selenium layer by means of a lens system. This semiconductor layer rests against a translucent metal coating that has been applied to the outside of the glass envelope. This thin, conductive coating, which acts as an electrode, is known as the signal plate.

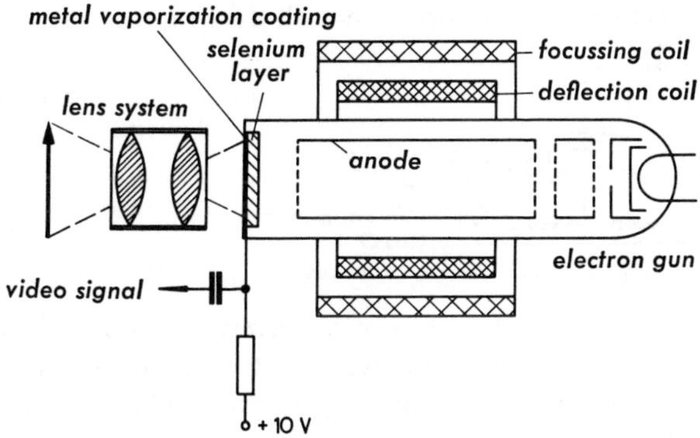

FIG. 183. Semiconductor television camera tube, the so-called resistron or vidicon. The semiconductor photolayer (selenium or antimony trisulfide or the like) on the transparent metal vaporization coating converts optical images into latent, electric charge images.

A positive potential of about 10 volts is applied to this signal plate. The selenium coating, which rests against the metal coating and which is not conductive in the absence of illumination, can be imagined as consisting of numerous, microscopically small cells. Each surface portion of these cells forms a capacitor with the thin metal coating next to it. The metal coating acts as the common counter-electrode for all the cells. The dielectric formed by the photosensitive selenium semiconductor layer in between. The semiconductor surface facing the cathode, and consequently the sum of all the cell surfaces, is subjected to interlaced scanning by a focussed electron beam or, more accurately, subjected to electron bombardment. The beam is guided in the interlace pattern by means of a magnetic or electric deflection system as in a cathode-ray tube or television receiver. The electrons of the scanning beam obtained from the electron gun are initially accelerated to a potential of 300 volts; slightly in front of the selenium layer, however, they are again retarded as a result of a low potential. The velocity of incidence of the electrons must be kept very low to prevent secondary electrons from being liberated from the semiconductor layer. The excess of electrons at the surface elements of the semiconductor layer over which the electron beam passes in the course of scanning creates there a negative potential with respect to the metal coating. Consequently, the scanning creates a definite charge on the capacitor storage element in relation to the initial positive voltage of the signal plate. A "virtual image" resistance distribution is formed in accordance with the brightness values of the

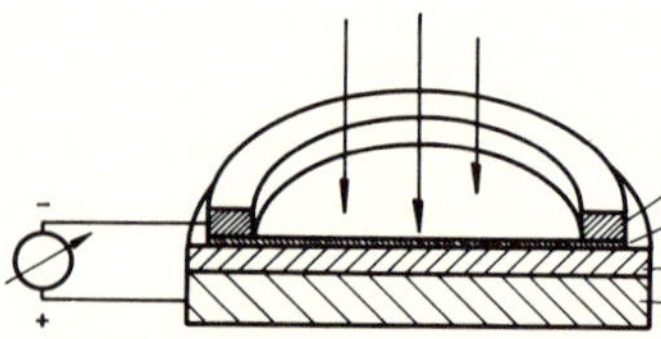

FIG. 184. Cuprous oxide photoelement.

object projected upon the absorption layer. At the points where the light falls upon the selenium layer, electron–hole pairs are created and the selenium becomes locally conductive. The negative charge created by the scanning beam flows off in the interval between successive scans in proportion to the conductivity, which is modulated each time by the brightness values. When the beam passes once more, the loss in charge between successive scans, which is a function solely of the local brightness, is restored. At the points upon which no light has fallen the selenium remains an insulator, and there is no need for replacement of charge because the semiconductor capacitor is still charged to capacity. Current pulses are produced at the external resistance R, which is connected in series with the signal circuit, only when charges must be supplied during a scan; these required charges are proportional to the brightness values of the picture. The current pulses thus obtained are called picture signals or video signals. These signals are finally passed on by the receiver to the picture tube. Here the electron beam is modulated in accordance with the intensity of the receiver signal and passed in an interlaced scan over the phosphor screen of the picture tube. The phosphor layer is activated with varied intensity in accordance with the modulation and thus, necessarily, reproduces the original image.

Resistrons operate at an extremely low illumination intensity of approximately 100 lux and give a picture of television broadcast quality. However, resistrons can be made with other semiconductor materials, such as antimony trisulfide, and these resistrons possess the advantage of less lag with moving pictures, though at the expense of higher illumination intensity requirements.

Wherever illumination results in liberation of charge carriers within a semiconductor and gives rise in the absence of a supplementary voltage to a directed flow of current through an externally closed circuit there is an internal photoeffect. Light converters of this type are called photoelements. For light meters photoelements of this type, made of selenium on iron, cuprous oxide (Cu_2O) on copper, i.e., materials similar to those used in rectifier elements, are preferred. The manner of production in accordance with Fig. 184 is the following: An atomically thin, transparent layer of copper is vaporized onto a semiconductor layer of, for example, cuprous oxide (Cu_2O), which is situated on a contact plate (copper). On top of this thin layer is placed a ring that acts as a lead. If light falls through the thin copper layer onto the semiconductor boundary surface, quasi-free electrons are liberated, which as a result of the rectifier effect reach the thin copper layer but do not pass through the Cu_2O to the contact plate. The thin copper layer acquires a negative charge with respect to the bottom plate. This element is capable of feeding a moving coil instrument or a relay. The potential per element on intensive illumination is of the order of approximately 1 volt. Modern photoelements are made with germanium or silicon. Whole "solar batteries" can be constructed by connecting four such elements in series. These are used to charge small batteries powering a radio or portable transmitter. The commercial silicon photoelements exhibit an efficiency above 10% on solar irradiation. Thus, solar irradiation of 100 mW/cm^2 yields an electric output of 10 mW/cm^2, i.e., 100 watts per m^2. Approximately 30% of the radiation is normally reflected, and the remaining loss is to be attributed to the poor quantum yield of the energy converter. The photoelectric efficiency, i.e., the quantity of photons converted into electrical energy, is primarily determined by the energy gap between the conduction and valence bands of the material used. The band gap of germanium is smaller than that of silicon, so that a higher percentage of electrons pass over into the conduction band, but the resulting voltage is lower. It follows by definition that the output voltage increases with the gap width. Thus, we arrive at two opposed dependencies. For long-wave "reddish" light, germanium would be preferable, whereas for short-wave "bluish" light, silicon would be the material of choice. For electrons to be activated into the conduction band silicon requires an energy of 1.2 e.v., and germanium 0.7 e.v. At sea level sunlight provides an energy converted into electron volts of 1.3–1.5 e.v. Thus, it is correct for silicon batteries to be preferred for solar batteries such as are used in rockets. A silicon photoelement is doped with trivalent boron atoms at the photosensitive surface in order to create a *p–n* junction. The layer created by diffusing-in boron is less than 0.002 mm thick in this case. On intense illumination the open circuit voltage is necessarily of the order of magnitude of 1 volt. As in all photoelements it is the current (not the voltage) that increases in proportion to the intensity of the light.

Illumination has the same effect upon the junction of a germanium transistor, with the result that the transistor then acts as a photoelement and photoconductor. In most transistors this is an undesirable effect. These transistors are sealed in airtight and lighttight capsules to protect them from the effects of light as well as moisture and dirt.

Recently, controlled semiconductor rectifiers have been constructed that are capable of thyratron-like action for heavy current uses and are consequently capable of replacing tubes in this area as well. These are special four-layer *p–n–p–n* "avalanche"-transistors with a middle tap at the second *p*-section. The highest reverse voltage of this element at the present time is 400 volts; it can carry a continuous forward current of up to 20 amps. The control current required for opening the rectifier is very low, lying below .1 milliamp with a switching time of less than 1 μsec.

If a reverse operated *p–n* junction is irradiated with light, the current that flows consists partially of the normal reverse current and partially of the charge carriers (electron–hole pairs) created by the light. As long as the reverse voltage is kept within moderate bounds, the current arising from the photoeffect is independent of the biasing voltage. But if the reverse voltage is raised beyond this point, the charge carriers are accelerated and collide with other particles or the atomic lattice; this results in charge carrier multiplication, which causes an increase in current and, at a critical voltage, a kind of breakthrough during which the current reaches the optimal short-circuit value. As soon as the biasing voltage exceeds the ionization point we are confronted with a multiplication factor of the sort encountered in gas-filled photocells. The ratio of the amplified current I thus obtained to the current I_{OL} flowing in the absence of the breakthrough is called the multiplication factor $M = I/I_{OL}$. For the transistor the current amplification factor is defined as follows:

$$\alpha_T = \frac{I_c}{I_l} \quad \text{or} \quad \frac{\Delta I_c}{\Delta I_l}$$

It is that portion of the emitter current that flows through the reverse biased *p–n* junction into the collector which in practice is less than 1 because of the recombination losses. Even in the initial stages of semiconductor research in studies of point-contact transistors it had been established that α can become apparently greater than 1 if the field strength at the collector is relatively high. Even then an avalanche effect was suspected. Under these conditions the new value α is not a constant, but is a function of the applied voltage or the multiplication factor. It is calculated from the equation

$$\alpha = \alpha_T \cdot M$$

in which α_T is the undisturbed value at low voltage. At the breakthrough voltage or ignition voltage U_z, α becomes infinitely great. More precisely, the original amplification factor is still a function of the emitter current and drops with increasing emitter current. With silicon, for example, a drop from the operating value of 0.95 to 0.25 and lower is observed when the emitter current drops to about 1/10 of the normal operating current. This double dependency on both α_T and M (parameter) makes the construction of semiconductor thyratrons possible. In the prescribed operating arrangement the *a* and *b* junctions are operated in the forward direction, and the middle transition is operated in the reverse direction. A current $I = I_0 + \alpha_a I + \alpha_b I$ flows through this barrier transition *c*; in the formula, $\alpha_a = \alpha_{Ta} \cdot M$ and $\alpha_b = \alpha_{Tb} \cdot M$. I_0 = normal reverse current; $\alpha_a I$ = injected hole current in system *a*; $\alpha_b I$ = injected hole current in system *b*. This formula can be transformed into:

$$I = \frac{I_0}{1 - (\alpha_a + \alpha_b)} \quad \text{or} \quad \frac{I_0}{1 - (\alpha_{Ta} \cdot M + \alpha_{Tb} \cdot M)}$$

Since for a low collector voltage $M = 1$ and is even less than 1 for a low emitter current $(\alpha_{Ta} + \alpha_{Tb})$, the current flowing through the four-layer system is:

$$I = \frac{I_0}{1 - (<1)}$$

In the event that $\alpha_{Ta} + \alpha_{Tb}$ is, for example, equal to 0.5, $I = 2I_0$. Since the reverse current I_0 is quite small, the current flowing through the diode is also very small. If the reverse voltage is increased, α_{Ta} and α_{Tb} remain constant, since no emitter current can flow as a result of this current direction. But the situation is very different for the factor M. Above the critical field strength a breakthrough occurs as a result of field emission and M increases. The value $\alpha_{Ta} \cdot M + \alpha_{Tb} \cdot M$ which appears in the parenthesis of the denominator of the equation above, quickly reaches the value of 1. This makes the denominator equal to zero, and the current I then becomes infinitely

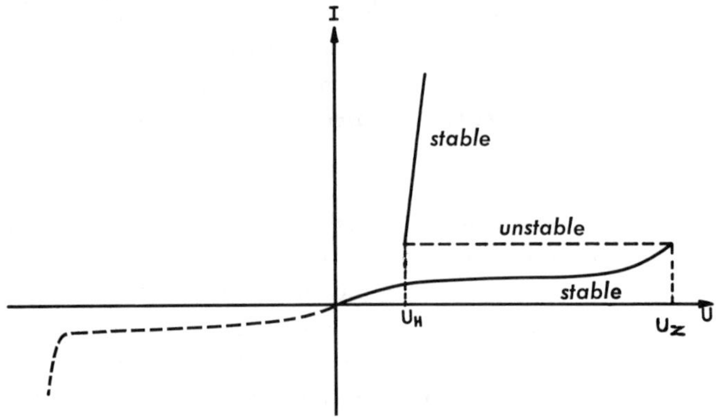

FIG. 185. Normal curve of the characteristic in four-layer semiconductor elements.

great. At the point of the current rise when both current carriers are created, the values α_{Ta} and α_{Tb} also increase. This mutual reinforcement of the spontaneous increase in conductivity results in a sharp characteristic, as shown in Fig. 185. As a result of the formation of the two types of charge carriers, the middle transition (in Fig. 186), which previously operated in the reverse direction, is now biased in the forward direction. This is another reason why the voltage drop becomes very small. The characteristic also shows how such an element can be fired, in the terminology of glow-discharge tubes, by an increase in the reverse voltage.

On interruption of the circuit, i.e., the holding circuit I_H or holding voltage U_H, the element again becomes controllable, i.e., as soon as the reverse voltage again reaches U_z, the diode is once again "fired." The ignition voltage for a particular four-layer diode can be very precisely determined. It depends upon the doping concentration and the type of semiconductor transitions. As already mentioned in the preceding section, the value α_T is a function of the magnitude of the emitter current. This phenomenon can be utilized to construct a thyratron, i.e., a circuit element in which the ignition voltage can be varied as desired by the application of external voltage.

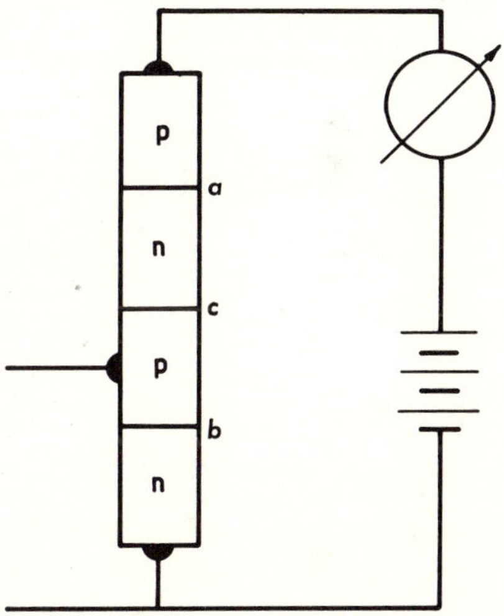

FIG. 186. Control circuit in which the ignition voltage can be varied by the application of external voltage. Transitions *a* and *b* are poled in the forward direction, *c* in the reverse direction.

The control circuit is illustrated in Fig. 186. The three lower layers of the circuit element act as an *n–p–n* transistor. If the emitter current is increased by the external control voltage, the current amplification factor is increased. Under these circumstances a small multiplication factor M, i.e., a smaller reverse voltage, is sufficient to bring the denominator of the formula to the zero value. The ignition voltage can be lowered in this way, as the dotted characteristic in Fig. 187 shows. Such semiconductor thyratrons, like their gas-filled counterparts, are in practice ignited within a desired time with current pulses or by selection of an ignition angle corresponding to a phase displacement between 0 and 180 degrees (see p. 102). In this manner one can obtain a relay or supply to a working resistance a voltage that is a function of the phase of the ignition angle. These new semiconductor elements possess a considerably lower voltage drop and are nearly lag-free in comparison to the gas-discharge thyratron. The deionization time of a gas-discharge tube is higher by several powers of ten.

In the development of the radio the triode was followed by the tetrode and pentode, and at present there is a similar tendency in transistor research. Figure 188 illustrates the principle of a surface tetrode. Here there are two diametrical contacts to the base, one of which is biased with about 6 volts during operation. The blocking field thereby created deflects the effective cross section, i.e., the

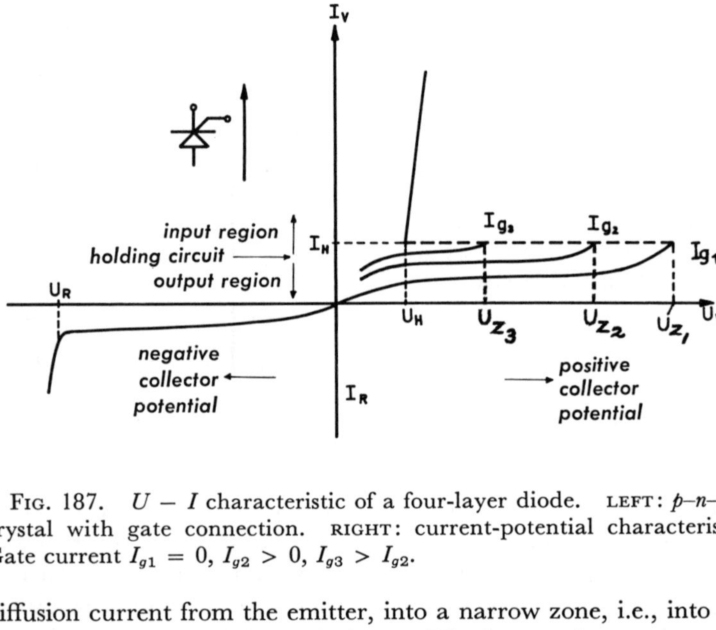

FIG. 187. $U - I$ characteristic of a four-layer diode. LEFT: *p–n–p–n* crystal with gate connection. RIGHT: current-potential characteristic. Gate current $I_{g1} = 0$, $I_{g2} > 0$, $I_{g3} > I_{g2}$.

diffusion current from the emitter, into a narrow zone, i.e., into the immediate vicinity of the main base. The base resistance is consequently reduced considerably. However, the purpose is also to reduce the capacitance between the individual layers, which is disturbing at high frequencies, and thus to obtain a fairly high upper frequency limit. The point-contact tetrode was developed elsewhere. Three point contacts are placed upon the crystal supporting layer. Two of these act as the emitter and the third as the collector. Two frequencies are supplied to the emitter pair, and the whole arrangement functions like a mixer tube. H. Statz developed the so-called spacistor, the principle of which is illustrated in Fig. 189. A *p–n* junction separated by a narrow space charge zone is operated in the reverse direction at a relatively high field strength of 200 volts. With a space charge zone of approximately 0.1 mm a relatively high field strength of approximately 20,000 volts/cm is obtained. An injection

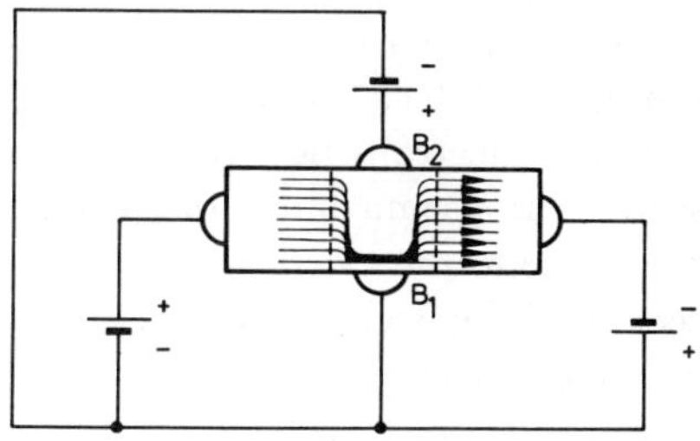

FIG. 188. Transistor tetrode. The drawn-in lines of force indicate the virtual narrowing of the effective conduction cross section.

electrode *I*, which is negative with respect to the collector *C* and is situated in the vicinity of the *p*-zone, is used to inject the electrons. The electrons consequently flow toward *C*. The intensity of the injector electron current can be influenced by an additional contact *M*, the so-called modulator, which makes this element a type of

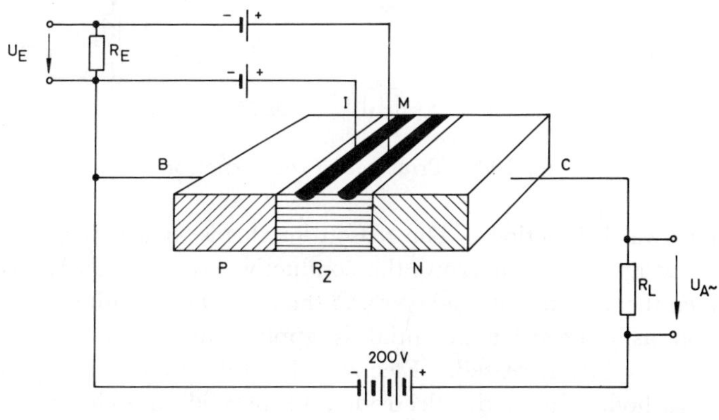

FIG. 189. Principle of the spacistor tetrode.

tetrode. The modulator is comparable to the control grid of a conventional electron tube. The spacistor is suited for very high frequency use; values of up to 10,000 Mc/s have been cited. Also, the input and output possess high ohmic resistances. The output resistance is of the order of magnitude of 10–30 megohms. The properties of this element therefore resemble those of a pentode.

The teknetron is a simple, controllable circuit element that was described by the semiconductor scientist W. Shockley. Figure 190 illustrates the principle of the teknetron. This system is directly analogous to a classical radio tube. For this reason we have chosen to close this section with a description of this element, though in view of certain deficiencies it is hardly likely to survive. The teknetron is a cylindrical bar of *n*-germanium approximately 2 mm long and 0.5 mm in diameter. As a result of the unipolar field between electrodes 1 and 3, electrons flow from left to right. In the narrower middle section the *n*-semiconductor bar is enclosed in an indium casing. The resistance to the electron current is essentially a function of the cross section of the neck of the bar. This can be varied by the application of a negative control voltage to the indium layer, which is operated in

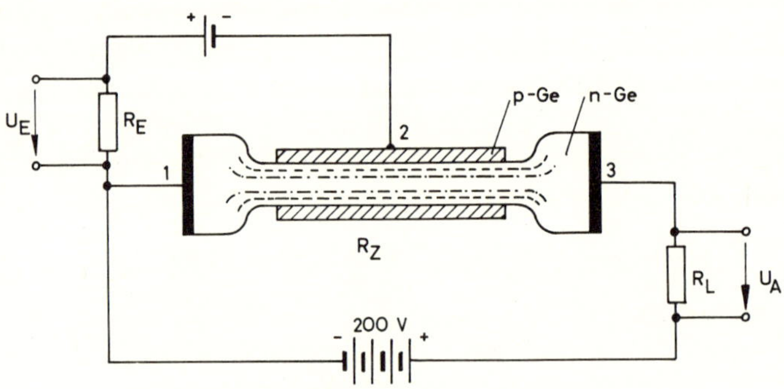

FIG. 190. Principle of the teknetron.

the backward direction. The reason for this is that a space charge zone that apparently narrows the conductive cross section is formed between the indium outer *p*-layer and the inner *n*-germanium cylinder as soon as a negative potential is applied to electrode 2. The resistance is thus increased. Electrode 1 of this system is comparable to the cathode of a triode, electrode 2 to the grid, and electrode 3 to the anode. A negative bias is applied to electrode 2 as in a radio tube. The input impedance is relatively high and amounts to several Mc/s. However, the frequency range barely exceeds 500 Mc/s.

The most recent developments are in the direction of power transistors in which the active boundary surfaces are increased by the use of annular or serrated forms. This ensures effective heat dissipation for the boundary layers.

3. Use of semiconductors as resistors

When an iron wire is heated its resistance increases. It is a cold conductor. Its temperature coefficient is therefore positive. In metallic conductors the atoms of the crystal lattice are caused to vibrate with increasing intensity by increased heating. This reduces the effective conductive cross section and thereby increases the resistance. One might also say that the conduction electrons in the course of their movements collide continually with metal atoms and thereby encounter resistance in the manner of a man who must push his way through a crowd. The energy obtained by the electrons by traversing the applied electric field is partially transmitted to the atoms of the conductor in each collision, with the result that the conductor becomes heated as the current passes through. This is why every metallic conductor exhibits resistance at room temperature. As the temperature drops, the resistance of a pure metal drops progressively. In the neighborhood of absolute zero in most cases there is an abrupt drop to zero. When the current can flow continuously without resistance and a permanent magnetic field is created around the conductor, we speak of superconductivity. If a thin iron filament is sealed into a hydrogen-filled glass envelope in such a manner as to prevent oxidation or burning out of the filament, we obtain a so-called iron-hydrogen resistor. It is used as a regulator or for stabilization of currents. Certain semiconductors, on the other hand, possess a negative temperature coefficient of specific resistance. They are consequently used to advantage in measuring instruments to compensate filament resistances or, also, for regulatory purposes. In the nomenclature these semiconductor resistances are known by the most diverse names. The most common is the NTC resistors (negative temperature coefficient), but they are also known as thermistors, hot conductors, warm conductors, Urdox resistors. Resistors of this type are produced from solid semiconductor solutions of Fe_3O_4 or Zr_2TiO_4, $MgCr_2O_4$, or Fe_2O_3 containing slight admixtures of TiO_2 or NiO, or CoO with slight quantities of Li_2O. The properties of these resistors vary in accordance with the composition. Generally, very high purities such as those demanded in transistor technology are required; a maximum of one foreign atom to 10^8 semiconductor atoms is permitted. Figure 191 gives the resistance as a function of temperature for an NTC resistor and for a copper wire. When the two are connected in series their opposed temperature coefficients compensate each other and the common resistance is no longer temperature-dependent. From Fig. 192 it can be seen that the characteristic of a thermistor exhibits a negative resistance over a certain range. Such "damping reduction" arrangements can be used in the generation of oscillations or time-base oscillations. The frequency is a function of the time constant of heating. An NTC resistor connected in series with a measuring instrument and a voltage source can be used for temperature measurements. If these resistors are of sufficiently small dimensions they can be used effectively in radiation measurements to determine quantum values $(h \cdot f)$. Thermistors in the form of thin plates or coatings are used in the measurement of infrared light. The principle of operation is based upon the fact that the electrons and

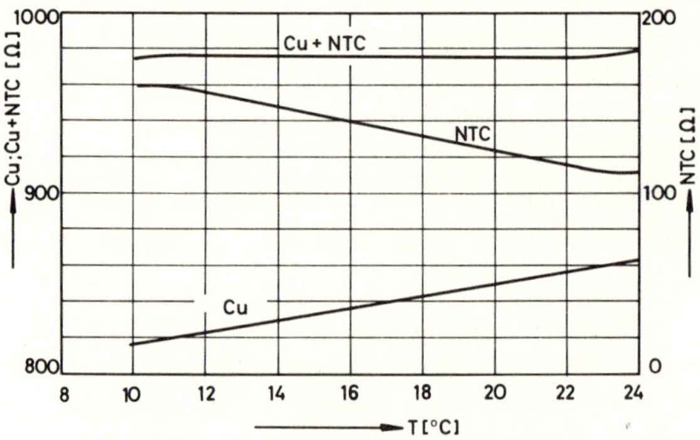

FIG. 191. Resistance as a function of temperature for a hot conductor (NTC) and for a cold conductor (copper wire). When both are combined (CU + NTC), the resistance, as may be seen from the linear characteristic at the top, is no longer temperature-dependent.

molecules are excited by the incident light and consequently are subjected to more intense elastic collisions with each other, so that the material becomes heated by definition. This increase in energy allows more electrons to enter the conduction band, where they are available for current carrying. Since the resistance of this material decreases over a wide range on being heated, the possibilities of application are very great. Aside from their electronic uses, they can be used as starting resistances in motors or to provide continuous regulation when connected in series with a lighting switch. A third type of special resistor is the voltage-dependent so-called varistor VDR. This type of resistor consists of sintered silicon carbide grains. The contact resistance between the carbide crystals is diminished as the voltage is increased. A type of breakthrough effect is utilized. Among other applications varistors are used to protect smoothing capacitors, and also as lightning arrestors for high tension plants. If the resistor is struck by lightning the voltage rises, and the resistance decreases until the high voltage has entered the ground, after which the varistor once again becomes an insulator.

In this case when the voltage is doubled the current is not doubled in accordance with Ohm's law, but is multiplied approximately twenty-fold. Varistors must be selected to allow the passage of little current at the normal rated voltage. They are often called over-voltage limiters.

FIG. 192. Basic thermistor characteristic.

The thermoelectric effect, which was first discovered by Seebeck in 1821, is observed when the site of a free junction between two different metals is subjected to heating. In principle there is not one junction site, but at least two junction sites V_1 and V_2, in accordance with Fig. 193, if the circuit is closed to demonstrate the presence of a current. From this it follows that the current disappears when the temperature is the same at V_1 and at V_2. A relatively high efficiency is obtained when one site is heated and the other cooled. The ideal condition is attained at $0° = -273.16°C$. If the cooling temperature is room temperature an efficiency of approximately 15–20% is obtained. Theoretically the efficiency cannot exceed the Carnot efficiency, which corresponds to $\frac{T_1 - T_2}{T_1}$, where T_2 is the higher temperature and T_1 the lower temperature.

The origin of the voltage in a thermoelement is attributed to the increased intensity of the zigzag motions of the charge carriers at the site of heating, i.e., to the greater kinetic energies of the charge carriers at the site of heating. If, for example, this occurs in an *n*-conductor, where the majority of carriers are electrons, the electrons flow from the hot end to the cold end and thereby create an electric potential and an electric field that will repel electrons. In this manner an equilibrium condition is created that is proportional to the temperature. All metals differ with respect to the intensity of the electric field created

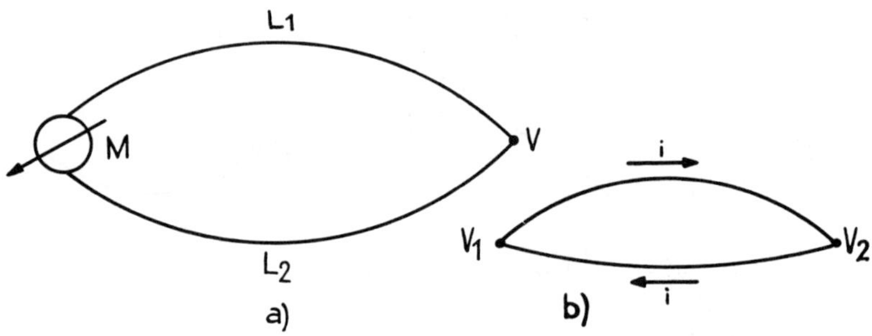

FIG. 193. Two different conductors are heated at the junction site V. At the measuring apparatus M a current can be ascertained. In practice, the junction sites V_1 and V_2 are not discernible, and generally only one is heated, while the other is at room temperature. If both sites are at the same temperature, the currents cancel each other out.

and are characterized by a property similar to the work function (Fig. 194), which we discussed in connection with electron emission from a cathode; like the work function this property varies with the material.[1] Metals with higher concentrations of charge carriers yield low thermoelectric potentials, and substances with low concentrations such as insulators and semiconductors yield high thermoelectric potentials. Two different materials (metals) result, as previously stated, in two different potentials, the difference between which can be put to use. If the material of one part is an *n*-conductor, while that of the other is a *p*-conductor, the charge carriers of the two parts are of

[1] The thermoelectric effect may also be interpreted as a difference in the work functions of the two contact materials, where one junction is heated and the other remains cool.

opposite sign and a high thermoelectric voltage is thereby obtained. Since charge carriers transport thermal energy as well as the charge (thermal conductivity is proportional to electrical conductivity) metals possess the great disadvantage of a high thermal conductivity. Consequently, from the standpoint of thermal conductivity doped semiconductor materials are particularly effective materials. The crystal lattices are disturbed by the admixture of atoms producing impurity spots and the thermal conductivity is thereby enormously decreased. In this way the thermal conductivity can be reduced to levels several thousand times lower than those of good conductors like silver or copper. At the same time, the electrical conductivity increases as the concentration of impurities is increased, so that suitable compositions can be obtained.

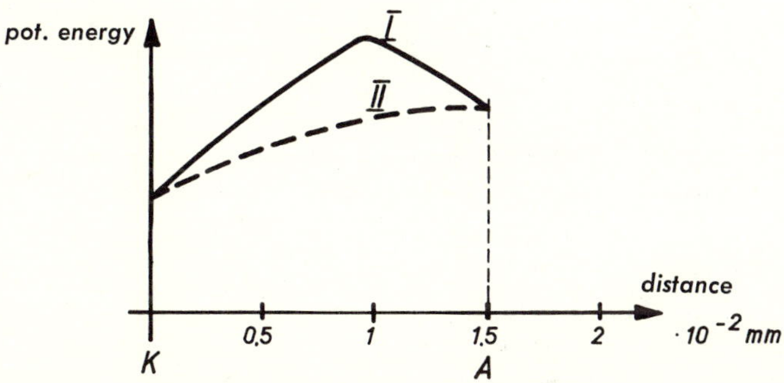

FIG. 194. Potential curve of the work function. Its course is comparable to that of a rocket, which must have a minimum velocity in order to escape the attraction of the earth. Here, too, the electrons must overcome the opposed potential of the space charge.

Appropriate compounds are: bismuth telluride Bi_2Te_3, antimony telluride Sb_2Te_3, or bismuth selenide Bi_2Se_3. Other additives can be included to ensure that conduction is primarily accomplished by electrons.

The voltage can be increased as desired by connecting several such units in series as in a battery. If instead of two different metals two different types of semiconductors, one *n*-type and one *p*-type, are used, the thermoelectric voltage per degree of temperature difference between V_1 and V_2 is fairly high and amounts to about 5 millivolts/°C. It is also possible to reduce the specific electrical resistance by doping, in which case the current is correspondingly raised. Figure 195 illustrates the principle of a three-element generator of this type for generation of electricity from heat. For the ohmic resistance to remain small the length l must be small. Also, the thermal conductivity must be low, so that the heat supplied at the lower junction does not immediately flow to the upper one.

Today generators supplying up to 4500 watts (110^{40} amp) are produced. Generators of this type are composed of many thermoelectric elements and selected semiconductor materials. The heat that generates the current is provided from charcoal, but can also be obtained from solar radiation. This means that it is possible to use such

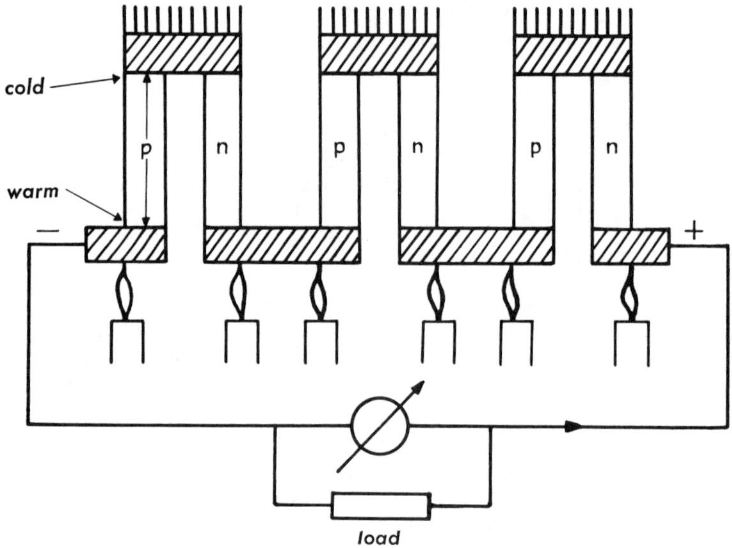

FIG. 195. Thermoelectric generator.

elements in space ships and satellites. Whether it will eventually be possible to obtain from them power capable of replacing steam turbines and rotary generators remains to be seen. If a.c. voltage is desired rather than d.c., appropriate converters are introduced into the circuits. The elimination of moving parts naturally constitutes an advantage.

The weight requirements of such elements are figured at 5 kg/kilowatt, which is very important in space ships. Smaller elements have been exhibited here and there that employ oil heating to generate the current for such uses as radio operation (2 volts, 3 amps).

The heat is dissipated from the second junction site by means of cooling fins or cooling ribs. Such ribs are also placed on the side where the heat is absorbed.

If a thermoelement of this type, which consists in principle of two different metals, is used in the opposite way, i.e., instead of heat being supplied, a current is passed through, it turns out that one junction, V_1, is cooled while the other, V_2, is heated. If the poles of the direct current source are reversed, not only does the current flow in the opposite direction, but the temperature variation is reversed. The side that was previously cooled is now heated, and the side that was heated is now cooled. This so-called Peltier effect, which is only observed with the use of d.c. current and not with a.c., makes the use of such elements possible in small refrigeration units. Here also *n*- and *p*-doped semiconductor materials are used to obtain a high yield.

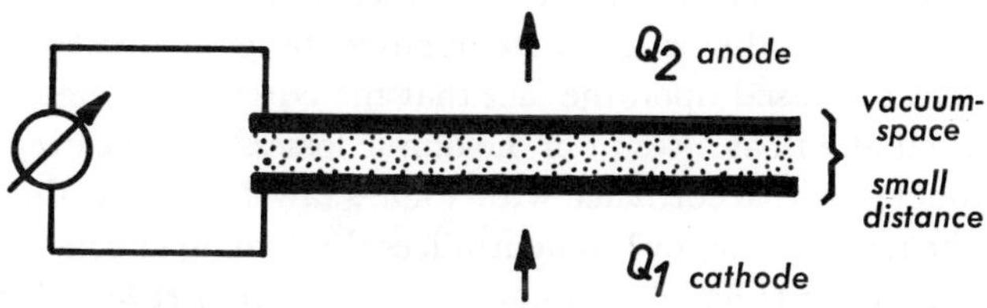

FIG. 196. Operation of the vacuum diode. For the space charge effect to be small, the distance between cathode and anode must be small (about 0.01 mm).

This phenomenon can be explained physically if we remember that the electrons, like the vibrating molecules, are carriers of thermal energy as a consequence of their kinetic energy and are then subjected to an electric field. The physical behavior can be explained figuratively in terms of a potential field between two transitions. It can also be said that the electrons absorb heat from the surroundings and the surroundings are consequently cooled. The phenomenon can be compared to that occurring at the cathode of a tube. It has been found that when electrons flow to the anode it is the fastest electrons that are drawn away. In the meantime the average kinetic energy of the electrons drops and the temperature of the heated filament decreases, i.e., the filament is cooled when electrons are drawn away; and if the filament is incandescent, it exhibits a duller glow.

Peltier cooling elements are usually made of 10–20 individual elements connected in series. In a Peltier cooling element the *n*- and *p*-arms are not welded directly together. The two parts lead to an ohmic contact piece of such a material as aluminum. This contact piece possesses a high thermal conductivity, by means of which the cold generated by the Peltier element is supplied to the object to be cooled. But metal contacts are also employed at the warm side to ensure effective transference or dissipation of the heat over cooling fins. From the standpoint of the electronics involved the use of a metal contact simply divides the transition into two stages. From an energy viewpoint, however, the result is the same.

The Peltier cooling elements are superior to conventional refrigeration units in that they eliminate the necessity for ammonia and other similar coolants as well as the rotary compressors, etc. Chemical or mechanical wear and tear is thus eliminated. The cold would increase almost in proportion to the current supplied if not for the presence of another effect that opposes the generation of the cold; this effect is based upon the fact that the current per element cannot be raised at will as a result of the ohmic losses, since heating occurs at the junctions in accordance with Joule's law $I^2 \cdot R$, and this heat at a definite temperature or loss neutralizes the cooling effect. Up to now cooling elements have been made in the form of reversed immersion heaters for cooling beverages up to 50 watts, with the heat being removed from the wall contact sites by water. In this way temperatures as low as $-40°$ have been attained. However, the cost of such high power units is still quite high; it would appear that we have not as yet found the right semiconductors for this purpose and that semiconductor combinations will have to be subjected to systematic study.

4. MHD generator, a new application for free charge carriers

The principle of the plasma generator (MHD generator) is illustrated in Figs. 197 and 198. If a conductor is passed across a magnetic field, the flux lines are cut and an electric potential is created at the ends of the conductor. This principle can also be applied when, instead of a conductor, charge carriers are passed across a magnetic field. A gas is completely ionized in a combustion chamber, so that it consists entirely of electric charge carriers. Such a gas is called a plasma. The gas is passed through a transverse magnetic field and a potential difference between the two electrodes is thereby created.

Because of the high temperature, several thousand degrees, the metallurgical problems are extremely difficult to solve. Because the temperature of this gas must be very high to ensure a high velocity,

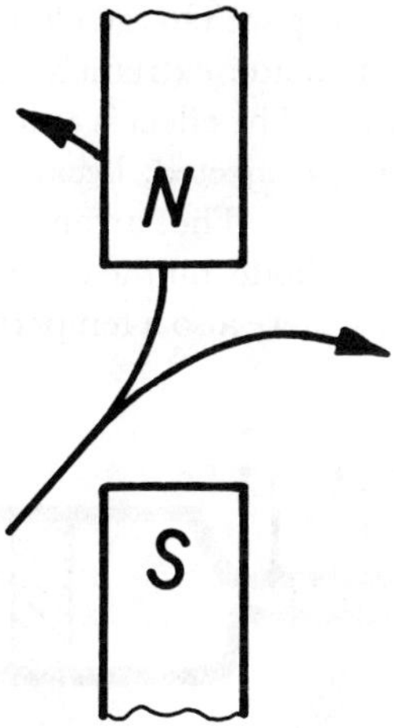

FIG. 197. Separation of positive and negative ions to the left and right as a consequence of the Lorentz force.

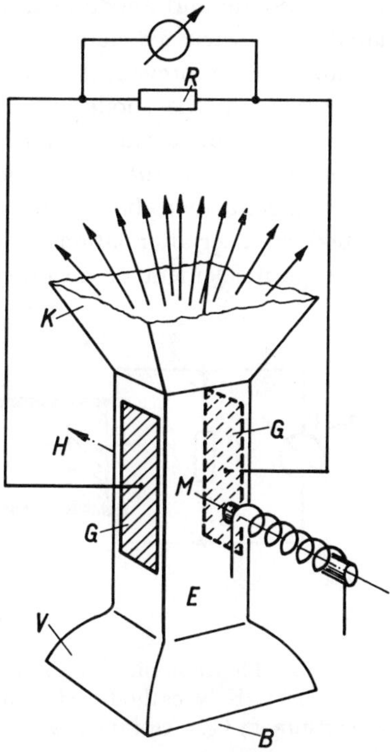

FIG. 198. Direct transformation of heat into electrical energy by means of the Magneto-hydro-system generator (MHD). It is hoped that in the future the exhaust gas from rocket mechanisms can be used for this purpose. M = large transverse magnetic field, through which the hot (3000°) ionized gas is passed; V = combustion chamber; K = chimney; E = place for the expanding ionized gas; G = graphite (or zirconium) electrodes.

a high efficiency of 60–70% can also be expected. It is hoped that generators of this type will be built that will be capable of delivering from a few to several hundred kilowatts of power.

5. Plasma cells

Researches in the use of ions alone to obtain a generator effect have led to an attempt to devise thermoionic generators, so-called plasma cells. Figure 199 illustrates the principle of a thermoionic generator, which consists of two metal electrode plates, one of which is maintained at a low temperature while the other is heated; the result of this is to create a cathode and anode as in a radio tube. If both plates are situated in an evacuated chamber, electrons are formed and pass over to the anode. To prevent the formation of a space charge cloud the distance between the anode and cathode is made extremely small, lying between approximately 0.1 and 0.2 mm. The effect is amplified if an ionizing vapor, such as cesium vapor, is injected, because the electric current then flows through the gas. The arrangement described has a certain relationship to a radio diode, on the one hand, and to a thermoelement, on the other, because here also a temperature

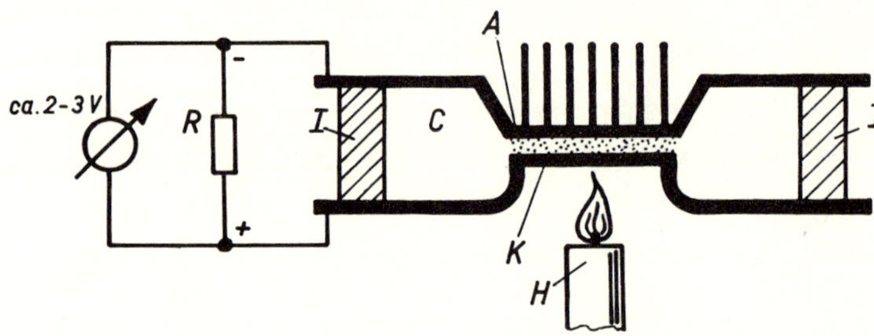

FIG. 199. Thermoionic generator. A = anode; I = insulation (heat + electricity); K = cathode; H = heating (possibly by the sun's rays); R = consumer; C = cesium vapor area.

gradient is utilized to accelerate the charge carriers in accordance with their increased kinetic energy toward the anode, which is maintained at a lower temperature.

The cathode could as well be heated by the sun's rays with the help of lenses or concave mirrors. Today units of from 3–200 watts with an efficiency of up to 14% are produced. Here also it is hoped that power sources yielding many kilowatts with an efficiency of approximately 30–40% will be created in the future. However, as the efficiency is increased, the required temperatures become higher and higher, and temperatures of approximately 2000° are anticipated.

6. Electrons in chemistry

Electrons are not only of fundamental importance in electrical science, but also in chemistry. The stabilizer cells that are of such great importance in electronics, and that will be discussed shortly, belong to a borderline area of chemistry.

The conduction of electricity in electrolytes cannot be considered as belonging to the domain of electronics: the electrons in this case are not free. The excess or deficit of these electrons in atoms turns the atoms into negative or positive ions, respectively, which are in turn available for current conduction. The absorption of electrons is called reduction, and the surrender of electrons, oxidation, in an extended sense of the term. When both mechanisms occur simultaneously, the phenomenon is known as a redox process. Redox processes occur in all electrolytic cells; a cell of this type exhibits a definite electromotive force that is equal to the difference between the redox potentials. A strongly interfering accessory phenomenon is the liberation of gas at the electrodes; this makes a tight cell construction impossible. G. Neumann has now succeeded in constructing a cell with a cadmium cathode and a nickel anode in which reduction at the negative electrode, resulting from the passage of current, is accomplished regeneratively by the oxygen liberated at the positive electrode.

The chemical process can be described as follows. The application of a constant potential to the nickel anode results, after the formation of a thin oxide coating (the polarization coating), in additional oxygen at the anode; this oxygen is obtained from the aqueous liquid (H_2O) of the separator. The oxygen passes to the cadmium cathode, which it partially oxidizes to cadmium oxide, but at the same time the current reduces the cadmium oxide to cadmium once more. The requisite hydrogen is likewise obtained from the aqueous (H_2O) solution of the separator; it combines with the oxygen of the cadmium oxide to form cadmium and water. This cyclical process enables a constant current to flow through the cell without the formation of more O_2 or H_2 gas than is reabsorbed by the cell. The redox potential at the cathode, on the one hand, and at the anode (oxygen liberation potential), on the other, is very stable and nearly independent of the current flowing through. The cell, which is called a stabilyte, can be constructed to be absolutely tight and is very well suited, because of its small dimensions, for electronic use in stabilizer circuits, for example, and wherever standard potentials are required. Furthermore, these cells, because of their storage properties, are not only capable of stabilizing the amplitude of an applied voltage but are also capable of smoothing them out like an extremely large capacitor. They are consequently used for smoothing and stabilizing filament voltages in mains-supplied instruments.

The a.c. resistance lies between 1.5 ohms and 20 milliohms, depending upon the type. At a frequency of 50 cps this corresponds to the filter effect of a condenser of 2100 μf and 160,000 μf, respectively.

The stabilization cell voltage is approximately 1.5 volts; for higher voltages several cells are connected in series. This fills a gap for the lower voltages (transistor technology) above which the task of stabilization is taken over by the zener diode and, at still higher voltages, by the glow-discharge tube.

7. Piezoelectric transducers

If a crystal structure consisting of positive and negative ions is deformed (under the influence of external pressure), assuming that there is a definite crystal structure on the sides to which the pressure is applied, an electric charge is created. This phenomenon is known as the piezoelectric effect and is exhibited by a number of substances such as quartz, tourmaline, Rochelle salt, etc. For the phenomenon to occur the lattice structure must be of such a nature that the point of concentration of positive and negative electrical charges will no longer coincide after the crystal is subjected to deformation. This property has found ready application in the construction of crystal microphones and sound pickups. Conversely, an external electric field by means of a reciprocal effect produces a deformation of the crystal. These elements are consequently suited for use in loudspeakers and ultrasonic transmitters (supersonic echo sounding, ultrasound therapy). For these purposes the relatively soft Rochelle salt crystal ($KNaC_4H_4O_6 + 4H_2O$) is quite suitable. On the other hand, quartz crystal slabs (SiO_2) are better suited as stabilizing crystals, i.e., for frequency stabilization of transmitters or for electrical quartz clocks, because of their high elasticity and lower internal damping. Here the characteristic frequency corresponding to the sharp mechanical resonance frequency of quartz, which is determined by the dimensions, and its mechanico-electrical transducer properties are used as a frequency standard as well as for regenerative amplification (damping reduction).

The number of piezoelectric materials has recently been enriched by the addition of the synthetically produced barium titanate crystal $BaTiO_3$. This is moulded, burnt, or sintered from ceramic material in a manner usual for ceramics. Then the two opposite surfaces are silver coated. In this way a disk condenser is formed. This raw material is then heated above the Curie point and slowly cooled over a long period of time under the influence of an intense transverse d.c. field of more than 3000 v/cm, so that the desired crystal structure is created. These transducers exhibit much lower ohmic resistance than quartz as a result of the particularly low impedance, so that in addition to possessing a great piezoelectric effect they also permit the use of longer leads to the amplifier. $BaTiO_3$ is also favored as a dielectric for miniature capacitors in transistor instruments because of its particularly high dielectric constant $DK(\varepsilon)$ (see Fig. 28).

Voltage-dependent condensers have acquired great importance in the so-called parametric amplifiers. An amplifier of this type is one where in the related oscillatory system, consisting perhaps of the above-mentioned capacitance C and an inductance L, one of the parameters, for example, ε, is varied periodically. Let us imagine a condenser consisting of two opposed plates, the distance between which is increased at the moment of the voltage maximum (the antinode of the wave). This would require additional work. When the voltage again passes through zero the situation returns to normal, a process which is not reflected in the work balance as a result of the zero voltage. The previously supplied work can, therefore, only benefit the oscillatory circuit. In practice the additional energy is supplied in the doubled oscillatory circuit frequency, i.e., in the so-called pumping frequency. Instead of a mechanical variation of the distance between the condenser plates, use is made of a voltage-dependent capacitor, which is fed in the correct phase relationship by the pumping frequency. In the last chapter of this book we will consider among other subjects the already quite advanced solid body maser amplifier, in which a nonconducting diamagnetic crystal is doped with low concentrations of paramagnetic ions. An amplification effect is obtained here also by means of a pumping effect and stimulation effect.

In order to explain these modern, atomic-type amplifiers we will have to go back to Chapter I and introduce several additional refinements into our model of the atom. In anticipation of the next chapter it may be pointed out that the name "maser" was coined by Townes, the inventor, and is an acronym for the descriptive phrase "*m*icrowave *a*mplification by *s*timulated *e*mission of *r*adiation." The practical importance of this new electronic element is based upon its extremely low intrinsic noise.

8. Magnetic amplifiers (transducers)

Electronic technicians and engineers are often called upon for the installation and maintenance of *magnetic amplifiers*. These control and regulating elements do not belong strictly to electronics. However, since controlled machines frequently contain electronic instruments alongside these magnetic amplifiers, it seems justifiable to describe these very widespread magnetic amplifiers or transducers in this supplement.

Magnetic amplifiers are very popular in industry because of their ruggedness and are manufactured for magnitudes ranging from several watts to several kilowatts. They are externally similar to conventional transformers and require no special attention. Several fundamentals of electrical engineering will have to be considered before we can discuss their manner of operation. The inductive reactance of a coil is a function equally of the frequency f and the inductance L.

$$R_L = 2\pi \cdot f \cdot L$$

The inductance of a coil with a good iron core (a choke coil) is, because of the high permeability μ, several hundred times as great as that of an air coil. The inductance can be obtained from the formula:

$$L = \mu\mu_0 \cdot \frac{F}{S} \cdot W^2$$

where W represents the number of windings, F the cross section (surface of the magnetic circuit), S the length of the flux lines, μ the relative permeability (for air this is equal to 1, for iron $\mu \approx 1000$), μ_0 the absolute permeability constant. The relative permeability not only varies from material to material, but is also a function of the degree of magnetization. For a high degree of magnetization, such as that resulting from a large d.c. current flowing through a large number of windings, μ drops considerably and gradually approaches the value 1 of the air coil. The inductance, and consequently the inductive reactance R_L as well, is thus a function of the magnetization as shown in Fig. 200.

If the choke coil, which is connected in series with a load, is initially magnetized by a separate coil, the inductive reactance will be smaller than in the absence of the initial magnetization. The magnitude of the current flowing through the load can therefore be controlled over a wide range by means of the magnetizing current. An arrangement such as that illustrated in Fig. 201 would not be advisable since an a.c. current is already being transformed in the magnetization coil S.

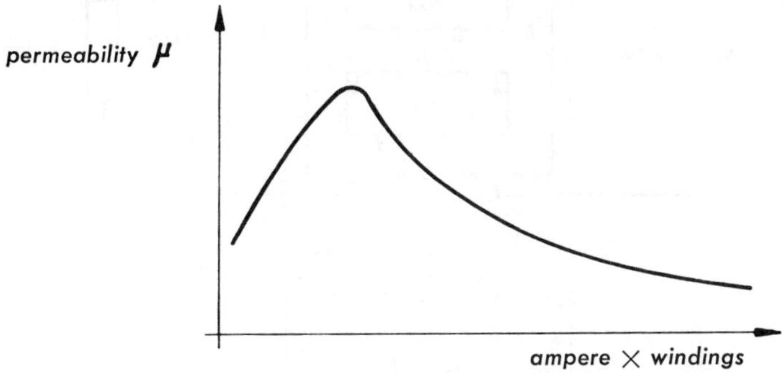

FIG. 200. Permeability μ or inductive reactance as a function of the coil current A in a coil with a closed iron core, for given frequency.

It is known that bridge connections are used to compensate magnetic or electrical circuits. A three-legged core such as that shown in Fig. 202, the two outside legs of which are enclosed in the choke coils,

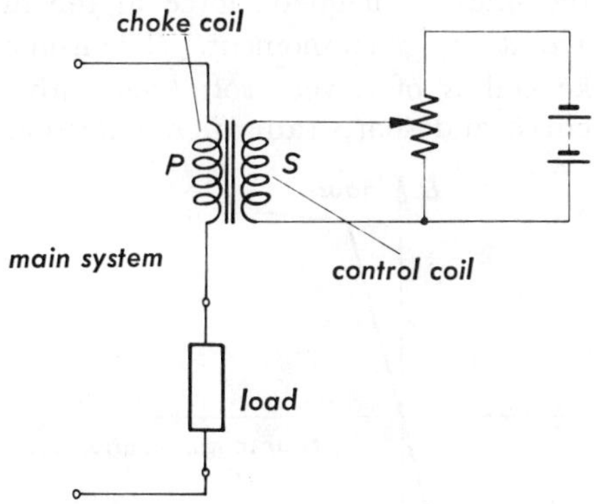

FIG. 201. The choke coil represents an a.c. resistance, the value of which is dependent on the preliminary magnetization; this is produced with the control or secondary coil. A backward effect of the alternating current would be harmful.

together with the center leg, which is provided with the d.c. coil, acts as such a bridge. In this arrangement the undesirable transformer current in the middle leg disappears. The arrows in Fig. 202

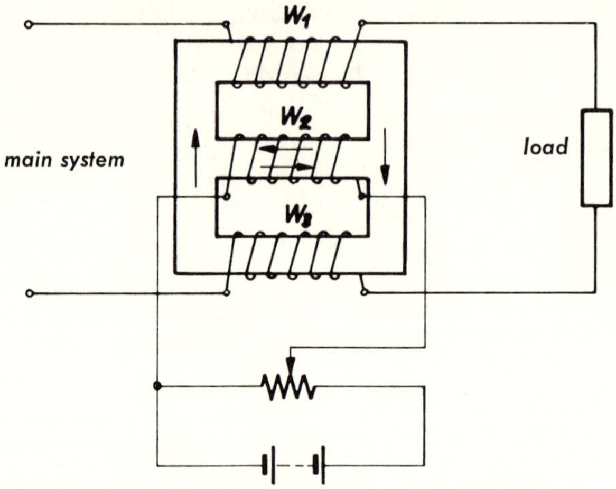

FIG. 202. Symmetric three-legged core of a transducer (magnetic amplifier). In the center leg, where the control S is located, the a.c. magnetic field is canceled as in a bridge circuit. The d.c. field, coming from the coil S, influences the two outer legs. D_1 and D_2 can be arranged in series or parallel depending on the course of the current.

illustrate how the lines of magnetic force in the middle core are oppositely directed at any given moment. The iron core used in the air-gapless choke coil is of a very soft type with a steep, linear magnetization curve and sharp saturation inflections, as shown in

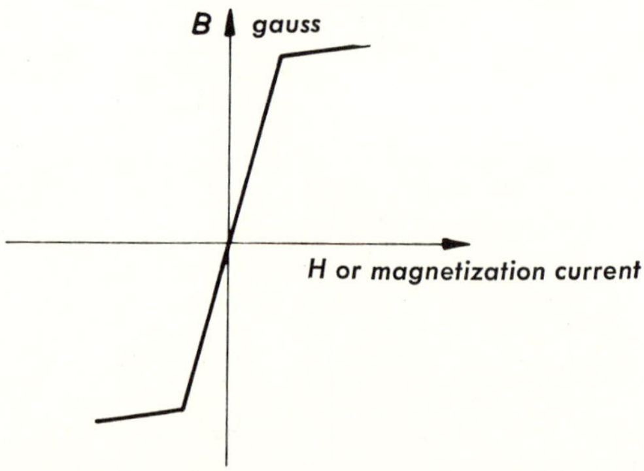

FIG. 203. Magnetization as a function of coil current, in iron. The magnetization curve for magnetic amplifiers should be steep, with sharp saturation inflections, B = induction or magnetic force flow; H = magnetic field strength $= \frac{\text{amp} \times \text{windings}}{\text{length}}$.

Fig. 203. Since the input control power required is lower than the controlled power in the load circuit, it is justifiable to call this an amplifier. The two primary windings are arranged in series or parallel, depending upon the desired current magnitude. Such transducers (as in Fig. 204) possess the property of permitting the passage of a nearly constant current at a given control voltage. If the load to be controlled uses rotary current, three units are used, each of which is arranged in one of the three phases. The control current in this case flows through all three of the control coils, which are arranged in series. When there is no control current the inductive

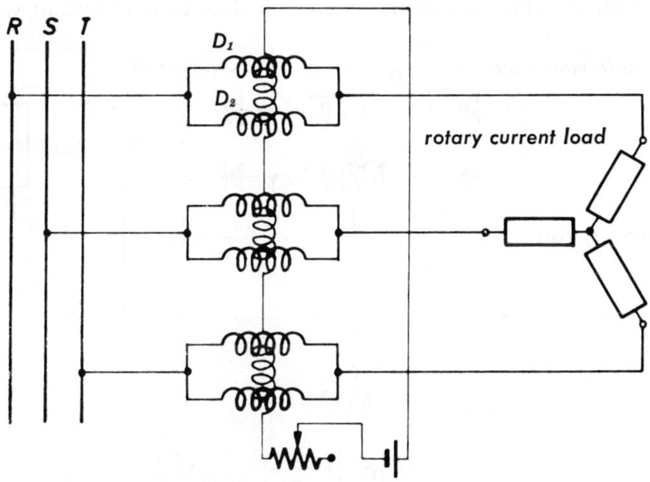

FIG. 204. Regulation of a rotary current load $V_{R,\,S,\,T}$ by means of three "current-controlling" transducers, in which the magnetization windings S are arranged in series. D_1 and D_2 are arranged in series (high ohmage) or parallel (low ohmage, for larger currents) depending on the size of the load.

reactance of all three control coils is at its highest, and the current to the load is at its lowest. It is possible to rectify a part of the output current and supply it to the control circuit. In this way it is possible to obtain regenerative coupling or a negative feedback, depending upon the polarity. Regenerative coupling intensifies the control. These are known as *current controlled transducers with separate excitation.*

As a result of the symmetrical course of the magnetization curve as shown in Fig. 203, the direction of the d.c. current is of no importance. The power consumed by the d.c. current control is of the order of several hundredths to several thousandths of the load power. It is also possible to connect a rectifier to the output of a transducer and thus ultimately use the transducer to regulate the intensity of a d.c. current. In industry it is frequently necessary to control d.c. motors and other d.c. using loads. However, it is also possible to use an internally generated d.c. current as an auxiliary source for the control arrangement, as shown in Fig. 205.

Since a load current flows through each of the two coils, the load current also flows through each of the rectifiers. This results in a superimposed d.c. current, which magnetizes the iron core in accordance with the intensity of the current. The control circuit with its

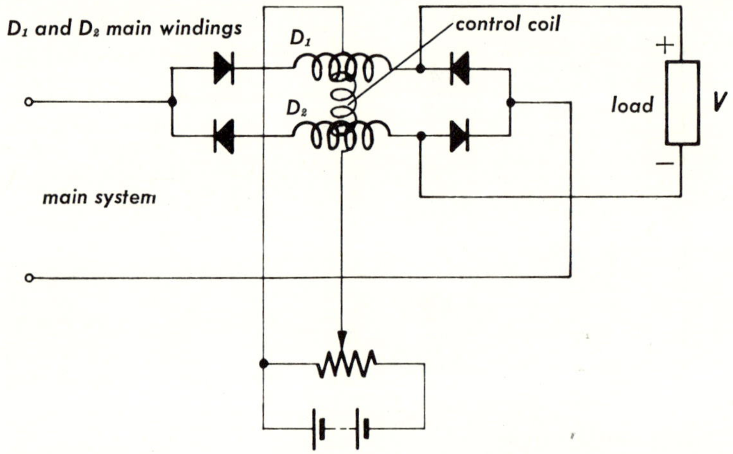

FIG. 205. Transducer circuit for d.c. loads (with voltage control).

separate winding can have the poles so arranged that an increase in the control current is able to reduce the degree of magnetization (even from the saturated state) and bring it into the unsaturated state, as a result of which the inductive reactance is reduced on the increase in control current. If a d.c. consuming load is used the arrangement can be combined with the necessary rectifier (see Fig. 206). In contrast to a control transformer, where the output voltage may vary, the no-load voltage in transducers is constant and has the value of the power supply. Essentially, in this case it is the inductive series reactance that is controlled and with it the current. The voltage at the load is then the result of the voltage drop in accordance with Ohm's law. It is precisely this property that makes these units so particularly well suited for controlling the brightness of fluorescent tubes. If a gas-discharge tube is connected the whole no-load voltage is available for ignition, for which the choke coil from the primary circuit of the transducer forms a nearly ideal current limiter after ignition. However, care must be taken that the filament voltage of the fluorescent lamp remains unchanged. A separate transformer can be used for this purpose (Fig. 207).

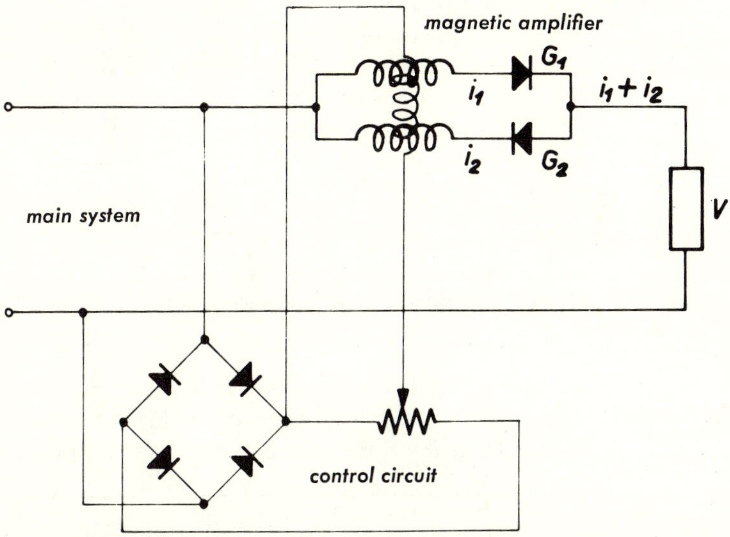

FIG. 206. Rectifiers as auxiliary elements of the transducer. Alternating current flows at the load. Through each of the choke coils D_1 and D_2 flows a half-wave "direct current," which independently saturates the iron core. By this arrangement the voltage is maintained nearly constant even when the load is changed (voltage-controlled transducer).

The control current is very low, so that remote control over great line lengths of small cross section is possible. A transducer can not only be used for control purposes, but also for automatic regulation. When it is used in this way the measured magnitudes are used to adjust the control resistance to the appropriate value. The adjustment can be made mechanically or directly. A popular way of using the transducer for such purposes is to provide it with two or more separate control coils (see Fig. 208). One coil is used in the manner previously described to adjust to the so-called theoretical value in accordance with the potentiometer setting, while the second is supplied

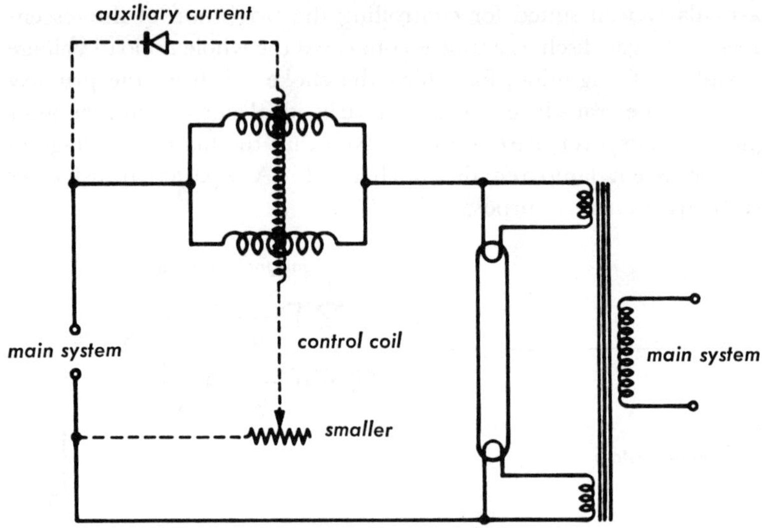

FIG. 207. Brightness control of a fluorescent lamp by means of a transducer.

from the measured magnitude with a second current corresponding to the actual value. The control current operating the magnetic amplifier is derived from the difference between these two currents,

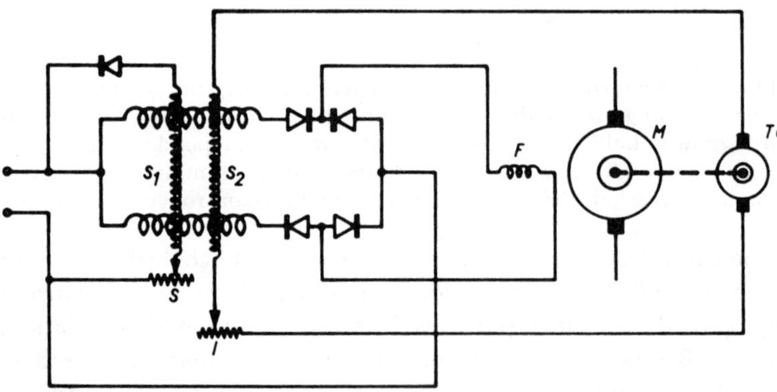

FIG. 208. Solution of a regulation problem by means of a transducer with two control coils, S_1 for the theoretical value setting and S_2 for the actual value setting. The measured magnitude (number of rotations of the motor M) is supplied to the control coil S_2 by the tachometer generator TG. The number of rotations of the motor being regulated is kept constant by altering the field current at F.

which are connected to opposite poles. The current that determines the magnetization is therefore always the difference between the actual value current and the theoretical value current. When the theoretical value has been adjusted, an actual value may be formed that partially upsets once more the effect of the former. An equilibrium develops in accordance with the transducer characteristic; this equilibrium, which is indispensable in regulatory systems, continually readjusts itself to changes in the load circuit such as those arising from a new load. In r.p.m. or speed regulation the measured magnitude, i.e., the number of rotations, is converted by a small tachometer generator into a voltage whose value increases linearly with increasing velocity. This voltage or current, which corresponds to the actual value, is fed into the second control coil of the magnetic amplifier. Usually this magnitude is also passed through an adjustable potentiometer. In this type of speed regulation the magnetic amplifier controls the exciter coil of a d.c. motor through a rectifier. A decrease in the field results in an increase in the number of r.p.m., and an increase results in a decrease, in accordance with the law of inductance. Nowadays many different problems in regulation are solved with magnetic amplifiers. To be sure, compared to electronic amplifiers they exhibit a very great time lag. The response time, i.e., the time required for the current in the output circuit to adjust to the new value, calculated against the variation in control current, is several tenths of a second. In many cases such a response time is desirable. When it is undesirable electronic amplifiers are used.

VII. MOLECULAR AMPLIFIER ELECTRONICS

1. Introduction

Electronics and radio engineering have gone through three different significant periods of development. Several decades ago the position of prime importance was held by electron tubes; approximately fifteen years ago W. Shockley discovered semiconductor transistors; and about five years ago C. H. Townes invented the so-called maser (an acronym for "*m*icrowave *a*mplification by *s*timulated *e*mission of *r*adiation"). This is a high sensitivity molecular amplification principle which permits operation with a much lower noise factor than electron tubes and transistors and consequently makes possible the amplification of still weaker signals.

New as this area of research may appear at first sight, it nevertheless is based upon long familiar fundamental principles of atomic structure and electronics. An understanding of electromagnetic radiation forms the basis for this modern branch of electronics which is finding ever increasing significance and opening new perspectives in such areas as radar, space communications, radio astronomy, medical therapy and other applications.

Since this new technique requires a certain understanding of atomic theory and atomic electronics we will begin by introducing certain refinements into the fundamentals of the Bohr model of the atom described in Chapter I. Here we will have to make use of figurative models to aid our understanding. However, practical experience, in particular the results of experimentally obtained measurements, are in good agreement with the energy values and frequencies deduced by calculations from our models.

2. Various types of radiation

Democritus and Aristotle, the great philosophers of antiquity, saw far beyond their time. There is a passage attributed to Democritus that says: "Only apparently does a thing possess color, only apparently is it sweet or bitter. In reality there are only atoms in a void." That was 400 B.C. Two thousand years later this intuition became fact, when H. Becquerel demonstrated in 1896 that uraniferous substances emit rays or particles that are invisible to the human eye. He was able to prove this by means of dark spots produced on photographic plates and by means of light phenomena (fluorescence) produced by these rays when they fell upon zinc sulfide crystals. Two years later Madame Curie discovered radium, which exhibited the same property known as radioactivity, to an even greater degree. If these rays are allowed to pass through a strong transverse magnetic field, a separation into three different components with three different directions is observed. The three different types of radiation are called α, β, and γ rays. From the observed orbits, which are determined by the mass and energy charges in accordance with mechanical laws, it was possible to derive the following conclusions regarding these rays:

(1) The α rays are positive ions. From their mass, which has been found to be four times that of a hydrogen atom, and from the fact that they possess double the positive unit charge, it was inferred that alpha rays must consist of helium atoms without the electrons, i.e., so-called helium nuclei, consisting of two protons and two neutrons.

(2) The β rays are fast-traveling, i.e., high energy electrons with the familiar negative unit charge of $1.6 \cdot 10^{-19}$ ampsec. Because of their opposite sign, the β rays travel in the opposite direction from the α rays, and because of their very much smaller mass (1/1830) they are subjected to incomparably greater deflection in the magnetic field.

(3) The γ rays are (hard) x-rays and consequently possessed of great penetrating power. These roentgen rays can also be described as a form of light invisible to the human eye and of extremely high frequency or, in view of the relation $\lambda = c/f$, as electromagnetic waves of very short wavelength. (In the formula, c represents the speed of light and f the frequency.) In contrast to the α and β rays, which consist of particles possessing mass, x-rays are not deflected in a magnetic field. They travel in a straight path through the magnetic field. The trajectories of the above described rays can be traced easily in a quantitative and qualitative fashion in a so-called Wilson cloud chamber (C. T. R. Wilson, 1912). The direction and length of trajectory, as well as any eventual collisions are observed. This makes it possible to draw additional conclusions concerning the

particles. The Wilson cloud chamber operates in the following manner:

The air is supersaturated with water vapor. The excess vapor changes into drops of water as soon as condensation nuclei occur. The passage of small charged particles through the cloud chamber causes the formation of gas ions, and the latter act as condensation nuclei and consequently form a cloud path.[1] We are familiar with the appearance of a cloud created by a series of such extremely small water droplets from the passage of jet aircraft at high altitudes. The trajectories of charged particles and photons can be reproduced on the photosensitive layer of photographic plates (a thick emulsion is desirable) and appear as sharp black lines after developing.

The typical wave characteristics—so-called interference phenomena, intensity amplification when waves are in the same phase, and intensity amplification when waves of opposite phase occur at a single location—are not only observed for light rays, x-rays and electromagnetic radio waves. The same phenomena have also been observed on the passage of corpuscular α and β rays through thin metal foils. From this it can be inferred that these particles are also characterized by wave motion (matter waves). From observations of this type it is possible to determine very exactly the wavelength λ. From the energy equation

$$E = m \cdot v^2 = h \cdot f$$

where

$$f = \frac{v}{c}$$

de Broglie derived the relation

$$\lambda = \frac{h}{m \cdot v}$$

in which h = Planck's constant. Planck proved that radiant energy can be transmitted only in multiples of h, and thus, $h = E \cdot f$, in which relation h is a constant independent of all material properties, whose value is obtained directly from the radiation formula. Experimentally h is encountered in the measurement of the photoelectric

[1] Nowadays a supersaturated layer is maintained constantly by means of a temperature gradient between the roof of the chamber, which is electrically heated to 20°C, and the bottom of the chamber, which is cooled to approx. −55°C by means of dry ice (solid carbon dioxide). The medium used is an 80% water-alcohol mixture which evaporates under the relatively warm roof, diffuses toward the bottom and recondenses to a liquid. A high d.c. voltage is applied to the chamber between the roof and floor to "blow away" quickly the easily visible cloud tracks.

effect on a zinc plate. The light quanta (photons) are not capable of liberating electrons from a zinc plate until the ultraviolet range of the spectrum ($f = 7 \cdot 10^{14}\ \text{sec}^{-1}$ or $7 \cdot 10^{8}$ Mc/s) is reached; thus, they are unable to liberate electrons in the red portion of the spectrum. The work function for liberation of electrons from a zinc plate is known from voltaic elements; it is equal to 1 e.v. Thus, the value of h is found as:

$$h = \frac{E}{f} = \frac{eU}{f} = \frac{1.6 \cdot 10^{-19}\ \text{ampsec} \cdot 1\ \text{V}}{7 \cdot 10^{14}\ \text{sec}^{-1}}$$

$$= \text{about } 10^{-34}\ \text{wattsec}^2\ \text{(or Joulesec)}$$

m = the mass of the particle, the mass of the electron in this case

v = the velocity of the particle, in this case of the electron

If the velocity v of the electron is known from other measurements and the wavelength is known from the previously mentioned interference

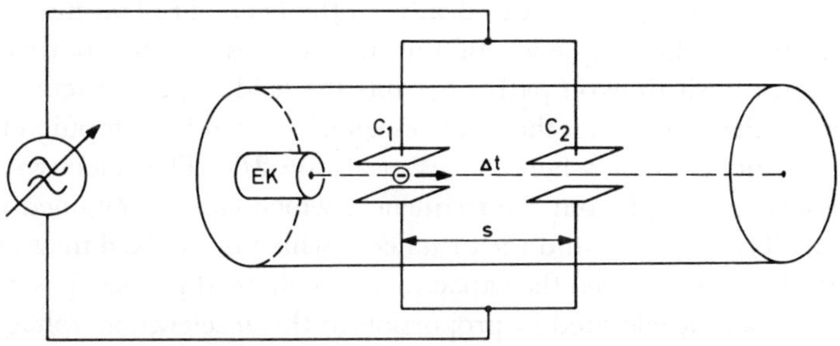

FIG. 209. Measurement of the velocity of electrons.

measurements, the formula above can be used to calculate the mass m of the electron. The velocity of electrons can be measured with the use of a measuring arrangement such as that illustrated in Fig. 209. The apparatus resembles a cathode-ray tube except that the deflection plates are placed parallel in the vacuum chamber and are also arranged in an electrically parallel circuit.

An electron will then travel through the two capacitors C_1 and C_2 which are supplied with high frequency a.c. current in such a manner that their electric field is perpendicular to the path of the electron. The electron is generally deflected to a lesser or greater extent upward or downward in accordance with the magnitude and direction of the electric field. The electron can, however, pass through the plates of the capacitors in a straight line if the time of traversal coincides exactly with the zero value of the alternating field. Let us assume that this occurs, for example, in the first capacitor C_1. As a consequence of the final transit time Δt on path s (the distance between the two sets of capacitor plates) an electron will be delayed in reaching capacitor field C_2, where it is then deflected if there is an electric field. It is also possible for an electron to pass undeflected through the second capacitor C_2 if the phase of the a.c. current passes through the zero value at exactly the same time during which the electron passes between C_1 and C_2. This occurs for all transit times at the oscillation phases 1/2, 2/2, 3/2, ... In all other cases the electron beam is readily deflected from a straight path, and the vertical deflection is immediately seen as a light trace on the picture screen.

Naturally, we cannot use this method to measure the path of an individual electron; we make use instead of a beam pulse composed of an enormous number of electrons. However, for the purpose of illustration it is convenient to deal with the behavior of an individual electron. If the frequency of the a.c. field is so chosen that the electron transit time on path s permits the field to pass exactly twice through the zero value, the electron velocity v can be computed from the equation $v = s/t$, where in this case $s/t = 2f$. The velocity v may therefore be found from the frequency, which can be obtained from the oscillator reading and the distance s, which is the fixed magnitude of the distance between the capacitors. A charged particle possessing a charge e is accelerated in proportion to the acceleration voltage U. From the equation for the conservation of energy it follows that

$$e \cdot U = \frac{m \cdot v^2}{2} \qquad v = \sqrt{2 \frac{e}{m} \cdot U}$$

Since, however, the velocity can never exceed the propagation velocity of an electromagnetic wave, i.e., 300,000 km/sec, it is to be understood that no matter how high the voltage U the velocity of light c is only approached and never quite attained. As the velocity of light is approached the electric and magnetic deflection changes in the same way as when the mass is increased.

According to Einstein's relativity theory, the original mass m_0 (the rest mass) increases to m in accordance with the following relation:

$$m = \frac{m_0}{\sqrt{1 - (v^2/c^2)}}$$

At low anode or acceleration voltages, where the electron velocity v is considerably smaller than c, the apparently increased or inertial mass m is the same as the rest mass m_0, i.e., $\sim 10^{-30}$ kg. If, however, the velocity v is raised to 0.9 c, i.e., to 270,000 km/sec, the inertial mass is already 2.3 times as high as the rest mass. At 0.995 c it should be ten times as high and, in the case of the speed of light, should attain an infinitely high value. F. C. Champion in 1932 confirmed the correctness of the relativistic "law of mass increase" from information provided by the visible orbits in the Wilson cloud chamber.

Here an electron accelerated in an electric field is fired against an electron at rest. According to the laws of classical mechanics, which govern macroscopic bodies such as, for example, billiard balls, if two balls of equal mass collide they must move apart along paths separated by a 90° angle on the condition that they do not collide head on.

Observations in the Wilson cloud chamber confirmed this law for slow traveling electrons as well. However, if the electron is accelerated by a very high acceleration voltage to a velocity v that is no longer negligibly small with respect to the speed of light c, the angle between the two paths after collision is smaller than the anticipated 90°. It must, therefore, be assumed in accordance with the previously cited laws of mechanics, that we are concerned here with an increase in mass.

If the laws governing the conservation of energy and momentum in elastic collision apply also to the microcosmic domain, an increase in voltage should no longer produce a linear increase in the velocity of particles whose velocity approaches the speed of light; on the other hand, the mass m of such particles should increase for energetic reasons. The literature gives the values for the radius of an electron as approximately 10^{-13} cm. These radii are thus approximately 1/10,000 the size of atomic radii. These findings can be checked by scattering and absorption measurements. Beta rays are passed through metal foils of known atomic weight. The greater the atomic weight, the more electrons contained and the greater the scatter angle resulting from the electrical interaction between the electrons. Charges q of the electrons, possessing the same sign, repel each other in accordance with Coulomb's law

$$K = \text{prop}\,\frac{q_1 \cdot q_2}{r^2}$$

in which K represents the force and r the distance between the charges.

After the electrons have been passed through such a foil, the electrons scattered and absorbed can be determined with a Geiger-Müller counter. The calculations are based on the fact that the smaller the radii of the electrons, the lower the probability of collision. It turns out that the number of deflected atoms is proportional to the atomic number (see periodic system). From our present understanding of the "structure of the atom" with its large empty spaces, it can be established that electrons that pass through hundreds of thousands of atoms hardly undergo any loss in velocity.

If alpha rays, which are positive particles, are substituted for the electrons and fired in a similar manner through a material of known atomic weight, values are obtained here, too, for the scatter and absorption resulting from interaction with the nucleus. Since the mass of these particles is considerably greater, i.e., since these particles exhibit greater inertia, they undergo, as might be expected, less deflection. Backward scattering and deflections of more than 90° are very rarely observed in the Wilson cloud chamber because the positive charges very rarely collide with charges of the same sign, i.e., the atomic nuclei. Here also the deflections are elastic collisions resulting from the repelling electric field. When collisions involving direct backward reflection occasionally do occur, they are very intense and consequently could not arise from the atomic shells. It is precisely the rarity of backward scattering that indicates how extremely small and concentrated the atomic nucleus must be. Rutherford in 1908, using the laws of classical mechanics and taking into consideration the coulombic forces involved, derived the scattering formulas for atomic particles.

The Millikan experiment (see p. 41) provides the most accurate measurement of the electron charge. But the charge can also be determined electrolytically in the manner which enabled Faraday (1833) to study it in greater detail. In this method one makes use of a glass container filled with an aqueous solution of an acid and containing two separated electrodes. In such a solution the atomic structure of the molecule becomes weakened (dissociated). Thus, in dissociated hydrochloric acid, for example, the hydrogen atom acts as if it is separated from its electron, whereas the latter acts as if it belonged to the Cl atom. The H atom is, therefore, positively charged (a so-called ion) and the Cl atom possesses one electron too many and is consequently negatively charged ($HCl \rightarrow H^+ + Cl^-$).

When a d.c. voltage is applied to the electrodes the positively ionized atoms are attracted by the negative electrode, the so-called cathode, and repelled by the positive electrode (anode). At the same time the negatively charged atomic groups travel to the anode.

Thus, the flow of electricity in an aqueous solution consists of a current of ionized atoms, in contrast to conduction in metals, where the current is carried by electrons. The H ions, thus, travel to the cathode, where free electrons are obtained from the current source to make up for the electron deficit. If the solution consists, for example, of copper sulfate ($CuSO_4$), the copper sulfate undergoes the usual dissociation into the ion groups Cu^{++} and SO_4^{--}. In this case also the ionized copper molecules travel to the cathode. At the cathode they precipitate out of solution, giving rise to the familiar electrolytic copper plating phenomenon. The atomic groups possessing the double negative charge travel to the anode, where the free electrons pass into the metallic conductor and finally travel to the positive pole of the current source.

To determine the charge of an electron the following considerations are taken as a starting point. It is known from experience that the quantity of material precipitated m is independent of the concentration of the solution and the temperature, and a function of the current intensity I, the time t, and a material proportionality factor P.

Consequently, $m = I \cdot t \cdot P$. P gives the mass in milligrams of the material precipitated out of the electrolyte per second at a current of one ampere. This proportionality factor P (the electrochemical equivalent) varies from material to material and is determined by measurement. Thus, the electrochemical equivalent for hydrogen $P_H = 0.01045$ milligram/ampsec. For the precipitation of silver from silver nitrate $P_{Ag} = 1.118$ milligram/ampsec. These values are proportional to the atomic weights and, consequently, to the mass numbers. The value of P for copper is found to be 0.3294 milligram/ampsec. This figure corresponds to only one-half the anticipated value. Faraday found that the valence W (which provides a picture of the number of electrons occupying the outer, so-called valence shell) of the substance also had to be taken into account, and copper is commonly known to be divalent in copper sulfate. (Hydrogen, on the other hand, with its single electron, is monovalent.) The electrochemical equivalents are consequently proportional to the atomic weight A divided by the valence W. The ratio $A/W:P$ has been found to be equal to 96,521. This is the so-called Faraday constant F and is valid for all electrochemical conversions. The Faraday constant gives the quantity of electricity in ampsec required to precipitate 1 gram equivalent of any material. To obtain 1 g of hydrogen 96,496 ampsec are required. The charge of an electron can be computed from the above figure F and the Losschmidt number L. The latter gives the number of molecules or atoms in a gram molecule and gram atom of a substance; this number is always equal to $6.02 \cdot 10^{23}$. The same number of atoms must therefore be present in a gram equivalent of a monovalent substance. Since the electrochemical equivalents of all substances is proportional, on the one hand, to their chemical equivalent weights, and since, on the other hand, all substances require 96,496 ampsec for the precipitation of one gram equivalent of material, the charge on a single atom of a monovalent substance is

$$q = \frac{F}{L} = \frac{96{,}496}{6.02 \cdot 10^{23}} = 1.602 \cdot 10^{-19} \text{ ampsec}$$

This is the smallest unit of electrical charge and is equivalent to an elementary quantum and, thus, the charge of an electron. In electrolytic phenomena, as in electronics and atomic physics, all charges must be whole multiples of q. An ion of a trivalent substance is consequently associated with three times the elementary charge, i.e., each ion in electrolytic conduction transports as many elementary quanta as indicated by the valence.

3. Refinement of the atomic model

The conceptual model of the Bohr atom, which is easily treated from a formal point of view, consists of an atomic nucleus about which electrons rotate in fixed, circular, and planetary "Kepler orbits."

Niels Bohr was the first to recognize the quantic graduations in the structure of the atom, in accordance with which the electrons were not able to revolve in just any orbits, but only in certain orbits prescribed by definite quantic conditions; moreover, while revolving about the nucleus in these prescribed orbits, the electrons, in defiance of classical conceptions, did not emit radiation and thus contributed no energy to the field. Each orbit (or the electrons occupying the orbit) has an assigned energy. Each atomic nucleus possesses a radially symmetrical electric field corresponding to its positive electric charge (the atomic number Z times the unit charge plus e).

An electron in one of the orbits about the atom is subject to this field and possesses a potential energy corresponding to its distance from the nucleus

$$E_p = -\frac{(Ze) \cdot e}{r}$$

The negative sign means that the potential energy decreases as the proximity to the nucleus increases, i.e., as the radius r decreases. If the radius $r = \infty$, $E_p = 0$, so that the energy value for all existing radii is negative.

If the orbits are numbered progressively from inside outward ($n = 1, 2, 3, \ldots$) the potential energy increases as the value of n increases and, consequently, as the radius increases. As the potential energy increases the negative value becomes progressively smaller and smaller and terminates at the nominal value zero, which expresses the fact that an electron that is situated at an infinite distance from the nucleus and is consequently not attracted by the nucleus possesses a zero energy.

In accordance with Coulomb's law, the electrostatic attraction force is expressed by the relation

$$K_c = \frac{(Ze)\cdot e}{r}$$

and in accordance with the planetary motion of the electrons about the nucleus the electrostatic attraction force must be in equilibrium with the centrifugal force

$$K_Z = \frac{m\cdot v^2}{r}$$

By setting K_c equal to K_Z and solving for $(m\cdot v^2)/2$ the kinetic energy can be calculated. Thus, for the kinetic energy we find

$$E_K = \frac{1}{2}\cdot\frac{(Ze)\cdot e}{r}$$

The factor 1/2 in this formula means that the kinetic energy of an electron moving about a nucleus is equal to only one-half the previously given potential energy.

The total energy of an electron in the atomic shell of an atom of atomic number Z is equal to the sum of the potential and kinetic energies

$$E_{\text{total}} = E_p + E_K = -\frac{1}{2}\cdot\frac{(Ze)\cdot e}{r}$$

This equation with r in the denominator indicates that the energy of an electron moving about an atomic nucleus increases as its distance from the nucleus increases.

The lowest energy level, consequently, lies close to the nucleus. This elementary arrangement of electrons is simple. All the electrons tend to occupy the lowest possible energy level (compare with the situation of a weight near a floor). If the inner levels are already filled, the additional electrons (in elements of higher atomic number) must occupy precisely fixed shells and positions at progressively greater distances from the nucleus. An "inner electron" can be raised to a higher level E_n by means of externally supplied energy, perhaps resulting from impact. When this occurs, however, the vacancy at E_m is immediately reoccupied by an electron from a higher level. The duration of the vacancy or the so-called relaxation time is consequently very short (approximately 10^{-8} sec). The resulting energy yield, which corresponds to the prescribed orbit and consequently occurs in quantized units, is called emission. The previous absorption of energy, which also occurs in quantized units, is called absorption. The difference between the two energy levels gives an electromagnetic radiation of the frequency

$$f = \frac{1}{h}(E_n - E_m) = \frac{1}{h}\Delta E$$

where $E_n > E_m$.

The greater the quantic level difference, the higher the frequency f of the emitted radiation. Quantization is to be understood to mean that all energy values and, consequently, also all frequencies occur in accordance with discrete, fixed quantities. Planck's universal constant h is equal to $6.628 \cdot 10^{-34}$ Ws^2. The range of choice of discrete frequencies in an atom is very great; it extends from the frequencies of radio waves to that of hard x-rays. If an electron drops from permissible quantic orbit n to the permissible orbit m, which is closer to the nucleus, in accordance with the electrostatic formulas the electron loses the potential energy

$$(Ze) \cdot e\left(\frac{1}{r_m} - \frac{1}{r_n}\right)$$

in which r represents the radius of the orbital path about the nucleus of the atom.

Half of this serves to increase the kinetic energy $(m \cdot v^2)/2$ since the speed of revolution must increase in the vicinity of the nucleus in accordance with the law of equilibrium between the centrifugal and coulombic attraction forces, because

$$\frac{(Ze) \cdot e}{r^2} = \frac{m \cdot v^2}{r}$$

(as in the case of the planets, where the centrifugal force and the gravitational force are in equilibrium). The other half is emitted as radiation in accordance with the relation $\Delta E = h \cdot f$ or $f = \Delta E/h$. Other considerations result in equations from which the frequency f can be very precisely computed from the known data of an atom. A very important contribution was made by Sommerfeld, who demonstrated that the action of the revolving electrons, i.e., the product of the momentum $(m \cdot v)$ and the length of the orbit (the circumference $2\pi r$) is always a multiple of Planck's action quantum h. As n increases (n represents the orbit number, which increases with increasing distance from the nucleus, i.e., n is an integral multiple $1, 2, 3, 4, \ldots$) the circumference of the orbits increases by a series of leaps, which is to say that we are concerned here with orbits of discontinuously increasing magnitude. In the hydrogen atom, which is the simplest case, in which there is only one electron and in which the orbits are assumed, the charge of the electron is equal to $-e$ and the charge of the nucleus (proton) $+e$. In accordance with the Bohr-Sommerfeld quantum theory, $2\pi r \cdot m \cdot v = n \cdot h$, in which m represents the electron mass and r the radius of orbit. When this is taken together with the equation representing the equilibrium between the coulombic attraction force and the centrifugal force

$$\frac{e^2}{r^2} = \frac{m \cdot v^2}{r}$$

the solution for the velocity gives

$$v_n = \frac{2\pi \cdot e^2}{n \cdot h}$$

i.e., the velocity of revolution in the nth quantum orbit decreases, as expected, as n increases and the orbit is further and further removed from the nucleus. If v_n is introduced into the quantum equation the radius of the nth quantum orbit is obtained directly:

$$r_n = \frac{n^2 \cdot h^2}{4\pi^2 \cdot m \cdot e^2}$$

The atomic radius r is consequently proportional to the square of the quantum number n. For a single quantum orbit, r is found to be equal to $5.29 \cdot 10^{-11}$ m, which is in good agreement with the experimentally obtained values.

As previously mentioned, when an electron passes from radius r_m to radius r_n precisely

$$\frac{1}{2}\cdot e^2\left(\frac{1}{r_m} - \frac{1}{r_n}\right)$$

is emitted as radiation. If the previously given equation is substituted for r_m and r_n, then

$$E = \frac{1}{2}\cdot e^2\cdot\frac{4\pi^2\cdot m\cdot e}{h^2}\left(\frac{1}{m^2} - \frac{1}{n^2}\right)$$

or

$$\frac{E}{h} = f = \frac{2\pi^2\cdot m\cdot e^4}{h^3}\left(\frac{1}{m^2} - \frac{1}{n^2}\right)$$

$$f = K\left(\frac{1}{m^2} - \frac{1}{n^2}\right)$$

The constant K in the preceding equation was first worked out by Rydberg and today the constant, in the form of a wave number constant, is named for him.

The equation from which the frequencies of the radiation can be calculated has the form

$$f = R_y\cdot c\left(\frac{1}{m^2} - \frac{1}{n^2}\right)$$

where $n = 1, 2, 3, 4, \ldots$ and c is the speed of light. The Rydberg constant R_y is equal to 10,967,758 m^{-1}. All spectral lines, which can be demonstrated by optical spectroanalysis with a prism (Fig. 231), have been arranged in accordance with a system of nomenclature. The lines corresponding to the leaps from the energy levels n of varying height to the lowest level m form a so-called series. The transitions from the higher levels $n = 2, 3, 4, 5, \ldots$ to the ground level, i.e., to the orbit closest to the nucleus $m = 1$ (i.e., the series number or series energy level = 1) are known as the Lyman series (1914). The "large leaps" from the outer shells with high orbit numbers or orbit levels give rise to the higher frequencies of electromagnetic waves, in particular to ultraviolet light. If the leaps lead from the orbit levels $n = 3, 4, 5, 6, 7, \ldots$ to the second lowest level or the series term $m = 2$, they are known as the Balmer series (Johann Jakob Balmer, 1885); these frequencies lie in the range of visible light. If the electron leaps go from the various height orbits $n = 4, 5, 6, 7, \ldots$ to the third lowest level, they are known as the Paschen series (Friedrich Paschen); these frequencies lie in the infrared or thermal radiation range. If we leave the hydrogen atom with its atomic number $Z = 1$ (nuclear charge $= 1$) and consider the possible spectral lines appearing in heavier atoms, the formula can be expressed in the form

$$f = Z^2 \cdot R_y \cdot c\left(\frac{1}{m^2} - \frac{1}{n_2}\right)$$

As a result of the increasing magnitude of Z^2, the frequencies or lines in the infrared, visible, and ultraviolet range are now created only by leaps in the outer electron orbits. Quantum leaps to inner electron shells give rise to x-ray spectra because the energy differences or the frequencies here are very high. For an electron to leap to one of the lower shells a vacancy must first be created by such means as an intense collision with a foreign electron (as in an x-ray tube, or electron acceleration by means of high voltage).

It should be mentioned in regard to the outermost electron shells that the series term n cannot be infinitely large; the n-term attains its maximal value at the so-called series-limit continuum. This series-limit continuum begins at the outermost possible orbit, where the already weak attraction force of the nucleus is barely able to satisfy the condition of equilibrium with the centrifugal force. The continuum, which begins at this point, arises because already freely flying electrons are occasionally trapped. These electrons bring with them a continuously distributed kinetic energy that is added to the energy that corresponds to a leap from the outermost orbit to an inner orbit.

When the substances are subjected to heating the electron orbits are disturbed by the thermal commotion, with the result that the continuous spectrum becomes very marked. Generally, the interest of physicists and electronic engineers is centered on the discrete, quantized transitions, of which, as we will see, there is a greater multiplicity than would appear from the foregoing. Special terms are used to describe these atomic energy levels and transitions; the first of these we are already familiar with, but it will be included to help provide an overall view.

1. *The principal quantum number* (n). It begins at the lowest value 1 and the shell closest to the nucleus. The latter has the lowest energy level. All subsequent n-shells are numbered progressively. As the number n increases ($n = 2, 3, 4, \ldots$) the diameter of the permissible orbit, or the major axis a of the elliptical orbit, increases. The radii do not increase linearly, but as the square of the principal quantum number, i.e., 1, 4, 9, 16, etc. The orbits lie within spatial shells.

The principal quantum number n characterizes the eigenvalues of the energy reservoirs of the large orbits. Its so-called series number merely indicates the large relationships, thus, the macrostructure. Since the principal quantum number does not determine the form of the orbit and consequently does not set forth all the subdivisions of the energy levels, we still require:

2. *The azimuthal quantum number* (l). It begins at the value zero and is also always a whole number such as 0, 1, 2, 3, ... This number characterizes the form of the ellipse (Fig. 210). If we let a

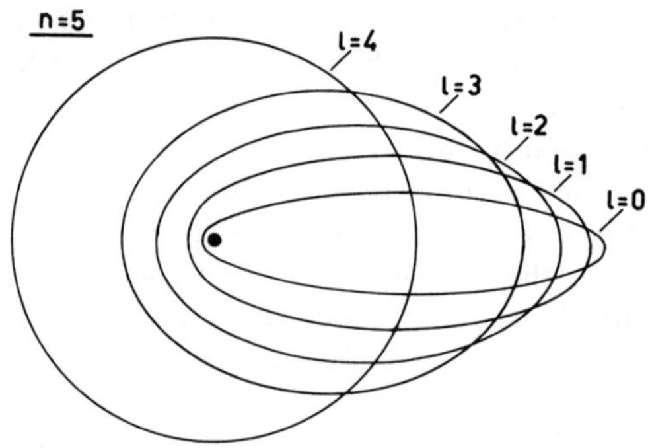

FIG. 210. Orbit forms as a function of the azimuthal quantum number. The nucleus is situated at one of the ellipse foci or at the center of the circle. The lower the number l as compared to n, the flatter the ellipse. The major axis, however, remains the same for all the orbit forms governed by the same principal quantum number n.

represent half the major axis and b half the minor axis of an ellipse, the following condition is observed:

$$\frac{a}{b} = \frac{n}{l + 1}$$

l cannot be smaller than zero, for that would give an infinite value, $n/0 = \infty$. But l cannot be larger than $n - 1$, either. When $l = n - 1$ then $a = b$. This corresponds to a circular orbit. For an orbit of $n = 4$ and the azimuthal quantum number $l = 3$, for example, the electron follows a circular orbit on a spherical shell. If in the same example l equalled 2, 1, or 0, the electrons would travel along elliptical orbits that become increasingly extended as l diminishes. In an unactivated hydrogen atom with only one shell, $n = 1$, the azimuthal quantum number, in accordance with the designated conditions, must correspond to $l = 0$, which confirms our previous statement: that when $n = 1$ a circular orbit ($a = b$) is the only orbit "energetically" possible.

There are as many permissible orbits for each principal quantum number as the number indicates. Thus, for example, if $n = 3$, l can have the values 0, 1, and 2. If $l = 2$, $a/b = 3/3$, i.e., $a = b$ and we have the classical Bohr circular orbit. The other orbits, $l = 1$ and $l = 0$, are increasingly flattened ellipses. Thus, when $n = 3$ there should be three orbit forms. Figure 210 shows an example with five types of orbits. Each of these orbit forms possesses a definite number of electron positions. When fully occupied one orbit form can contain a maximum of $2(2l + 1)$ electrons, a quantity that consequently depends only upon the azimuthal quantum number l and not upon n.

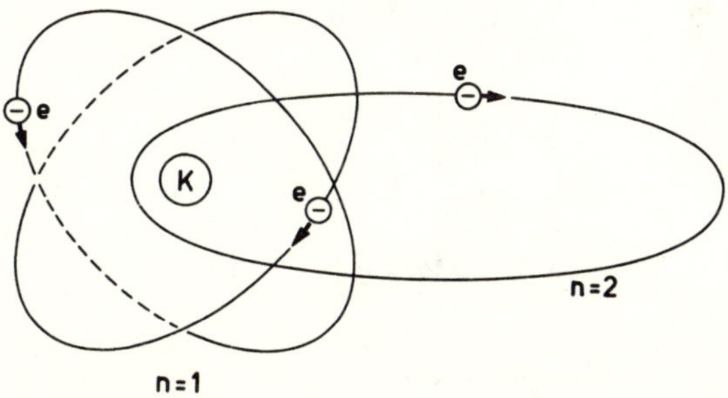

FIG. 211. Model of the lithium atom. Two electrons revolve along the circularly shaped inner orbit, one in a left-handed manner, the other in a right-handed manner. The third electron travels on the next orbit ($n = 2$), the lowest level of which exhibits an elliptical form. The atomic nucleus K is located as in all other cases at one of the foci. Lithium is the lightest of all metals; it is easily cast, moulded, alloyed, and welded. The silvery white metal is softer than lead and only half as heavy as water, upon which it is therefore able to float.

In contrast to our earlier atomic model we now find that every orbit with the same l always possesses the same number of available electron positions regardless of the magnitude of the principal quantum number.

In the first orbit $n = 1$ and $l = 0$, for example, there is never room for more than $2(2 \cdot 0 + 1) = 2$ electrons. Thus, we are able to obtain a picture of the way the electrons are arranged in any element. Figure 211 shows a lithium atom. A single shell has room for $2 \cdot n^2$

electrons. Elements of higher atomic number, which exhibit various completely filled n-shells repeat the l-states, for example, $l = 0$, in the various n-shells. In these l-states there are always only two electron positions. In accordance with the previously given formula, when $l = 1$ there are six electron positions, when $l = 2$ there are ten, etc. The azimuthal quantum number is a refinement of the concept of energy levels that takes into account the Kepler orbits (Fig. 212). If the electrons compared belong to a shell governed by the same principal quantum number, the energy of an electron with a lower l number is lower than the energy of an electron with a higher one.

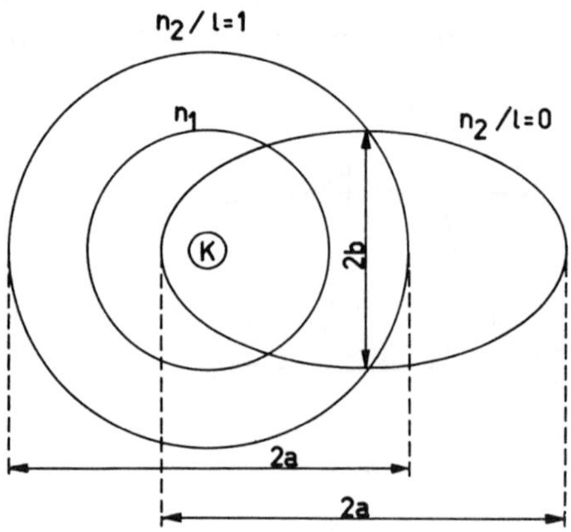

FIG. 212. Orbit forms for the quantum numbers $n = 1$ and $n = 2$.

Figure 213 shows an extract from an energy-level diagram that also takes into consideration the elliptical orbits (the azimuthal quantum numbers). For example, the energy of the electron on the shell $n = 4$ and $l = 1$ is greater than the energy of $n = 4$ and $l = 0$, but it is certainly lower than $n = 5$ and $l = 1$. An electron with a higher value for l possesses a circular orbit when it is the highest state governed by a principal quantum number, while the orbit of the next highest state is a rounded ellipse; and each successively lower value of l corresponds to an orbit of progressively more and more flattened elliptical shape. This is understandable inasmuch as an electron in an orbit with a high l-value comes less near to the nucleus. Its energy is consequently also higher than that of an electron in a much flattened ellipse whose path takes it into the vicinity of the nucleus. Similar conditions are observed for satellites in space.

The following abbreviations are commonly used in the literature:

$$l = 0 \rightarrow S\text{-state}$$
$$l = 1 \rightarrow P\text{-state}$$
$$l = 2 \rightarrow D\text{-state}$$
$$l = 3 \rightarrow F\text{-state}$$
$$l = 4 \rightarrow G\text{-state}$$

When used with respect to an individual electron the states are expressed by the small letters *s*, *p*, . . ., etc., rather than capitals.

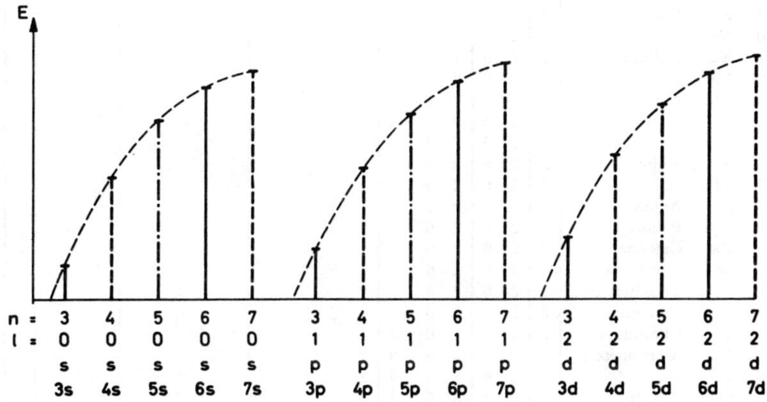

FIG. 213. Energy-level configuration taking into account the azimuthal quantum numbers. E = energy.

In front of these abbreviations the principal quantum number is placed. An electron whose $n = 4$ and $l = 2$ is described as 4*d*, and the reader who knows his physics will understand that the electron in the Bohr model belongs to the 4th energy shell and follows a flattened elliptical orbit ($l = 2$). The number of frequencies now possible (permissible energy levels) is much greater here than in the classical model of the hydrogen atom with its spherical shells. The azimuthal quantum number l provides us with clear information concerning the possible configurations of the Kepler orbits, or elliptical orbits, in this miniature solar system. When several electrons are present the azimuthal quantum number l is composed vectorially of the orbital angular momenta of the individual electrons with left-handed and right-handed revolutions l_1, l_2, l_3, . . ., and is a measure of the resulting orbital angular momentum of the electron system.

Electronic configuration of the elements. Distribution of the electrons among the energy levels or orbits of the atomic model

			Shell																	
			K	*L*		*M*			*N*				*O*				*P*			*Q*
No.	*Symbol*	*Element*	1*s*	2*s*	2*p*	3*s*	3*p*	3*d*	4*s*	4*p*	4*d*	4*f*	5*s*	5*p*	5*d*	5*f*	6*s*	6*p*	6*d*	7*s*
1	H	Hydrogen	1																	
2	He	Helium	2																	
3	Li	Lithium	2	1																
4	Be	Beryllium	2	2																
5	B	Boron	2	2	1															
6	C	Carbon	2	2	2															
7	N	Nitrogen	2	2	3															
8	O	Oxygen	2	2	4															
9	F	Fluorine	2	2	5															
10	Ne	Neon	2	2	6															
11	Na	Sodium	2	2	6	1														
12	Mg	Magnesium	2	2	6	2														
13	Al	Aluminium	2	2	6	2	1													
14	Si	Silicon	2	2	6	2	2													
15	P	Phosphorus	2	2	6	2	3													
16	S	Sulfur	2	2	6	2	4													
17	Cl	Chlorine	2	2	6	2	5													
18	A	Argon	2	2	6	2	6													
19	K	Potassium	2	2	6	2	6		1											
20	Ca	Calcium	2	2	6	2	6		2											
21	Sc	Scandium	2	2	6	2	6	1	2											
22	Ti	Titanium	2	2	6	2	6	2	2											
23	V	Vanadium	2	2	6	2	6	3	2											
24	Cr	Chromium	2	2	6	2	6	5	1											
25	Mn	Manganese	2	2	6	2	6	5	2											
26	Fe	Iron	2	2	6	2	6	6	2											
27	Co	Cobalt	2	2	6	2	6	7	2											
28	Ni	Nickel	2	2	6	2	6	8	2											
29	Cu	Copper	2	2	6	2	6	10	1											
30	Zn	Zinc	2	2	6	2	6	10	2											
31	Ga	Gallium	2	2	6	2	6	10	2	1										
32	Ge	Germanium	2	2	6	2	6	10	2	2										
33	As	Arsenic	2	2	6	2	6	10	2	3										
34	Se	Selenium	2	2	6	2	6	10	2	4										
35	Br	Bromine	2	2	6	2	6	10	2	5										
36	Kr	Krypton	2	2	6	2	6	10	2	6										
37	Rb	Rubidium	2	2	6	2	6	10	2	6			1							
38	Sr	Strontium	2	2	6	2	6	10	2	6			2							
39	Y	Yttrium	2	2	6	2	6	10	2	6	1		2							
40	Zr	Zirconium	2	2	6	2	6	10	2	6	2		2							
41	Nb	Niobium	2	2	6	2	6	10	2	6	4		1							
42	Mo	Molybdenum	2	2	6	2	6	10	2	6	5		1							
43	Tc	Technetium	2	2	6	2	6	10	2	6	6		1							
44	Ru	Ruthenium	2	2	6	2	6	10	2	6	7		1							
45	Rh	Rhodium	2	2	6	2	6	10	2	6	8		1							
46	Pd	Palladium	2	2	6	2	6	10	2	6	10									
47	Ag	Silver	2	2	6	2	6	10	2	6	10		1							
48	Cd	Cadmium	2	2	6	2	6	10	2	6	10		2							
49	In	Indium	2	2	6	2	6	10	2	6	10		2	1						
50	Sn	Tin	2	2	6	2	6	10	2	6	10		2	2						
51	Sb	Antimony	2	2	6	2	6	10	2	6	10		2	3						

			Shell														
			K	*L*	*M*	*N*				*O*				*P*			*Q*
No.	*Symbol*	*Element*	1*s*	2*s* + 2*p*	3*s* + 3*p* + 3*d*	4*s*	4*p*	4*d*	4*f*	5*s*	5*p*	5*d*	5*f*	6*s*	6*p*	6*d*	7*s*
52	Te	Tellurium	2	8	18	2	6	10		2	4						
53	I	Iodine	2	8	18	2	6	10		2	5						
54	Xe	Xenon	2	8	18	2	6	10		2	6						
55	Cs	Cesium	2	8	18	2	6	10		2	6			1			
56	Ba	Barium	2	8	18	2	6	10		2	6			2			
57	La	Lanthanum	2	8	18	2	6	10		2	6	1		2			
58	Ce	Cerium	2	8	18	2	6	10	1	2	6	1		2			
59	Pr	Praseodymium	2	8	18	2	6	10	2	2	6	1		2			
60	Nd	Neodymium	2	8	18	2	6	10	3	2	6	1		2			
61	Pm	Promethium	2	8	18	2	6	10	4	2	6	1		2			
62	Sm	Samarium	2	8	18	2	6	10	5	2	6	1		2			
63	Eu	Europium	2	8	18	2	6	10	6	2	6	1		2			
64	Gd	Gadolinium	2	8	18	2	6	10	7	2	6	1		2			
65	Tb	Terbium	2	8	18	2	6	10	8	2	6	1		2			
66	Dy	Dysprosium	2	8	18	2	6	10	9	2	2	1		2			
67	Ho	Holmium	2	8	18	2	6	10	10	2	6	1		2			
68	Er	Erbium	2	8	18	2	6	10	11	2	6	1		2			
69	Tm	Thulium	2	8	18	2	6	10	12	2	6	1		2			
70	Yb	Ytterbium	2	8	18	2	6	10	13	2	6	1		2			
71	Lu	Lutetium	2	8	18	2	6	10	14	2	6	1		2			
72	Hf	Hafnium	2	8	18	2	6	10	14	2	6	2		2			
73	Ta	Tantalum	2	8	18	2	6	10	14	2	6	3		2			
74	W	Tungsten	2	8	18	2	6	10	14	2	6	4		2			
75	Re	Rhenium	2	8	18	2	6	10	14	2	6	5		2			
76	Os	Osmium	2	8	18	2	6	10	14	2	6	6		2			
77	Ir	Iridium	2	8	18	2	6	10	14	2	6	9		0			
78	Pt	Platinum	2	8	18	2	6	10	14	2	6	9		1			
79	Au	Gold	2	8	18	2	6	10	14	2	6	10		1			
80	Hg	Mercury	2	8	18	2	6	10	14	2	6	10		2			
81	Tl	Thallium	2	8	18	2	6	10	14	2	6	10		2	1		
82	Pb	Lead	2	8	18	2	6	10	14	2	6	10		2	2		
83	Bi	Bismuth	2	8	18	2	6	10	14	2	6	10		2	3		
84	Po	Polonium	2	8	18	2	6	10	14	2	6	10		2	4		
85	At	Astatine	2	8	18	2	6	10	14	2	6	10		2	5		
86	Rn	Radon	2	8	18	2	6	10	14	2	6	10		2	6		
87	Fr	Francium	2	8	18	2	6	10	14	2	6	10		2	6		1
88	Ra	Radium	2	8	18	2	6	10	14	2	6	10		2	6		2
89	Ac	Actinium	2	8	18	2	6	10	14	2	6	10		2	6	1	2
90	Th	Thorium	2	8	18	2	6	10	14	2	6	10	1	2	6	1	2
91	Pa	Protoactinium	2	8	18	2	6	10	14	2	6	10	2	2	6	1	2
92	U	Uranium	2	8	18	2	6	10	14	2	6	10	3	2	6	1	2
93	Np	Neptunium	2	8	18	2	6	10	14	2	6	10	4	2	6	1	2
94	Pu	Plutonium	2	8	18	2	6	10	14	2	6	10	5	2	6	1	2
95	Am	Americium	2	8	18	2	6	10	14	2	6	10	6	2	6	1	2
96	Cm	Curium	2	8	18	2	6	10	14	2	6	10	7	2	6	1	2
97	Bk	Berkelium	2	8	18	2	6	10	14	2	6	10	8	2	6	1	2
98	Cf	Californium	2	8	18	2	6	10	14	2	6	10	9	2	6	1	2
99	E	Einsteinium	2	8	18	2	6	10	14	2	6	10	10	2	6	1	2
100	Fm	Fermium	2	8	18	2	6	10	14	2	6	10	11	2	6	1	2
101	Mv	Mendelevium	2	8	18	2	6	10	14	2	6	10	12	2	6	1	2
102	No	Nobelium	2	8	18	2	6	10	14	2	6	10	13	2	6	1	2
103	Lw	Lawrencium	2	8	18	2	6	10	14	2	6	10	14	2	6	1	2

Since we will have frequently to deal with the magnetic moment in this chapter, it is explained in terms of the illustration in Fig. 214. If an electric current i is passed through a coil, a magnetic field is created the directional sense of which can be indicated by an arrow. Outside the magnet the directional sense of the field is from the north pole to the south pole. Consequently, the arrow inside the "generating system" must point in the direction of the north pole. The same effect is obtained by the motion of an individual electron

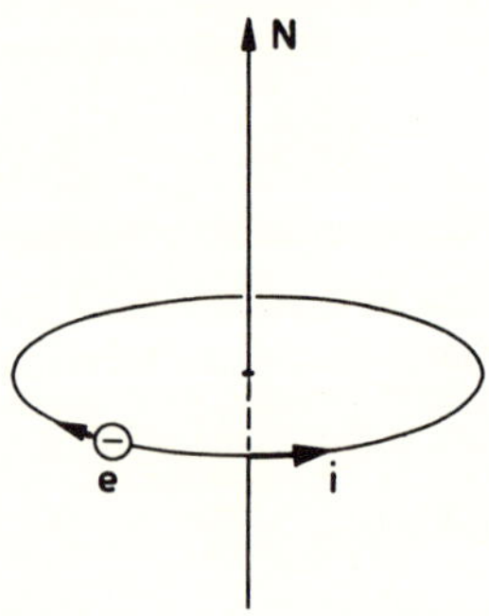

FIG. 214. The directional sense of the magnetic field lines H are found by means of the corkscrew rule. Since the charge of the electron is negative, in contrast to the defined directional sense of an electric current i, the arrow must point in the opposite direction.

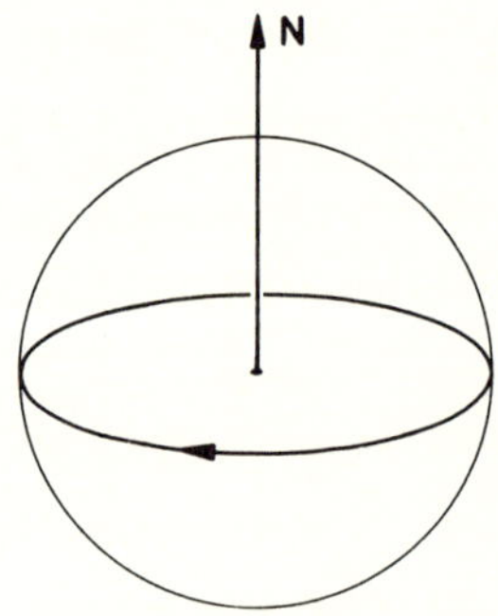

FIG. 215. An electron spinning around a nucleus corresponds to a circular current and similarly gives rise to a magnetic field.

with its charge of $1.6 \cdot 10^{-19}$ ampsec around a nucleus. In this case also the directional sense of the magnetic field can be expressed by an arrow (a vector) perpendicular to the plane of the circular orbit. If an electron were to rotate about its own axis (Fig. 215), this would similarly give rise to a magnetic field. In this case also a vector arrow gives the direction of the magnetic field. Here the arrow represents the axis of rotation at the same time.

3. *The quantum number of spin* (s). Under certain conditions, such as the use of an external magnetic field and the utilization of highly selective spectroanalytic instruments, there is observed a splitting of the spectral lines into separate components, revealing a fine structure.

The basis for this phenomenon was found in the fact that electrons, in addition to revolving about the nucleus, rotate about their own axis. The angular momentum associated with this rotation about their own axis is called spin. In accordance with the laws of electrodynamics this rotational motion of a charged particle, as we have seen previously, also gives rise to a magnetic moment (unit = the Bohr magneton $\mu = 9.27 \cdot 10^{-21}$ gauss/cm^3),

$$\mu = \frac{e}{m} \cdot \frac{h}{4\pi}$$

On the basis of these and other measurements Pauli (1925) concluded that an atom cannot exist in a state in which two of its electrons possess the same value for all the quantum numbers. In the simplest model of the Bohr atom, as we know, there are two electrons revolving about the nucleus in the *K*-shell, which is closest to the nucleus. They are, however, distinguished by the direction in which they rotate about their own axis. For the same quantum number $n = 1$ there are the spins $s = +\frac{1}{2}$ and $-\frac{1}{2}$.

The splitting of one spectral line into two neighboring lines, i.e., into two somewhat different frequencies, leads to the following consequences. An electron that is only lightly bound to the atomic nucleus in the presence of an external magnetic field H possesses two oppositely directed equilibrium states whose energy difference ΔE, expressed by the directional quantization, is proportional to $\mu \cdot H$. This difference between the two energy stages, which is a function of the external magnetic field, is so slight that the spectral lines are in the microwave range.[1]

The frequency $f = \Delta E/h$, which corresponds to the quantic transition between the two possible energy values, is

$$f = \frac{g \cdot \mu}{h} \cdot H = \gamma \cdot H$$

For free electron spin the so-called spectroscopic splitting factor is $g = 2$, from which it follows that the magnetic field-dependent frequency $f = 2.8 \cdot H$(Mc/s).

[1] Microwaves are defined as very short electromagnetic waves with wavelengths between 10 cm and 1 mm. Very often these waves are described by frequency rather than wavelength; in this event the maximum frequency lies between 3 gigacycles and 300 gigacycles (1 gigacycle = 1000 megacycles). The microwave range forms a bridge between the range of radio waves and the region of the optical spectrum.

Since the electron is normally subject to interaction with the neighboring molecules and its own atomic nucleus, which also gives rise to a magnetic field as a result of its own rotation, as we will see, it is possible for a certain splitting of energy levels to occur even in the absence of an external magnetic field.

In atoms with several electrons the individual characteristic spins of $s = +\frac{1}{2}$ and $s = -\frac{1}{2}$ add up vectorially to a total spin sum, the so-called quantum number of spin S. The resulting number is either a whole number or a half, depending upon whether the total number of electrons is even or odd.

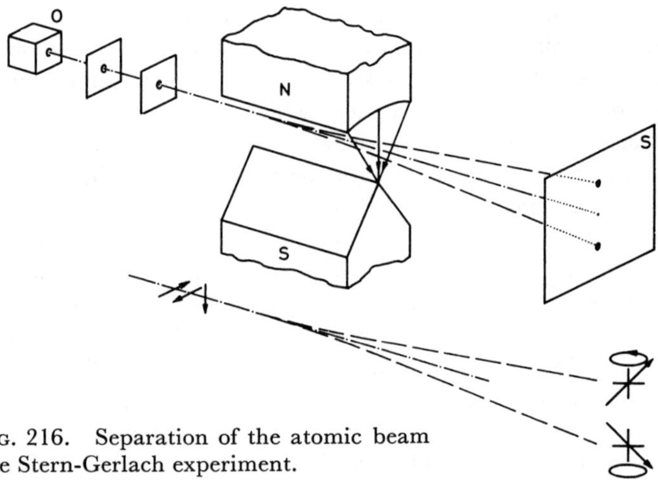

FIG. 216. Separation of the atomic beam in the Stern-Gerlach experiment.

This increasing multiplicity of the energy transitions makes necessary another refinement of the atomic model. It is generally true that if there is a favored orientation in space, created, for example, by an external magnetic or electric field, the energy in the individual electron orbits is also a function of the relative position of the orbital planes with respect to the direction and strength of the field. The mechanically possible adjustments to the direction of the field are limited by the quantic conditions to a finite number of definite angles.

Since an electron describes an orbit about the nucleus of the atom at the same time that it rotates about its own axis and is subject to precessional motion, the strength of the applied external field is of the greatest importance, particularly in relation to the field created by the motion of the atomic charges within the atom. If the external field is large in relation to the internal field, the precessional motion of l and s is about the direction of the field. We therefore have the behavior of two small magnets to consider, specifically the one resulting from the circular current of the large orbital motion and the other from the spin of the electron. If the external field is small in relation to the internal field, the precessional motion of l and s about each other is much more rapid than the precessional motion about the direction of the external field. Thus it is extremely helpful in this case to combine l and s into a total momentum J, whose precession about the direction of the external field can be considered.

4. *The internal quantum number* J. The directed magnitudes (vectors) of the orbital moment L and the spin moment S, which we just discussed, produce as a result of their interaction (so-called coupling)

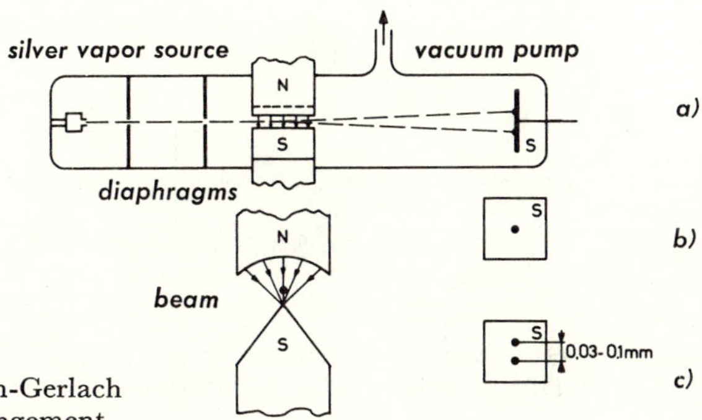

FIG. 217. Stern-Gerlach arrangement.

the total magnetic moment or the total angular momentum J of the atom. The system of all the so-conceived spinning circular currents forms the magnetic moment. In fact, on splitting of the S-, P-, D-states the J-values are characteristic. $J = 0$ means that an equal number of electrons rotate to the left and to the right. The circular currents consequently neutralize each other. This symmetry, which is particularly characteristic of the rare gases, with their closed groups of electron arrangements, results in a total angular momentum of zero, as a result of which the total magnetic angular momentum is zero. Atoms such as this therefore cannot be paramagnetic; they are diamagnetic, as we will see in the next section. Stern and Gerlach in 1921 were the first to investigate the manner in which free atoms (gases) possessing a magnetic moment behaved in an external magnetic field. In an evacuated chamber they passed a sharply defined atomic beam (silver vapor) through a strong, inhomogeneous field (Figs. 216 and 217). It turned out that this beam separated into $2J + 1$ component beams. For silver the relationships were particularly simple, since J was 1/2. Thus, only two component beams occurred deviating symmetrically from the original and no longer observable pencil of beams (Figs. 217*c* and 218*a*). The separation can be explained in a simplified way as follows. If atoms possessing a

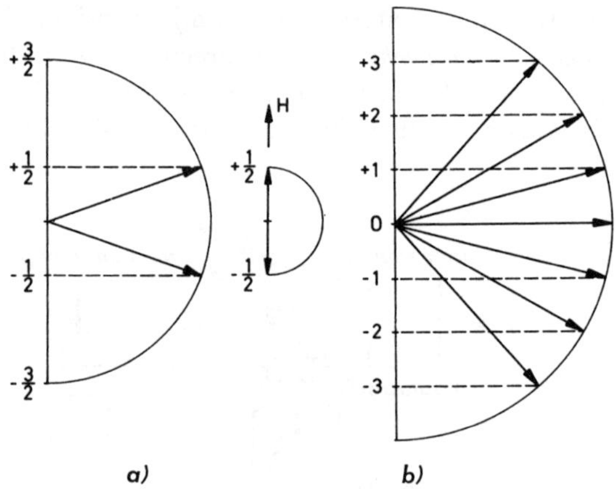

FIG. 218. (*a*) Angle adjustments of silver atoms (no zero line); (*b*) energetically permissible angle adjustments in accordance with $2J + 1$.

magnetic moment—whose circular currents therefore do not neutralize each other—enter a homogeneous field, the originally arbitrary positions are altered by the magnetic forces. In Fig. 219*b* the orbit planes tend to orient themselves in the direction of the magnetic field. The orbit planes could be described in the simple case of a rotating electron moving around a nucleus. In Fig. 219*c* the flux lines are neutralized in relation to the external field, whereas in Fig. 219*b* an angular moment is created. If we consider an atom with several circular electron currents, this phenomenon is conceived as possessing a resultant that cannot be equal to zero. Since lateral forces (or angular moments) are exercised on the rotating circular currents, the familiar mechanical phenomenon of precessional motion occurs.

The axis of the circular current describes a cone-shaped shell (Fig. 220). The axis of the cone is parallel to the direction of the lines of force of the external magnetic field, with the result that the

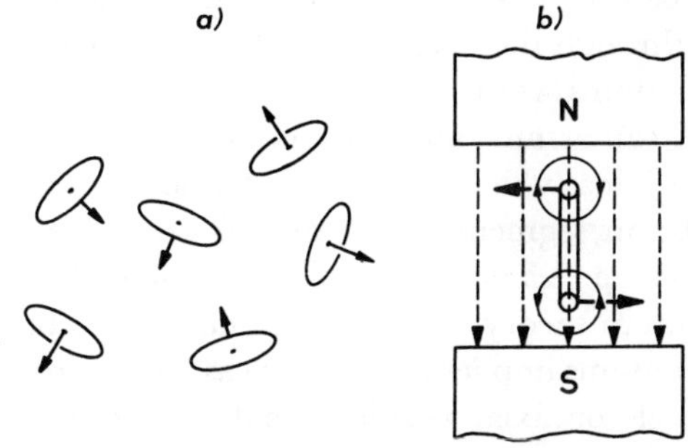

FIG. 219. (*a*) Free atoms (e.g., silver vapor) in a stationary distribution; (*b*) Orbit axis perpendicular to the field direction, appearance of an angular moment; (*c*) Orbit axis parallel to field direction, no angular moment.

precessional motion is about the direction of field H. The experimentally established Stern-Gerlach separation indicates that an atomic system can only assume certain quite definite angular positions in a magnetic field. In quantum theory one speaks of a directional

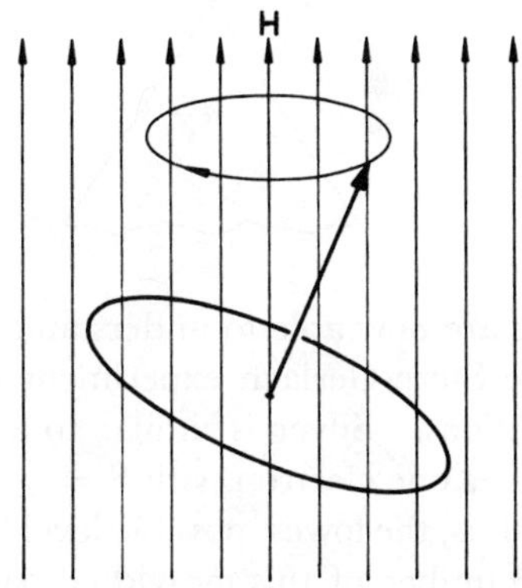

FIG. 220. Precession of a circular current in a magnetic field.

quantization with a family of $2J + 1$ discrete situations or energy stages, the deflections of which are bilaterally symmetrical. The transition from the various discrete positions of the rotational axes to the visible separation of the atomic beam on the screen S is accomplished by means of the inhomogeneous field (Fig. 221). This exercises upon the atom as a whole a force that is positive or negative in accordance with the rotational sense of the magnetic moment. If there were atoms whose vectors were antiparallel to the field, they would pass undeflected through the magnetic field.

But what about the atoms with other axial positions? If we imagined that the atoms can assume all arbitrary angle orientations, it would follow that the originally sharply defined atomic beam would widen, i.e., a blurred beam segment would appear on the screen. In the experiment, however, a clear separation of the beam is observed in accordance with quantum theory. The occurrence of two beams as in the case of the silver atoms implies that in this case there are only two separate energy levels or axial positions that are possible, as shown in Fig. 218.

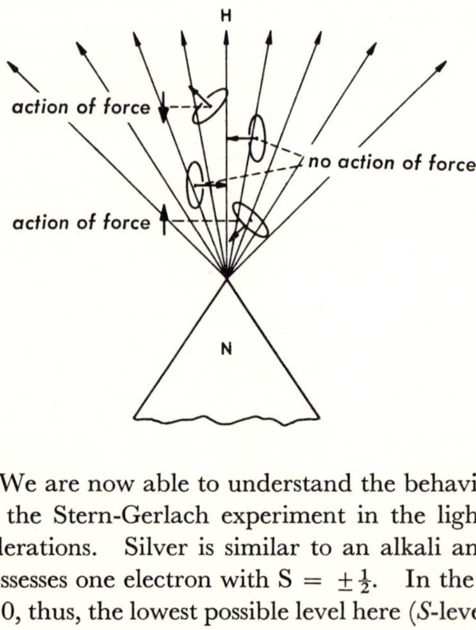

FIG. 221. Action of force in an inhomogeneous magnetic field.

We are now able to understand the behavior of the silver particles in the Stern-Gerlach experiment in the light of the foregoing considerations. Silver is similar to an alkali and the last valence shell possesses one electron with $S = \pm\frac{1}{2}$. In the ground state L is equal to 0, thus, the lowest possible level here (S-level). The internal quantum number of this individual particle j is $L + S$ and consequently possesses the value $\frac{1}{2}$. Consequently we can expect the orientations to the magnetic field, corresponding to a J-value of $\frac{1}{2}$. The orientations of the vector J in the magnetic field are limited to $2J + 1$. The projections of the vector J on the field direction are expressed by the quantum number m (see Fig. 218). As the experiment shows, in the case of silver atoms there are two lateral beams and there is no null line (undeflected atoms). The experiment is thus in agreement with the atomic model. The separation of the optical spectrum lines[1] effected by an external magnetic field was discovered by P. Zeemann as early as 1896. His experiments also showed that the orbit planes could only assume certain definite angles with respect to the spatially favored axis (magnetic field direction). In his experiments he placed a light source in a strong magnetic field and found that the simple lines observed in the absence of a field split into several adjacent lines. J. Stark in 1913 studied the similar symmetrical splitting of energy levels, produced by means of strong electric fields. He found that the light emitted by atoms split into separate components of sharply distinguished frequency when the light source, i.e., the atoms, were placed in a strong electric field. He found that the component number increased with the series number n. The intervals between the numbers were found to be multiples of a definite wave number interval. This is to be attributed to the fact that in electric fields an atom becomes a complicated dipole structure. Even if we restrict ourselves to the behavior of a single electron in an elliptical path, it will be seen that an external electric field would be superimposed upon and reinforce the effect of the attractive force of the nucleus on one side while weakening it on the other. The atom consequently becomes a dipole and itself describes elliptical orbits about the undisturbed center of gravity. The configuration of these orbits leads to superimposed motions and consequently to a splitting of the energy levels. In accordance with Stark's interpretation the atomic model contains at least one marked major axis that is related to the axes of the electrons (spin), with particular importance being attached to axes of the outer, valence electrons. Other spectroanalytic measuring methods of even greater selectivity (microwave analysis) have shown that the lines can be still further separated. The intervals between the lines are approximately 40,000 times smaller than those of the fine structure and are beyond the limiting resolving power of optical spectrographs. This so-called hyperfine structure is based on the magnetic moment of the atomic nucleus, which consists of protons and neutrons. These nucleons must, consequently, also possess characteristic rotations. Thus, we come to the fifth characteristic quantum datum.

[1] By means of the different colors obtained on passing the light rays through a prism.

5. *The nuclear spin* I. From the nuclear spin we calculate the angular momentum of the nucleus $Y = h/2\pi\sqrt{I(I + 1)}$; here I is a whole number (0, 1, 2, 3, 4, . . .) when and only when the protons and neutrons add up to an even number. If both protons and neutrons are even-numbered, I is equal to 0. If the protons and neutrons form an odd number, I is equal to 1/2, 3/2, 5/2, . . ., i.e., it is always an odd multiple of 1/2. Since the nucleus possesses a mass 1836 times as large as the electron, its magnetic moment is 1/1836 as great. The selection of separate spectral lines has here been considerably increased. The atomic nuclei, which are charged particles with spin, are also subject to a directional quantization in a magnetic field just like that of the electrons in paramagnetic atoms. An atom of nuclear spin I possesses $2I + 1$ different angular orientations. Acting like a spinning top, the nucleus attempts to adjust itself to the direction of the magnetic field. This force, which leads to an angular momentum, is in this case also the cause of a precessional motion with a so-called Larmor frequency about the axis of the external magnetic field. This frequency is proportional to the strength of the magnetic field. For a magnetic field of 10,000 gauss the Larmor frequency lies between 1 and 50 Mc/s, depending upon the magnetic field and the nucleus. To study this frequency the optical methods heretofore employed must be abandoned in favor of microwave spectroscopy. If an external high frequency a.c. voltage with a resonance frequency is employed, the magnetic nuclear moments shift their angle in the direction of the magnetic field. The nucleus is thus raised to a higher energy level. This absorption of energy can be observed in a high frequency field. Since the nucleus rotates and yields an electric moment, it must in accordance with our view also exhibit a definite form. It can have the form of an elongated or flattened ellipse about its own axis of rotation, depending upon whether the moment is positive or negative. In the literature this is referred to as a quadrupole moment. The electrical coupling of this moment with the orbit electrons, which possess their own special configuration in a molecular structure, influences the molecular rotation, which we will discuss later. In fact, the so-called rotation lines are observed to undergo further splitting into values corresponding to the quadrupole frequencies. These can also be shown to exist in solid substances by placing them between two electrodes (or in a wave guide), whereupon they exhibit certain resonance properties if they are subjected to appropriate frequencies. Such quantum mechanically fixed atomic

frequencies of solid bodies give promise of being useful as frequency standards.

In an atomic clock, where the finest energy level transitions are required to remain within the electronically determinable microwave range, an understanding of electron spin and nuclear spin is of great importance. The orientation of an electron with respect to the nuclear spin, which may be parallel or antiparallel (opposed), gives rise to two alternative energy states of the atom, which are independent of external influences and can serve as frequency norms.

4. Paramagnetism and diamagnetism

Michael Faraday had already made the discovery that magnetic forces could be exercised upon all substances. On the basis of their behavior in a magnetic field, it is possible to distinguish two different groups of substances:

1. The paramagnetic substances, among which the ferromagnetic substances (previously discussed on p. 34) form a special case.
2. The diamagnetic substances, which we will examine in some detail in this section.

A bar of a paramagnetic material P assumes a position parallel (Fig. 222, top) to a homogeneous magnetic field H, while a diamagnetic material D assumes a position perpendicular to the field (Fig. 223, top). If, however, an inhomogeneous magnetic field is employed in accordance with Fig. 222, bottom, and Fig. 223, bottom, a sphere of paramagnetic material is attracted toward the sites of greater magnetic field strength, i.e., in the direction of the sharp magnetic cuts, whereas a diamagnetic sphere will tend toward the sites of lower field strength. As we saw in the previous section, paramagnetism is always encountered when the atoms or molecules possess a fixed magnetic moment, i.e., when the internal quantum number J differs from zero. The stronger the external magnetic field, the more it tries to adjust the atomic "magnets" to the direction of its own field. This ordering tendency succeeds only partially because of the interference of thermal molecular motion (thermal energy). In paramagnetism the magnitude that determines the magnetic condition (permeability μ) is consequently temperature dependent. We find here a certain analogy with the polar dipole molecules in a dielectric (as discussed on p. 37). To divert all the atomic magnetic moments of a paramagnetic material into the direction of field H at room temperature would require a magnetic field of nearly unobtainable strength.

Consequently, paramagnetic substances are much less strongly magnetizable than ferromagnetic materials. Ferromagnetic substances such as iron (*ferrum*) possess paramagnetic properties, only to a higher degree. Their permeability μ is a thousand times as high, or three orders of magnitude higher, i.e., the magnetic flux density $B = \mu H$, where H represents the magnetic field. The so-called Weiss zones are responsible for this excellent magnetic property. The existence of these zones is explained in the following way:

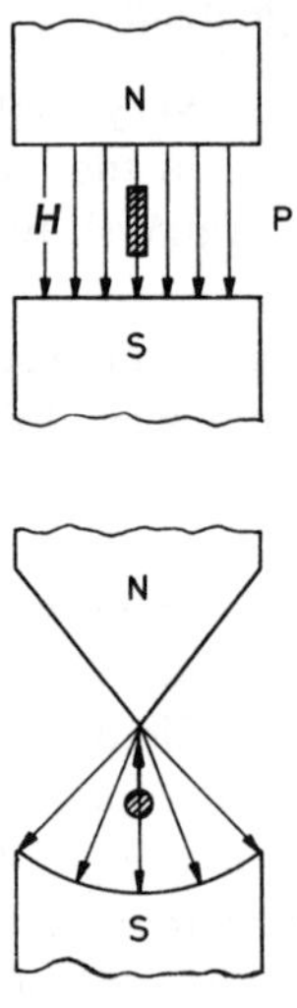

FIG. 222. Paramagnetism.

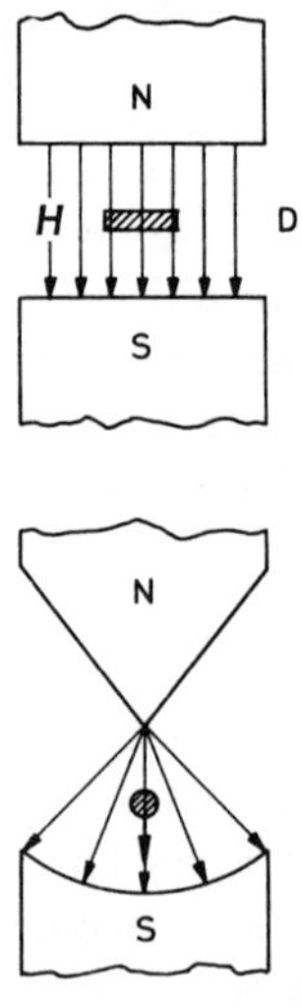

FIG. 223. Diamagnetism.

It is assumed that there is an electric or magnetic coupling between the electrons of neighboring ferromagnetic atoms. As a consequence of this coupling the magnetic moments produced by the spin of the electrons exhibit a tendency to assume the same parallel axial positions. This coupling effect extends no further than the individual microcrystal edges, which is the reason why unidirectional magnetic moments are encountered only within these restricted zones. The Weiss zones act somewhat like miniature parallel magnets that are oriented along the edges of the microcrystals (see Fig. 224). Although these are saturated magnets of "maximum intensity," the piece of iron as a whole is not outwardly magnetized. The reason for this is that the Weiss zones are statistically distributed in all directions, and thus neutralize each other's effect.

If the strength of a magnetic field H is progressively increased, an increasing number of the magnetic moments of the microcrystals swing into line with direction of the field, until all the zones have undergone optimal adjustment and magnetic saturation is reached. The nature of the microcrystal structure determines the size of the stages by which the turnabout is achieved. This also makes it clear why a magnetic parallel circuit can be attained at technically obtainable field strengths only with the help of the Weiss zones. Iron is magnetically saturated at approximately 20,000 gauss (magnetic induction/cm^2).

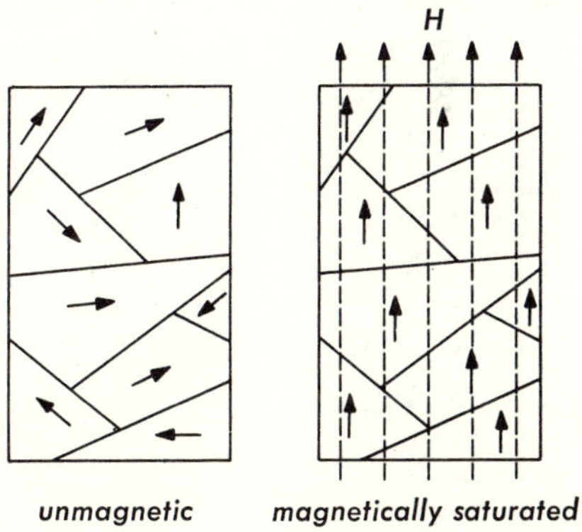

FIG. 224. Weiss zones.

If a saturated bar magnet, suspended lengthwise, is suddenly subjected to reverse magnetization by an external bar magnet, it rotates about its longitudinal axis. From this observation and from the mechanical laws of rotary motion it is inferred that a rotation does in fact occur in atoms, as assumed. Conversely, it has been observed that an unmagnetized ferromagnetic bar can be magnetized by acceleration about its longitudinal axis. These considerations lead to the conclusion that the total magnetic moment arises only from the spin of the electrons and not from their orbital motion.

Let us then designate as ferromagnetic those substances in which the directions of spin of the electrons do not compensate each other. In an atom of four electrons the angular momentum arising from the spin would be balanced out if two of the electrons possessed a left-handed spin and two a right-handed spin. A substance of this type cannot be magnetized. If, on the other hand, three of the electrons possessed a left-handed spin and one a right-handed spin, the magnetic equilibrium would be upset. We then speak of electrons with unpaired spin, where the angular moments and consequently the magnetic moments no longer compensate each other. Figure 225 illustrates the relationships in an iron atom and shows the direction of spin of the electrons in the individual shells. This represents an elaboration of the earlier energy level diagram.

On the basis of the foregoing discussion, atoms are diamagnetic only when they exhibit an internal symmetry, i.e., when the internal quantum number J is 0. Only then do they possess no external magnetic moment. All atoms and ions containing closed electron shells like the rare gases belong to this group. Many solid diamagnetic

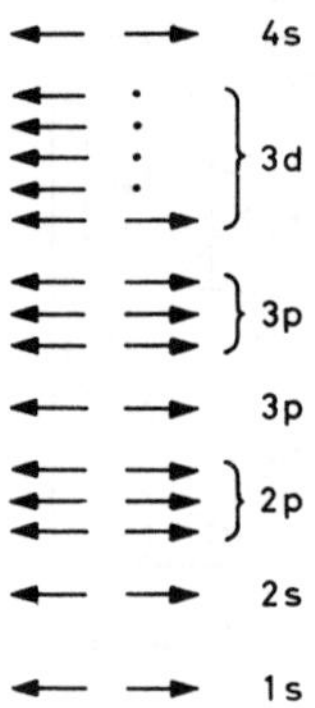

FIG. 225. Directions of spin in an iron atom (the bottom electrons lie nearer the nucleus).

substances possess ionic lattices whose electrons are arranged in the manner of rare gases. The fact that diamagnetic substances in an inhomogeneous field are repelled in the direction of lower field strengths indicates that here also there is an interaction between the electrons moving about the nucleus and the magnetic field. The direction of the repulsion indicates that what is here involved is the fulfillment of Lenz's electrical law, which states that an induction current is always directed in such a manner that it works in opposition to the phenomenon that gave rise to it. It is therefore a phenomenon to which all substances are subject, paramagnetic substances as well as diamagnetic ones. The electronic motions must consequently give rise to a moment the direction of which is in opposition to the external field. This leads to the conclusion that such a moment can only arise from the precessional motion to which the electron orbit is subjected as a result of influence of the external magnetic field. It is this relatively slow rotational motion described by all spinning bodies when they are subjected to a laterally acting force (angular momentum) such as that arising from the gravitational force of the earth and thus no longer rotate merely about their own axis. This circular motion, which was computed for the first time by Larmor, describes a path along the shell of a cone. The rotational sense on the shell of the cone, i.e., the precessional motion, is independent of the direction of spin of the individual electrons. The Larmor precessional rotation frequency f is calculated from the formula:

$$f_L = \frac{1}{2\pi} \frac{e}{2 \cdot m \cdot c} \cdot H$$

where H is the magnetic field, c the speed of light, e the charge of the electron, and m the mass of the electron. This precessional motion is an additional motion of the electrons about the direction of the field and is independent of the orbital motion about the nucleus. The current thereby created gives rise to a magnetic moment the direction of which is opposed to the magnetic field H, thus creating diamagnetism. Diamagnetism is a property that is independent of temperature. As previously mentioned, all substances are subject to these influences, the effect of which increases as the electrons of an atomic system become more numerous and especially as the atomic radii increase. Since it is the square of the radius that enters into the formula, the outermost electron orbitals play the dominant role. The question then remains: what type of moment is present when paramagnetic and diamagnetic influences are both present? From the previous discussion it follows that substances are diamagnetic when the paramagnetic influences do not prevail.

5. From the atom to the molecule

In many gases the molecules consist of two atoms, for example, H_2. In other substances the molecules are composed of a great many atoms, and in some instances many hundreds of atoms may enter into a compound if this will enable them to complete their valence shells (see Fig. 4, Chapter I). From an energetic viewpoint, this enables them to assume their lowest possible energy level. In a gas the motion of the individual molecules is random and the direction and magnitude of the individual molecules is completely independent of each other. The average kinetic energy of the molecules is proportional to the absolute temperature.

It follows from this that the absolute temperature is the measure of the average kinetic energy E of the molecules. The formula derived in thermodynamics for one mole of an ideal gas is

$$E = \frac{M \cdot v^2}{2} = \frac{3}{2} R \cdot T$$

in which T is the absolute temperature in °K and R is the so-called universal gas constant. The number 3 indicates three degrees of freedom corresponding to the possibilities of motion in three directions. The various aggregate states can also be explained. In solid bodies with the ideal structure of a pure crystal the equilibrium between the molecular and electric forces of attraction and repulsion orders the molecules in an essentially regular, geometrical "crystal" form. The transverse bonds connecting the molecules are conceived of as not being rigidly fixed at normal temperatures. They tend rather to vibrate elastically about their equilibrium positions; the mean kinetic energy of these vibrations is proportional to the absolute temperature. One proof of this is given by the Brownian movement of such bodies as the dust particles that are visible in microscopes. At absolute zero (-273°C = 0°K) these molecular particles are in a state of rest. The transition to the liquid and, finally, to the gaseous state is effected by an increase in temperature. The intermolecular bonds that maintain the order of the crystal are broken by the increasing intensity of the vibrations of the molecules. This also explains thermal expansion. The simplest model of the molecule is one consisting of two atoms, whose distribution of mass can be compared to that of a dumbbell. A structure of this sort is capable not only of propagational (translational) motions, but also of rotary motions (rotations), for example, about the axes perpendicular to the line connecting the two atoms. In polyatomic molecules the direction of rotational vibration is characterized by still more degrees of freedom; thus the molecules possess a greater ability to store heat, and this is reflected by their greater specific heat. A dumbbell-shaped molecule consisting of two oppositely "charged" ions can also be compared to a dipole. There are polar molecules that possess a fixed atomic center of gravity, and there are others that possess several different possible prescribed positions. When various atomic configurations are possible there are quantized energy transitions, since two energy states can never be the same in the same system. Consequently, there are quantized energy transitions that correspond to changes in the various molecular bonding states as well as to the electron transitions. The frequencies arising from molecular energy transitions are relatively low and in accordance with the molecule usually lie in the range of the long infrared waves. They are therefore long waves in the light spectrum or the shortest in the scale of radio waves. Heavy molecules exhibit quantum mechanical transitions with frequencies lying in the range of millimeter and centimeter waves and thereby open the way to measurement with high frequency instruments. These rotation lines are raised to higher energy states by means of externally supplied energy. When energy must be supplied in this way at definite frequencies by a measurement transmitter we speak of so-called absorption lines. There is a great advantage offered by the presence of lines or frequencies that can be reproduced by the generation of continuous a.c. currents. Resonance tuning to the molecular frequency with a wavelength-calibrated oscillator gives the characteristic frequency f from which the internal energy transition E can be calculated. The molecular resonance line can also be used as an oscillation frequency in the construction of molecular clocks.

6. The atom as a source of electromagnetic vibrations (atomic clocks)

The time scale is derived from astronomically observed periodicities and fixed by the daily rotation of the earth. In another respect, the time unit is given by the period of revolution of the earth about the sun. The length of one revolution of the earth is determined from the so-called star transit by means of a telescope. The conventional clock, which reproduces the time scale down to the unit of the second, is based on the periodic repetition of a vibrational motion. A pendulum possesses an oscillation period that is a function of its length

$$T = 2\pi\sqrt{\frac{l}{g}}$$

where l represents the length and g the acceleration due to gravity. The frequency is obtained from the reciprocal of the oscillation period.

Macroscopic bodies such as pendulums are subject to various external influences such as temperature. Even if variation in length can be nearly eliminated by the combination of various materials, the accuracy of motion of such a clock is limited by other influences.

Considerably greater accuracy than that of the pendulum is given by the quartz oscillator (the principle of which was described on p. 256). The quartz crystal is a piezoelectric transducer with a definite characteristic frequency

$$f = \frac{k}{2d}\sqrt{\frac{E}{\rho}} = \frac{286}{d}\cdot k \text{ (kc/s)}$$

where E represents the modulus of elasticity, ρ the density of the quartz crystal, $k = 1, 2, 3, \ldots$ = ordinal number of the oscillations,

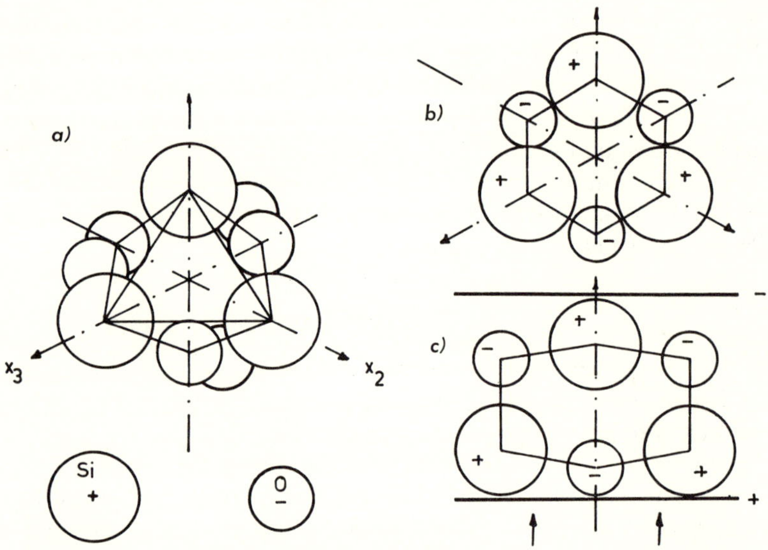

FIG. 226. (*a*) "Spatial" arrangement of the atoms in a quartz crystal lattice. (*b*) The same arrangement projected on a single picture plane. The large circles represent the silicon ions, and the small ones the oxygen ions. Under normal conditions the effects of the positive and negative ionic charges of a cell neutralize each other externally. (*c*) Displacement of the center of concentration of the charges on deformation of an SiO_2 crystal, resulting from external pressure applied to the upper and lower boundaries. The upper surface thereby acquires a negative charge and the lower one a positive charge.

and d = the thickness in centimeters. As Fig. 226 shows, the quartz crystal works as a generator (or microphone) when it is deformed by an external pressure. The centers of concentration of the charges of the ionic lattice are displaced in such a manner that the previously neutralized charges are no longer externally compensated because of the change in the ionic distances. Figure 226*c* shows schematically how two positive silicon ions are brought closer and closer to the lower boundary as a result of the deformation, while two negative oxygen ions are brought closer and closer to the upper boundary. This variation in charge can be taken off by means of two electrodes between which the quartz, situated as a dielectric, acts as a form of capacitor, and the signal obtained can be amplified by means of transistors or tubes. Operated in the reverse fashion, the crystal is deformed by an electric field and works like a motor (or loudspeaker). It is consequently capable of replacing the usual coil and capacitor resonant circuit. The mechanical characteristic frequency of the quartz oscillator is very sharp and well defined as a result of the low internal damping. Since the fundamental frequency is a function of the thickness, which is equal to exactly one half wavelength, $\lambda = 2d$ (mechanical), these oscillations are easily reproduced. Quartz oscillators of this type are favored for reference frequencies of 100,000 c/s (0.1 Mc/s). These values can be stepped down by means of series-connected frequency divider stages to drive a synchronous motor which drives—possibly by means of further mechanical transmission elements—a clock mechanism. The frequency of a quartz oscillator can also be extensively multiplied by amplification of the upper harmonics obtained by overmodulation to obtain frequencies of 1, 10, or 100 Mc/s. In the last case the time elapsing between individual oscillations is one hundred-millionth of a second.

A quartz oscillator is also a macroscopic oscillating body, whose characteristic frequency is a function of the density, the modulus of elasticity and mechanical values; it is subject to material aging. The resulting errors in motion are very small. These influences do not become significant until one desires to exceed the precision of the terrestrial time scale, which today appears to be quite possible. It is known that the earth's rotation is subject to periodic fluctuations brought about by such means as tidal ebb and flow, as well as other influences, and that the velocity of the rotation progressively decreases as a result of the damping losses due to friction. The loss in velocity of rotation over a thousand years is only a fraction of a second. This results in a secular change in astronomical time. To determine such changes clocks are required that are subject to no external influences.

Since the quantum mechanical transitions in atoms are independent of terrestrial influences and are limited only by Planck's natural constant h, they would appear to offer advantages in molecular, atomic, and nuclear clocks. The emitted or absorbed frequencies of these clocks are subject to the strict condition

$$f = \frac{E_n - E_m}{h}$$

where E_n represents the initial energy and E_m the final energy, which in turn represent an invariable, definite standard. Depending upon the energy values E_n and E_m, the electromagnetic radiation falls in the range of x-rays, light rays, infrared, or short radio waves. The last of these are the important ones in the transitions with which we are here concerned. To construct a clock utilizing this principle, the quantized changes of configuration of the electrons with respect to the atomic nucleus or of the individual atoms with respect to the molecule must be put to use. In the former case the result is an atomic clock, and in the latter a molecular clock. Both of the atomic oscillators, like the quartz oscillator, must yield an a.c. current with a frequency f, which in the case of the atomic oscillators is, of course, higher by several orders of magnitude. From the earlier discussion we are familiar with the numerous permissible and firmly fixed energy transitions in the atomic model. To construct an atomic clock the hyperfine structure arising from the magnetic interaction between atomic nucleus and valence electron would appear suitable. Conventional atomic clocks therefore make use of the cesium atom (Cs 133), because it contains only one valence electron; the spin of the electron is $s + 1/2\hbar$, and the nuclear spin is $I = 7/2\hbar$, where $\hbar = h/2\pi$. There are two energy forms corresponding to the two possibilities of parallel or antiparallel position.[1] For the parallel spin

$$F_p = I + s = 4\hbar$$

whereas for the antiparallel

$$F_a = I - s = 3\hbar$$

The energy levels are so close together that the frequency absorbed or emitted on a transition falls in the radio range. The frequency

[1] Cesium possesses 55 electrons, 54 of which fill all the inner shells as in a rare gas. The remaining electron lies in a new and practically empty shell. This so-called valence electron exhibits a rotational direction parallel to the nucleus, but either the same as or opposite to the rotation of the nucleus. These positions are described in shortened form as parallel or antiparallel.

is 9192.63183 Mc/s or 9.192. . . Gc/s (1 Gigacycle/s equals 1000 Mc/s), or in wavelength $\lambda = c/f \sim 3$ cm. Figure 227 illustrates the principle of such an atomic clock. During operation the entire envelope is kept constantly evacuated to 10^{-6}–10^{-7} mm Hg to prevent atomic collisions as much as possible. At one end there is a small heating chamber with a heating coil. Here the cesium is vaporized by heating to approximately 200°C. It might seem that this is a cathode system; this, however, is not the case. The electrically neutral cesium atoms travel as a result of thermal acceleration to a collimator diaphragm. Part of the gas passes through the diaphragm, is focussed to a beam by the aperture, and reaches an inhomogeneous magnetic field u. As a result of the thermal equilibrium the beam originally possesses as many

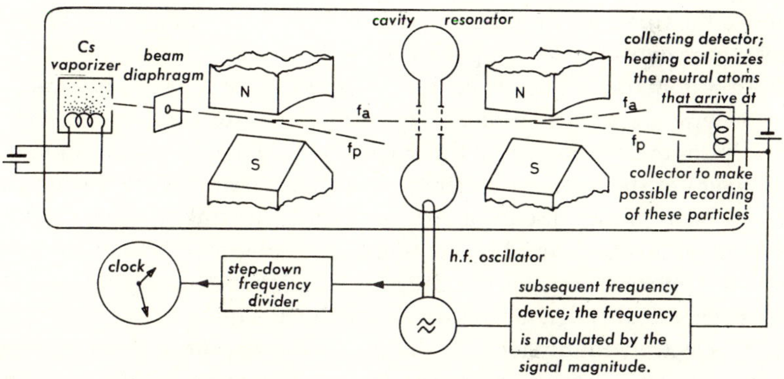

FIG. 227. Diagram of a cesium frequency standard.

parallel as antiparallel atoms (precisely considered in accordance with a statistical distribution). The inhomogeneous magnetic field forces the atoms in opposite directions—in accordance with the relationship of the direction of their magnetic moment to the direction of the external magnetic field in their path—and consequently divides them into two groups f_a and f_p. The atoms of the lower energy level f_p are denied entrance to the cavity resonator, while the higher energy atoms f_a are able to enter. Interaction occurs as the cavity resonator, which is supplied by a microwave oscillator, is in exact resonance with the atomic characteristic frequency of 9192 Mc/s. Consequently, the atoms, on transition to the lower level f_p, when they reattain thermal equilibrium, are able to emit the excess energy in radiation with the resonant frequency. The cesium beam passes through the cavity resonator (in practice there are usually two circuits), which is open on both sides, and once again consists of both types of atoms, i.e., parallel and antiparallel. As they continue their flight they come to a second inhomogeneous magnetic field *u*, where they are again separated into two groups. If all the atoms in the resonator were brought to the lower energy state, which is thermodynamically impossible, all of the atoms would have to ultimately land on the collector (Fig. 227, right). If, however, no interaction occurred at all, in the event, for example, that the search frequency of the h.f. generator was not tuned to the atomic frequency, then all of the atoms would necessarily fly past the collector. No so-called induced or stimulated transition would have occurred. For thermodynamic reasons neither of these ideal situations is possible. Consequently, the energy excess above the thermal equilibrium, accumulated by sorting at the cavity resonator entrance can be put to use.

An automatic frequency control is necessary here just as in a radio or television set with automatic frequency adjustment. In this case it adjusts the frequency of the microwave oscillator to the optimal midpoint of the atomic frequency resonance curve. Since the atoms of the beam under discussion are electrically neutral, at the collector they must first be ionized by heating with a tungsten filament (liberation of electrons, emission). Only in this way is a signal produced at the collector; this signal can then be amplified a million-fold by means of an electron multiplier. If the indicator current is too low, the microwave oscillator is trimmed to the resonance frequency by a reactance tube of the control system until the signal reaches the maximum. This is the position at which the highest possible percentage of the atoms effect the desired transition inside the cavity resonator. A sharply defined resonance frequency that corresponds to the atomic frequency can on principle be divided down to the usual unit of measure by means of electronic dividers. With the electronic tools now available it has, however, been found even more effective to multiply the frequency of a quartz clock that is being tested to the atomic frequency and then determine the difference in frequency. The accuracy of such a clock depends upon the sharpness of the resonance curve. The smaller the so-called half width of the resonance curve, the more accurately the control system tunes the search oscillator to the center of the spectral line. Figures of 100 cps are given for the half width, which in relation to the extremely high fundamental frequency corresponds to a figure of merit $Q = f/\Delta f$ of

10 million. Since the atomic spectral line represents a relatively sharp resonance curve, the auxiliary oscillator, which generates the centimeter waves, must be quite stable if it is not to continually fall behind during frequency comparison. The search frequency is consequently derived from a 5 Mc/s quartz oscillator and multiplied up to the characteristic frequency of the atomic clock, while the quartz is fairly steadily trimmed to the theoretical value by the control system. The final power is generated by a klystron oscillator that is pulled into tune by the above-mentioned search frequency. The comparison frequency, which is obtained by frequency multiplication and various superpositions (addition and subtraction of frequencies) in refined processes, is also frequency modulated with 100 cps. The use of such a wobble process as the integrating part of the control system makes it easier to determine the direction of deviation of the search frequency from the hyperfine structure frequency.

7. Microwave spectroscopy and molecular clocks

There are molecules in which several different stable atomic arrangements are possible. In going from one equilibrium condition to another, energy is absorbed or emitted in accordance with the direction of the energy difference. The differences are very small,

FIG. 228. NH_3 molecule in its two possible states. Ammonia exhibits a marked dipole moment.

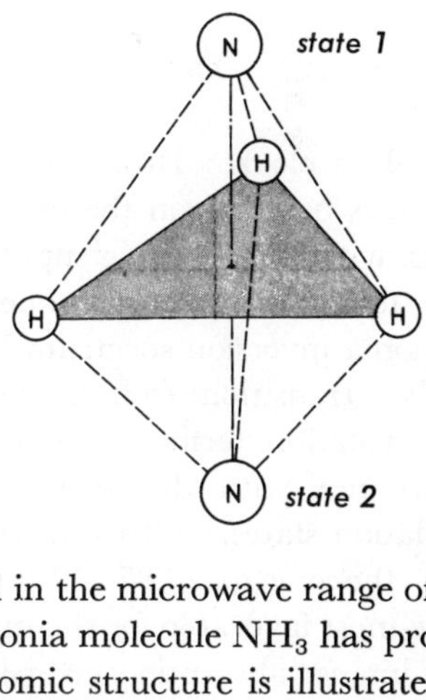

and the spectral lines consequently fall in the microwave range of the electromagnetic spectrum. The ammonia molecule NH_3 has proved advantageous as an oscillator. Its atomic structure is illustrated in Fig. 228. The hydrogen atoms H occupy the vertices of an equilateral triangle. The nitrogen atom N has two stable positions, one above the plane of the triangle (state or position 1) and the other below it (state 2). Since the absorption and liberation of energy are both accompanied by an electromagnetic radiation, a lower-energy nitrogen atom is capable of absorbing energy when subjected to radiation from an external h.f. field oscillating at a resonant frequency. This can be demonstrated by the absorption of microwaves in gases. If we disregard the tunnel and coupling effects, as in Fig. 229, we can obtain a very simplified idea by imagining the nitrogen molecule as a pendulum that oscillates between the two equilibrium states. The symmetrical alternation of the N atom with respect to the surface is called inversion. In this simple model we can find an analogy to a

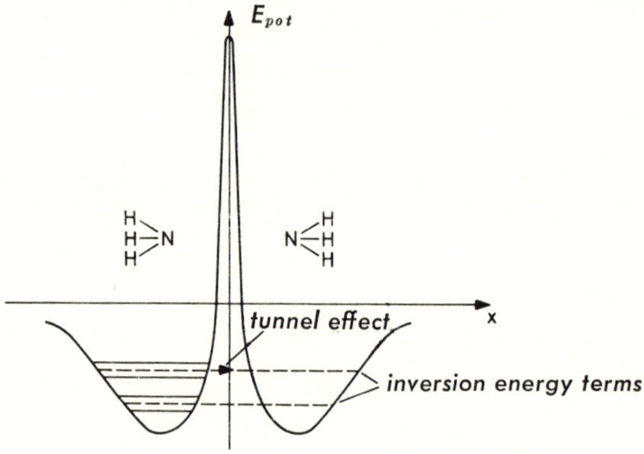

FIG. 229. The potential energy curve in the NH_3 molecule.

pendulum clock. However, the frequency with which we are concerned is based upon the energetic transition between the two final states, with energy being supplied for the N atom in the lower energy state to enter the upper one. To express it more precisely, the ammonia inversion spectrum lying in the microwave range arises as a result of transitions from the lower to the higher inversion level of the same rotation term. Several other inversion lines are created in accordance with the energetically possible molecular rotation or oscillation states. Of the various microwave absorption lines, however, the one at 23,870.12 ± 0.02 Mc/s (wavelength $\lambda = 1.26$ cm) is the most favorable for the manufacturer of molecular clocks. This very intense absorption inversion line 3.3 is relatively low in the frequency range, is easily found in the inversion spectrum, and is easily reproduced. Figure 230 shows the basic arrangement employed to determine the absorption lines of a gas or its molecules. A wave guide is evacuated and then filled with the gas to be studied till a low pressure of 0.01 mm Hg is attained. The wave guide is supplied at one end (Fig. 230, left) by a microwave oscillator of variable frequency. On the other side the microwaves are first rectified by a detector and then fed to the vertical deflection plates of a cathode-ray oscillograph. If the frequency of the testing transmitter is tuned to the molecular frequency, only a small portion of the h.f. energy arrives at the other end of the wave guide, since most of the energy is absorbed by the gas. This resonance effect indicates an intense interaction between the waves from the oscillator and the molecules. In order to locate the frequency, possibly for study of other frequencies,

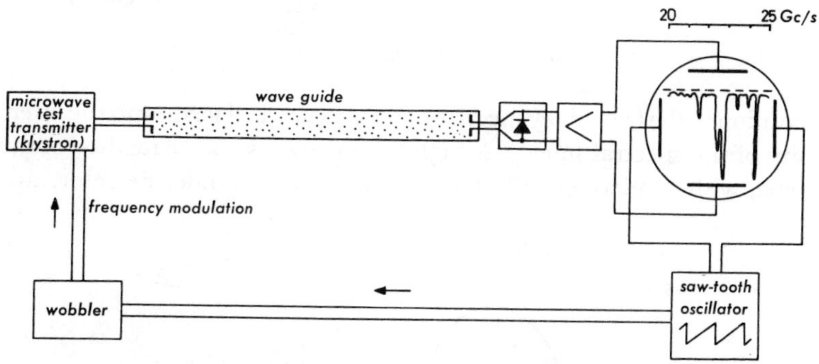

FIG. 230. Microwave spectroscopy arrangement.

the microwave oscillator is usually allowed to vary periodically over a certain frequency range in a wobble process. It is more efficient to employ the arrangement illustrated in Fig. 230, in which a saw-tooth generator is used and the cathode ray is started along the x-axis while simultaneously the frequency of the testing generator is linearly increased.

When the saw-tooth curve reaches its maximum amplitude the steep flank follows immediately, whereupon the luminescent point on the cathode-ray screen drops back to its starting position. The frequency of the microwave transmitter also returns to the starting position. Then the play of frequency increase and direction of the electron beam begins all over again. In this arrangement every point on the abscissa corresponds to a definite frequency. This allows a frequency or wave scale to be obtained. Since the absorption intensity is reflected by the vertical deflection of the cathode ray, the oscillogram represents a so-called frequency function or absorption frequency function curve. This makes it possible to determine the

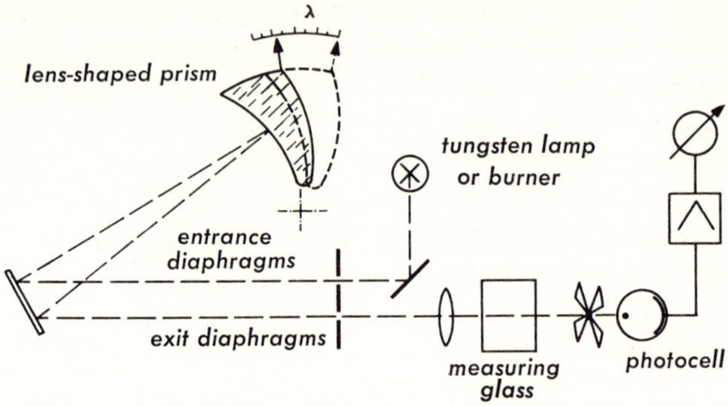

FIG. 231. Optical spectrometer.

position in the frequency spectrum, the geometrical form, and the intensity of the spectral lines. A high frequency resonance method of this type, with a microwave test transmitter, wave guide, detector, and

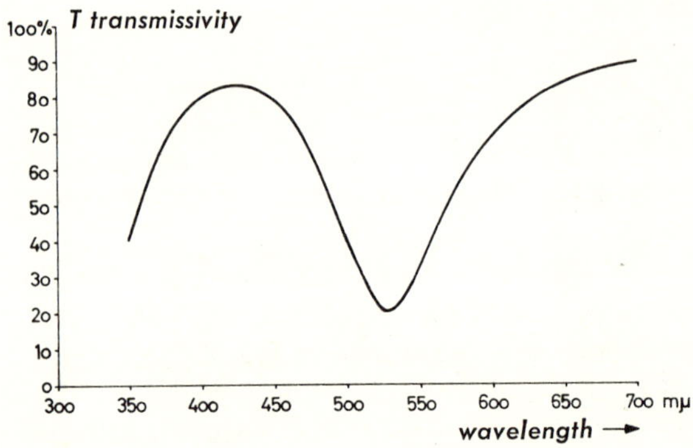

FIG. 232. (*a*) Transmission curve of potassium permanganate solution in an optical spectrometer (continuous transitions); (*b*) as intensity curve of a mercury vapor discharge (discrete transitions). The height indicates the positions of the corresponding frequencies.

oscillograph, makes possible the investigation of atomic and molecular spectrum lines in the millimeter and centimeter wave region. Microwave spectroscopy forms a supplement to optical spectroscopy (Figs. 231 and 232). From the spectra obtained conclusions can be drawn concerning the details of molecular structure and the interatomic distances in molecules. Spectroscopy is also employed to determine nuclear moments, moments of spin, quadrupole moments, dipole moments, and the internal electric fields within molecules. It can also be used for analysis if an index exists from previous measurement of the same material. Microwave spectroscopy is used in those areas that are beyond the discrimination power of even special optical equipment, where the distances separating the lines are only fractions of an Ångstrom.[1]

If a molecular clock is to be based on the absorption method the test transmitter must be automatically adjusted, as in an atomic clock, to the absorption maximum of the molecular frequency taken as a

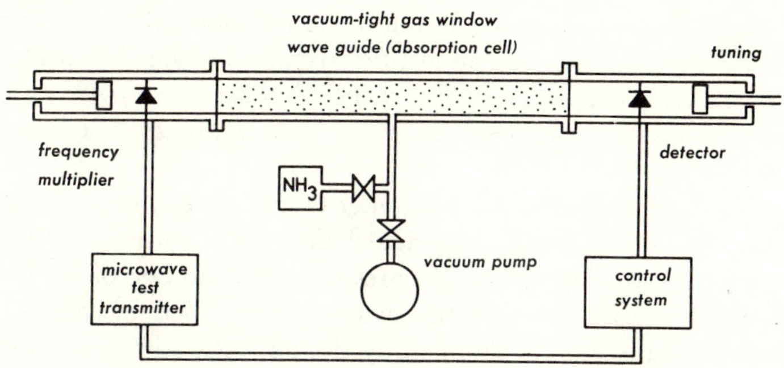

FIG. 233. NH_3 frequency standard based on the absorption method.

time norm (Fig. 233). In this case also the degree of accuracy depends upon the accuracy of adjustment to the absorption maximum. The half widths of the absorption curve for such ammonia absorption cells are given as 150 Kc/s and the adjustment accuracy, considering the control systems presently manufactured, is approximately 1/200, or ± 750 cps. The figure of merit $Q = f/\Delta f$ of about 150,000 is unfortunately lower by a power of 10 than that of the quartz oscillator. Still the molecular frequency standard possesses an advantage over the initially mentioned quartz oscillator in that only the width is affected, while the peak cannot be affected. The molecular resonance sharpness is limited by various factors:

[1] 1 Ångstrom equals 10^{-8} cm. Today wavelengths in the light range of mμ (millimicrometers) = 10^{-6} m or in nm (nanometers) = 10^{-9} m are given.

1. The oscillating molecules travel in all directions in accordance with a temperature-dependent statistical distribution. The rays consequently travel in relation to the observer or the detector at the exit. The relationships here are exactly the same as those in the case of a passing automobile, the frequency of whose pitch varies as a result of the Doppler effect. The decisive factor in the spreading of the frequency curve is the velocity components directed from the radiation to the detector. This source of error can be considerably reduced by reduction of the temperature.

2. In accordance with Fourier analysis it would be possible to obtain an ideal ("unspread") spectral line only with a sinusoidal oscillation of infinite length. If, however, the a.c. current is interrupted, the curve form is not uniformly broken and consequently yields additional frequencies. This also leads to spreading of the lines. The collisions among molecules and the collisions of the molecules with the walls of the absorption tube are also among the oscillatory disturbances. The molecules must therefore be able to vibrate with as little disturbance as possible, and to ensure this special measures are taken.

3. If in order to reduce the number of collisions among molecules the number of molecules introduced into the measuring system is reduced by lowering of the gas pressure, the number of oscillators available for absorption of the externally supplied energy is also reduced. If the energy supplied by the h.f. oscillator is too high, a saturation effect, which also results in spreading, is encountered. It has been found that at the recommended gas pressure of 10^{-2} mm the power density drops to 1 mW/cm^2. If the pressure is reduced further, the signal level drops lower and lower, and the proportion of noise level to signal level becomes correspondingly less favorable.

4. Just as any damping or any resistance connected to an oscillatory circuit (also an antenna with its characteristic radiation resistance) causes spreading, the unavoidable radiation damping in this case also leads to spreading of the curve.

8. The ammonia maser

It has been found that these unfavorable influences can to a great extent be eliminated by the so-called beam method. We are already familiar with the principle of this method from the atomic clock, which also dispenses with a homogeneous gas filling and uses instead a cesium beam. In this case a molecular beam (ammonia gas) instead of an atomic beam is passed into the vacuum. Since the molecules all assume the same propagation direction and velocity, and the indicator circuit is transverse to the direction of motion, the Doppler effect is reduced. Particles that undergo collision are deflected, fail to pass through the diaphragm, and are separated from the beam. The ammonia oscillators based on this method are also known as *masers* (*m*icrowave *a*mplification by *s*timulated *e*mission of *r*adiation). The principle is illustrated in Fig. 234. The NH_3 oscillator consists of three essential components: an NH_3 distributor (at 1 in the figure), a focussing electrode system (at 2), and a high frequency cavity resonator (at 3). These components are contained in an evacuated

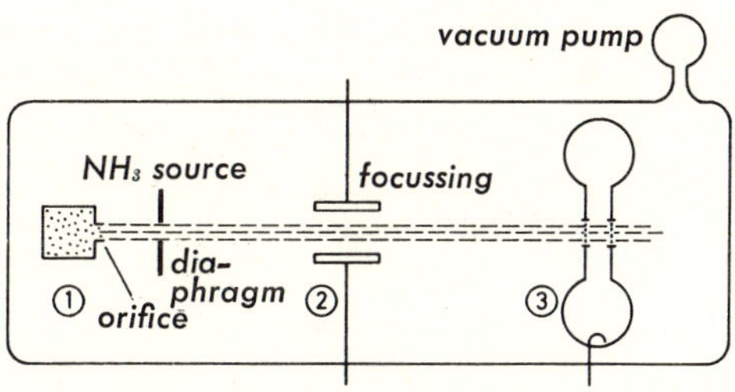

FIG. 234. Ammonia maser with a Stern-Gerlach field. Here the poles of the magnetic field or the electric field are very rapidly reversed, i.e., with the characteristic frequency.

vessel. The ammonia gas passes from the reservoir 1 through one or more hair-fine, parallel orifices at a high velocity through the diaphragm into the electric field of the focussing electrodes.

This electrode system in practice is composed of an even number of parallel metal wires that are completely insulated from each other (see Fig. 236). Between the adjacent wires there is a constant potential of 15,000 volts. The electric field thus created is at a maximum between them, i.e., at the periphery of the "circuit" and drops off axially toward the center. At the center of the axis the field gradient reaches a zero value.

As a consequence of the thermodynamic equilibrium the original ammonia beam contains a certain number of particles of higher energy state and a certain number of the lower state of the transition with which we are concerned (Fig. 235*a*). The statistical distribution is always determined physically by Boltzmann's law. Thermodynamically the higher energy state E_2 is less frequent than the lower energy state E_1. The following relation

$$\frac{n_2}{n_1} = e^{(E_2-E_1)/kT}$$

in which n represents the number of particles and k the Boltzmann constant $1.38 \cdot 10^{-33}$ ampsec/deg governs the statistical distribution.

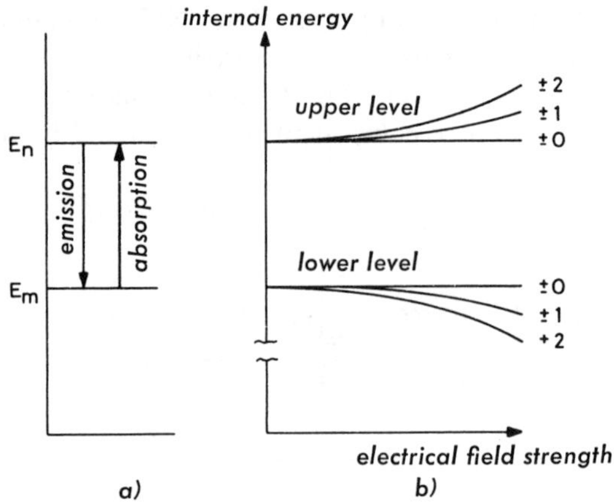

FIG. 235. Energy levels and their separation in an electric field.

To permit emission the statistical distribution must first be changed in such a manner that the concentration of higher energy state particles is increased. This can be accomplished by making use of measures that will separate the two types of particles and eliminate the lower energy particles. For this purpose, the atomic beam method makes use of the magnetic Stern-Gerlach field, which makes available a strong electric field of 15,000 volts. The uninfluenced molecular beam possessed a propagation direction perpendicular to the electrostatic quadrupole shown in Fig. 236. Here we are concerned with molecules that represent a type of electrical dipole. Stark already showed how energy levels can be separated by means of an electric field (as in Fig. 235*b*). An inhomogeneous field, moreover, separates the energetically quantized states into trajectories corresponding to the states. The molecules of the higher level are concentrated in the direction in which the field decreases, i.e., toward the center, and travel along the axis straight to the cavity resonator. The lower energy molecules, i.e., those on a lower energy level, are subjected to a radially outward-directed force, which consequently separates them from the central beam. The molecules that are separated out are frozen solid by means of a cooling coil through which liquid air is passed. All the ammonia molecules of the higher inversion level are collected and focussed along the beam axis by the electrostatic mechanism. This main beam, i.e., the beam consisting of the desired molecules of the higher energy level, travels to the cavity resonator, which is very precisely tuned to the molecular frequency of the appropriate quantized transition. The almost lossless resonator is caused to begin oscillating by natural thermal noise (at room temperature), with the result that

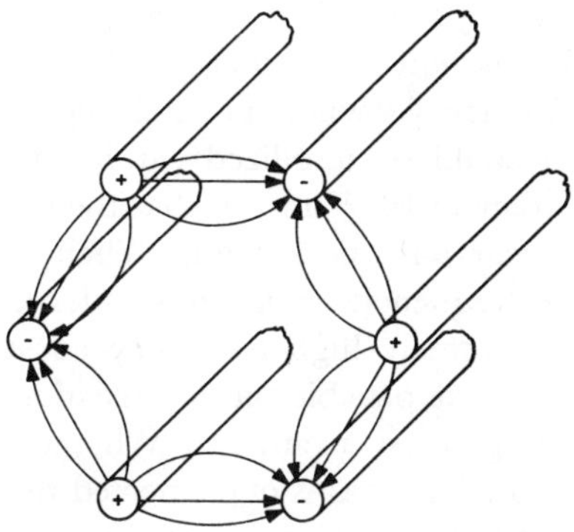

FIG. 236. Two types of focussing electrodes of a maser and their electric field.

the excited particles are subjected to an a.c. electric field. Since the molecules are in a state that does not correspond to the thermodynamic equilibrium distribution, the transition to the equilibrium condition is triggered by the resulting interaction with the high frequency field. Many molecules make the transition and consequently emit the excess energy. This power loss takes the form of electromagnetic radiation, and the oscillations of the resonator, which is tuned to the same frequency, are greatly amplified. This intensification of the field in turn stimulates cophasal emission from new molecules. If the power supplied by the molecules compensates the electrical resistances or losses of the resonator, the apparatus becomes a self-exciting oscillator. The generated frequency then corresponds to the inversion frequency 23,870 Mc/s ($\lambda = 1.26$ cm).

Since this type of molecular beam system exhibits the properties of both an amplifier and a stable feedback oscillator, it offers many new possibilities. The figure of merit is very high as a result of the self-actuated damping reduction, and the resonance lines are very narrow. Similarly, the adjustment accuracy, i.e., the stabilization, can be made very precise. In this way a very exact frequency standard can be constructed. To reduce the detuning effect of coupling between the cavity resonator frequency and the molecular frequency, in practice a system of two coupled identical cavity resonators is used. One of these is crossed by the molecular beam, and the other is easily coupled on by a small opening in a manner that is usual in wave guide technique. The resonance tuning, which moves over a range of only a few cycles per second, is effected by the variation in temperature in a calibrated thermostat. P. Kartaschoff (Neuenberg) indicates that an accuracy of $\pm 10^{-10}$, i.e., a frequency deviation of ± 2.4 cps, was obtained with a temperature that could be stabilized within the control range of ± 0.2°C. An accuracy of 10^{-10} means that the maximum deviation per 10^{10} sec (300 years) is one second. Their stable, quantically fixed, characteristic frequencies make molecules and atoms sources of a.c. current of extremely high frequency constancy; these very constant frequencies make possible the construction of molecular and atomic clocks of the greatest accuracy. Whether these frequencies should be stepped down by means of connected tube stages acting as frequency dividers in order to drive the synchronous motor of a clock mechanism, or used for time comparison measurement with the multiplied frequency of a quartz oscillator, is a question of the chosen electronic apparatus arrangement. The latter arrangement is the usual one. The quartz clock is maintained in continuous operation and only periodically calibrated and controlled by a molecular clock.

Light, thermal radiation, and radio waves were shown by Maxwell to belong to the same natural system of electromagnetic waves. Atomic energy transitions are encountered in all ranges, and con-consequently quite a wide range of frequencies is presented. Roentgen and light radiation are emitted only as a result of larger transitions, and the leaps of the individual particles resulting in emission are chronologically irregular. This so-called spontaneous emission is, therefore, not coherent. By incoherent radiation is meant that the emission from the excited electrons, atoms, or molecules occurs in a completely chaotic fashion and that there is no phase relationship between the individual emissions. As soon as the wavelength approaches the order of magnitude of one-tenth of a millimeter, however, it is observed that the radiation phenomena must be cophasal. Here the oscillation phenomenon has become long enough to speak of a wave, which cannot be said to exist, figuratively speaking, when only a piece of a backward and forward path is present, but only when there is, to some extent, a continuous wave train. Atomic emission of continuous waves is obtained by ordering the transition of the concentrated, high energy particles to the lower level by means of a resonator, i.e., by stimulation that is synchronized to the characteristic frequency. This gives rise to coherent radiation. With large transitions such as those characterizing the emission of light, such controlled stimulation is not possible without the use of special measures. The so-called relaxation time, i.e., the time between the entrance of the particle into the metastable state and its return to the lower energy level, is extremely short at light frequencies. The most provable ratio W of spontaneous to stimulated or so-called induced emission can be derived from the laws of the thermal equilibrium. The following relation is obtained:

$$W = \exp\left[-\frac{h \cdot f}{kT}\right] - 1$$

Compared to the optical frequencies f (of light) the radio frequencies f are low. The probability of storing here a concentrated energy level and subsequently stimulating its emission increases as the wavelength increases or as the frequency is diminished. The present state of research into optical molecular amplifiers (the optical masers) that are actually capable of amplifying light waves will be discussed at the end of the present chapter. A distinction must be made between stimulated molecular amplification and classical resonance phenomena. The typical characteristic of the transitions obtained in the ammonia and cesium masers was that a weak high frequency signal was capable of initiating radiations of considerably greater power at the same frequency. A molecular amplifier of the type described, which utilizes the interaction between an external electromagnetic field and the internally stored energy of matter is not tunable. The wavelength is determined by the fixed quantum mechanical transitions alone. This type of amplifier consequently demands the most precise agreement of the transmission frequency with the molecular frequency. This amplifier, like an intermediate frequency amplifier, is assigned only one prescribed frequency. The advantage of this molecular amplifier over conventional amplifiers lies in its freedom from fluctuations such as those exhibited by tubes as a result of the discontinuous manner in which they are known to generate electrons. This results in a sensitivity limit (determined by the ratio of signal level to noise level) that is theoretically about one thousandfold higher, which allows the amplification of even individual energy quanta. When used as an input stage an amplifier of this type is capable of picking up signals as weak as those required by radio-telescopes in radio-astronomy. Molecular amplifiers can also be used as input stages in radar receivers; this increases vastly the range at which the same output signal can be employed. Instruments of this type have been used successfully in satellite controls.

It is also possible to build tunable atomic amplifiers by making use of the precessional motions induced by an external magnetic field. As we saw in an earlier section, an external magnetic field H causes separation into $(2J + 1)$ energy stages. The intervals between the stages are functions of both the magnetic field H and the magnetic moment. Consequently, for every field strength there is a definite precessional frequency that represents the energy difference of the separation. An amplifier of this type can therefore be tuned by variation of the external field. In contrast to the ammonia maser or the cesium beam method, here the magnetic field is brought right to the site of the interrogation device (e.g., the resonator). This type of amplifier, also quantum mechanical, is known as the versitron. The noise in molecular amplifiers arises only from spontaneous emission-transitions. Care must therefore be taken to ensure long storage times. Consequently, such amplifiers are especially noise-free *when thermal motion is eliminated as much as possible by cooling to very low temperatures. Liquid helium is favored as a coolant* (4.2°K). As a result of the low noise the sensitivity is 20 to 30 db better than that of classical input stages. This allows reception of transmitter signals from a satellite at considerably greater distance.

9. Tunable masers and solid state masers

For several years attempts have been made to build masers using molecules in the solid state. Solid bodies allow the use of simpler equipment than gaseous materials, particularly since they eliminate the necessity of vacuum installations. Mastery of this new borderline area of electronics requires a thorough understanding of atomic and solid state physics, as well as microwave engineering. The purpose of this oscillator is also initially to raise the molecules to higher energy states in order to utilize them subsequently for stimulated emission. Individual paramagnetic ions, i.e., ions with unpaired electrons, are incorporated into a diamagnetic, nonconducting crystal. This crystal is then placed in an external magnetic field. The electrons responsible for the magnetic moment, which—as we have seen—are always the reason for paramagnetism, are subjected to the external magnetic force acting laterally on the axis of spin and give rise to a precessional motion. Aside from the Larmor rotational motions the rotational axes can assume various discrete positions determined only by the internal quantum number. Each of these positions corresponds to a definite energy value. Here also we are concerned with the absorption or emission of a radiation, accordingly as the transition is to the higher or lower energy level. As we already know, the frequencies are situated in the microwave range. Up to now we have always been concerned with two adjacent energy levels and the corresponding transition. However, when the internal quantum number permits it, a number of adjustments are possible. Consequently, it is possible to induce jumps past one or more positions by supplying an appropriate energy. Solid state molecular amplifiers making use of three, rather than only two, energy levels in the amplification process have gained a position of special importance. In these amplifiers separate frequencies can be used for the excitation or pumping frequency and the signal frequency, which is to be amplified and act as the stimulator. The statistical distribution of thermal equilibrium is governed once more by Boltzmann's law, according to which in the successive energy states $E_3 > E_2 > E_1$ the number of particles (the population numbers) is distributed in the following way $n_1 > n_2 > n_3$. This means that the higher the energy of the state (the higher level), the lower the population number. The lowest state E_1 consequently has the highest population. Figure 237 illustrates the energy level arrangement for the three different fixed energy levels; the horizontal line lengths illustrate the ratio of the particle numbers. On the left (Fig. 237*a*) the figure illustrates the initial statistical ratio at thermal equilibrium. Figure 237*b* shows a forced concentration of the higher states at the expense of the lower, induced by external causes. There is an equal probability that an external radiation field will raise a particle from the lowest state E_1 to the highest state E_3 or drop a particle from an upper state to a lower one. Since originally there were more particles "below," i.e., at thermal equilibrium there are more n_1 particles present than "above" (n_3), an intense interaction will result in a new statistical distribution in accordance with which there will be exactly as many particles "above" as "below." From a chronological viewpoint the sequence is in the direction of an increase in the population number of the upper level. In other words, there are more absorption than emission transitions. The external radiation field that does the so-called pumping work supplies the power in the energy balance.

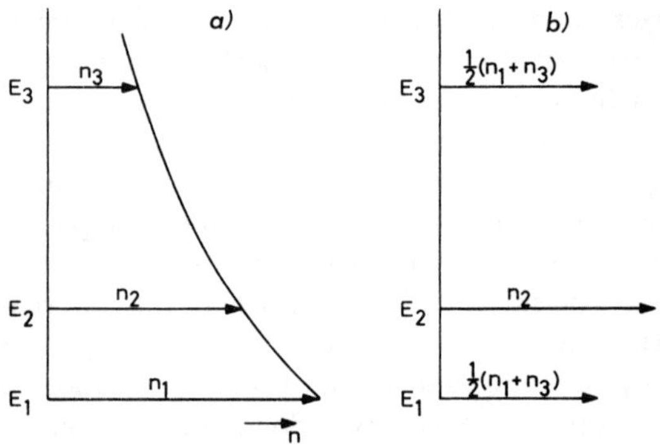

FIG. 237. Energy-level arrangement of a three-level solid state maser. Ordinate: energy; abscissa: population number. (*a*) The population numbers for the three energy levels are in energic equilibrium (stable). (*b*) The population number of the highest level E_3 has been raised by externally supplied radiation energy (so-called activation by pumping).

Since relatively many of the particles from the lower level E_1 are raised to the upper level E_3 or concentrated there, the population number of E_1 falls below that of E_2 also; this condition is a requirement for the subsequent stimulation of emission. But let us first consider the working operation of the "pump."

This concentration process is accomplished by an external energy supplied by a high frequency field at the frequency

$$f_p = \frac{E_3 - E_1}{h}$$

For this purpose the paramagnetic crystal is situated so that on one side it is in a constant magnetic field and on the other in a wave guide, where it is subjected to electromagnetic radiation at the pumping frequency f_p. There is here in fact a certain analogy with a mechanical pump, the action of which can, for example, raise the potential energy of a volume of water above the ground level. The origin of the energy-enriched (E_3) molecules may be conceived of by considering that the original quantized spin, which was more or less parallel to the direction of the constant external magnetic field, is shifted by the pumping frequency in such a manner that new, stable angles in the antiparallel position are assumed. These angles have higher energies than the parallel positions. Because this represents a disturbance of the equilibrium, on the disappearance of the pumping frequency the atoms tip back into the equilibrium position determined by the thermal influences and the external field. The relatively long

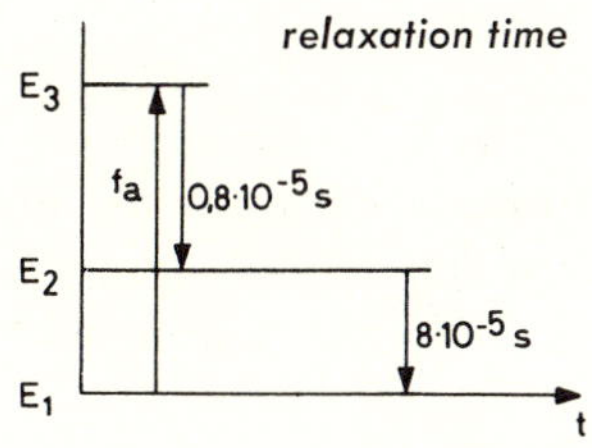

FIG. 238. Increase in efficiency by making use of different relaxation time values.

duration of the nonemissive state[1] is used, as we have seen, for stimulation or control by means of an externally supplied weak wave of the frequency used to effect the transition to the lower energy level E_1. Since the separation of energy levels induced in this solid state maser is, in contrast to the ammonia maser, determined by the external magnetic field, of which the separation is a function, it is possible to adjust the desired frequency by variation of the strength of the external magnetic field. Because this type of amplifier requires transitions between three levels for its functioning, it is known as a three-band solid state molecular amplifier. The distinguishing characteristic of this type of amplifier is that the pumping frequency used for concentration of the higher energy state does not coincide with the signal frequency that is to be amplified. The emission, which is associated with the transition that reestablishes the thermal equilibrium, can be stimulated or induced by the lowest control voltages of the frequency f_2.

[1] Also called the metastable state.

Figure 238 shows how the efficiency can be increased by making use of the different relaxation time values. The relaxation time for the transition E_3 to E_2 is 10 times as fast as the relaxation time for the transition E_2 to E_1. In the highest energy state the spins drop extremely quickly to the E_2 level, where they remain for a longer period of time before dropping back to E_1. This mechanism leads to an interim concentration of the E_2 spin states. If the transition to E_1 is stimulated during the relatively long relaxation period by an external signal of frequency f_2, a very strong amplification effect is obtained. The energy stored in level E_3 is then no longer used merely for frequency separation, but is also utilized for a considerable increase in the efficiency.

In practice the maser crystal is placed in a wave guide with a dividing filter (e.g., a magic T; see Fig. 132 on page 172) so that the pumping frequency and operating frequency, which are required for excitation and amplification, respectively, can be supplied separately. The three-band solid material consists of either a ruby crystal or chromium-potassium cyanide crystal[1] lightly doped with chromium ions. Values between 200...1000 are given for the attenuation factor.

10. Traveling-wave masers

In the so-called traveling-wave tubes (see p. 174) the electrons are passed through an entire system, the electrons being bunched into groups and interacting with the system, which is amplified progressively in an exponential fashion. In a similar manner it is possible to develop a cascade arrangement in the maser. Here the wave arising from coupling grows progressively and can be uncoupled in the form of intense radiation. Figure 239 illustrates the principle of such a so-called traveling-wave maser. An oblong-shaped paramagnetic crystal is incorporated into a wave guide, which serves as the reaction

[1] To obtain the three requisite levels in the microwave transition range, paramagnetic solid materials such as the following examples are used; the range of suitable materials should soon be made clear by further study and is presently the subject of research.

$$(Cr_{0.001\%}\ Al_{0.999\%})_2\ O_3 \text{ (ruby)}$$
$$Ni_{0.05\%}\ Zn_{0.95\%}\ SiF_6 \cdot 6\ H_2O$$
$$K_3\ Cr_{0.005\%}\ Co_{0.995\%}\text{–}(CN)_6$$

Whether the ultimate working materials will be solids, gases, or liquids is also the subject of present investigations.

chamber. A number of transverse metal bars are embedded as so-called coupling members in serrate fashion. This system is situated in an external magnetic field. The pumping power is supplied by means of the wave guide on the left and irradiates the entire reaction chamber. This power is used to continuously enrich the upper energy level E_3 with particles that had previously occupied the lower levels. A still weak input radiation is fed into the coupling system from the left; this wave has the task of stimulating emission, in the course of which it loses none of its energy. The resulting emission leads to amplification of the radiation, which results in further stimulation in the direction to the right. The interaction consequently continues

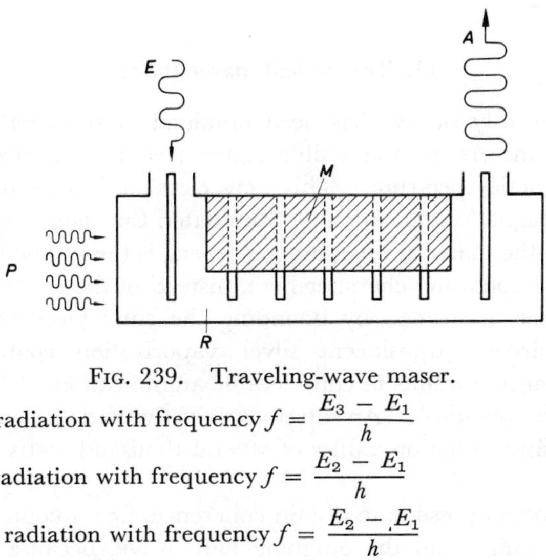

FIG. 239. Traveling-wave maser.

P—pump radiation with frequency $f = \dfrac{E_3 - E_1}{h}$

E—input radiation with frequency $f = \dfrac{E_2 - E_1}{h}$

A—output radiation with frequency $f = \dfrac{E_2 - E_1}{h}$

R—resonator
M—working material of the maser, e.g., ruby crystal or chromium-potassium cyanide crystal.

to increase as the radiation travels from left to right, so that the radiation field must increase progressively. It is certainly correct here to speak of an induced emission of radiation. Uncoupling is effected by a rod, thus, by means of an antenna or over a coaxial line, depending upon whether the atomic system operates at higher or lower frequencies; the pumping system here necessarily lies higher. Ruby traveling-wave masers can already be catalogued, and the following table will illustrate the presently obtainable technical values.

Mean operating frequency	5850 Mc/s
Tuning range	± 175 Mc/s
Band width	25 Mc/s
Power amplification factor	10^3
Maximum output power	$10 \cdot 10^{-6}$ watt
Magnetic field	3975 oersteds
Pump frequency	19,100 Mc/s
Pump power	$50 \cdot 10^{-3}$ watt
Operating temperature	1.5°K

11. The optical maser (laser)

Quite recently success has been obtained in the construction of solid state masers for the visible range (red and infrared) of the electromagnetic spectrum. This new element has been named a "Laser" (Light Amplification by Stimulated Emission of Radiation). Here, also, the material used, among others, is the synthetic pink-red ruby. The operating chamber here, instead of the resonator cavity of the maser, is formed by bounding the ruby piece with plane-parallel mirrors (translucent silver vaporization coating). The induced emission results in considerable amplification of the arriving light quanta (photons). An extremely selective and directed aperture angle $\frac{1}{20}°$ and radiation values of several thousand watts have been attained.

It has proved possible to obtain coherence, i.e., a cophasal course for the incoming and the outgoing light wave, because the entire stimulation is based on one resonance phenomenon. The coherence can be demonstrated by interference figures. In order to ensure that the emission resulting from stimulation is uninterrupted, i.e., to obtain a stationary oscillation, the population concentration of the higher energy level must be kept higher than the Boltzmann equilibrium value; this is accomplished by means of pumping radiation. Experiments have already shown that with too intense stimulation the population density of the upper level drops below the critical value, and that after this some time elapses before emission will begin again, i.e., until the pumping radiation can repopulate the upper level. This makes possible a relaxation oscillation.

Besides the optical solid state masers, optical gas masers (e.g., mixtures of helium and neon) have also been constructed. The activation here is supplied by electrical energy from a short-wave high frequency oscillator, which gives rise to the metastable state and to a gas discharge in which the photons resulting from the latter discharge act also as stimulators.

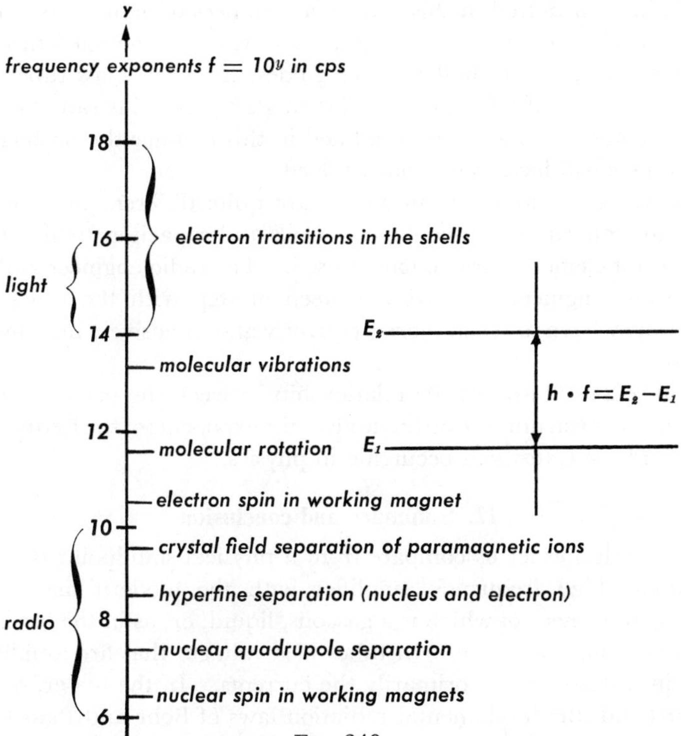

FIG. 240.

The amplified maser output radiation, especially in solid state masers, is reputed to reach radiation powers of over 10 kw. Focussing of such light intensities with lenses would have to yield illumination densities of 10^8 watts/mm^2, which is sufficiently high to immediately vaporize high temperature oxides. As a result of the coherence of such cophasal radiation, the light amplitudes[1] are so great that electric field strength values of more than 1 mio volt/cm have been

[1] Waves, i.e., rays of the smallest aperture angles (several minutes) have been generated without the underivable diffraction of the incoherent light system.

attained. Experiments have been made with projection objectives to bounce such an intense and nearly parallel-traveling beam off the moon in order to obtain a reflection visible from the earth. Besides, there is the possibility of modulating such a light beam to carry communication or even of using it as a carrier frequency for a multi-channel directional beam system; immense amounts of information could be transmitted in this way over distances of light years. Like the waves of the micro range, light waves can be conducted through equipment similar to hollow wave guides; however, glass tubes and rods must be used (cf. total reflection in glass plates) instead of metal.

If roentgen rays are ever amplified in this manner the problem of death rays will have been finally solved.

Atomic oscillators and amplifiers are quite different in structure from the conventional tube stages. Coils and capacitors, which were important elements, are no longer used. The radio engineer and the electronic engineer who wish to keep in step with the times will necessarily have to reach more frequently and intensively into atomic physics.

Figure 240 illustrates the relationship between the position in the frequency spectrum (the ordinate gives the exponents of the frequencies 10^y) and the transitions occurring in physics.

12. Summary and conclusion

In conclusion let us compare from a physical standpoint the conventional high frequency amplifier with the modern maser, the principal element of which is a gaseous, liquid, or, as in the last case, solid working substance. In tubes it is voltages that are amplified, and in transistors it is primarily the currents. In the maser, on the other hand, the fundamental radiation laws of Bohr and Planck are used to activate and then stimulate the emission of energy quanta as radiation. Activation occurs first and is effected by a pumping frequency source, which provides the requisite power to raise the atoms or molecules, most of which occupy the lower discrete energy states, to the higher states. This is described by saying that the working substance has thereby been activated to labile enriched states. The control signal which one wishes to amplify is an electromagnetic radiation that is fed into the apparatus and initiates or stimulates a transition from the activated state to the stable state. This control power is amplified if the radiation resulting from the emissive transition is synchronized, i.e., in the correct phase position. The resonator chamber or reaction chamber in which the maser material is situated plays an essential part in this sorting. The energy for the amplified radiation is provided by the pumping generator, which is analogous to the anode battery in tubes.

Its use of a quantic mechanism makes this type of amplifier or oscillator system in principle suitable for all types of electromagnetic radiations, particularly over an extremely wide spectrum range, including, for example, radio waves, thermal radiation, or light waves. Of particular importance are the possibilities it opens in communication engineering, where new channels are being sought continuously and where the answer can only lie in the direction of the shortest wavelengths.

The solid state circuit elements such as quartz, etc., which are generally nonconductors, are not directly electronic elements, but form a group of integrative components. We have touched upon this borderline area to show the connection between solid state nonconductors and the metallic conductors. The semiconductors fall into the area between, and they have been dealt with in greater detail.

Chapter VI looked briefly into the area of semiconductor engineering in order to give some idea of the profusion of interesting phenomena revealed by this new branch of electronics and of the new and as yet incompletely charted perspectives it has opened up.

No attempt has been made in this book to discuss electronic installations. The aim of this book has been to provide sufficient information about electronic elements for an understanding of the simple basic ideas underlying even complex instruments. For an understanding of the further development of electronics the elementary laws are of the greatest importance. Electronics has become an instrument for integrating many areas. The businessman uses electronic calculators; the physician, in addition to the x-ray machine, makes use of a host of electronic aids; even the sermon of the minister is heard over wireless hearing aids. Today all types of occupations come in contact with electronics.

A mastery of the fundamental principles is necessary not only to open the way to new solutions for those concerned with further developments. It is also necessary for all those concerned with maintenance of electronic equipment to be thoroughly familiar with the principles that allow correction and maintenance of operation.

We hope that this book will fulfill the purpose for which it was written: to provide a basic understanding of electronics.

INDEX

SOME DOVER SCIENCE BOOKS

SOME DOVER SCIENCE BOOKS

CHANCE, LUCK AND STATISTICS: THE SCIENCE OF CHANCE,
Horace C. Levinson

Theory of probability and science of statistics in simple, non-technical language. Part I deals with theory of probability, covering odd superstitions in regard to "luck," the meaning of betting odds, the law of mathematical expectation, gambling, and applications in poker, roulette, lotteries, dice, bridge, and other games of chance. Part II discusses the misuse of statistics, the concept of statistical probabilities, normal and skew frequency distributions, and statistics applied to various fields—birth rates, stock speculation, insurance rates, advertising, etc. "Presented in an easy humorous style which I consider the best kind of expository writing," Prof. A. C. Cohen, Industry Quality Control. Enlarged revised edition. Formerly titled *The Science of Chance.* Preface and two new appendices by the author. xiv + 365pp. 5⅜ x 8. 21007-3 Paperbound $2.00

BASIC ELECTRONICS,
prepared by the U.S. Navy Training Publications Center

A thorough and comprehensive manual on the fundamentals of electronics. Written clearly, it is equally useful for self-study or course work for those with a knowledge of the principles of basic electricity. Partial contents: Operating Principles of the Electron Tube; Introduction to Transistors; Power Supplies for Electronic Equipment; Tuned Circuits; Electron-Tube Amplifiers; Audio Power Amplifiers; Oscillators; Transmitters; Transmission Lines; Antennas and Propagation; Introduction to Computers; and related topics. Appendix. Index. Hundreds of illustrations and diagrams. vi + 471pp. 6½ x 9¼.
61076-4 Paperbound $2.95

BASIC THEORY AND APPLICATION OF TRANSISTORS,
prepared by the U.S. Department of the Army

An introductory manual prepared for an army training program. One of the finest available surveys of theory and application of transistor design and operation. Minimal knowledge of physics and theory of electron tubes required. Suitable for textbook use, course supplement, or home study. Chapters: Introduction; fundamental theory of transistors; transistor amplifier fundamentals; parameters, equivalent circuits, and characteristic curves; bias stabilization; transistor analysis and comparison using characteristic curves and charts; audio amplifiers; tuned amplifiers; wide-band amplifiers; oscillators; pulse and switching circuits; modulation, mixing, and demodulation; and additional semiconductor devices. Unabridged, corrected edition. 240 schematic drawings, photographs, wiring diagrams, etc. 2 Appendices. Glossary. Index. 263pp. 6½ x 9¼. 60380-6 Paperbound $1.75

GUIDE TO THE LITERATURE OF MATHEMATICS AND PHYSICS,
N. G. Parke III

Over 5000 entries included under approximately 120 major subject headings of selected most important books, monographs, periodicals, articles in English, plus important works in German, French, Italian, Spanish, Russian (many recently available works). Covers every branch of physics, math, related engineering. Includes author, title, edition, publisher, place, date, number of volumes, number of pages. A 40-page introduction on the basic problems of research and study provides useful information on the organization and use of libraries, the psychology of learning, etc. This reference work will save you hours of time. 2nd revised edition. Indices of authors, subjects, 464pp. 5⅜ x 8.
60447-0 Paperbound $2.75

THE RISE OF THE NEW PHYSICS (formerly THE DECLINE OF MECHANISM), *A. d'Abro*

This authoritative and comprehensive 2-volume exposition is unique in scientific publishing. Written for intelligent readers not familiar with higher mathematics, it is the only thorough explanation in non-technical language of modern mathematical-physical theory. Combining both history and exposition, it ranges from classical Newtonian concepts up through the electronic theories of Dirac and Heisenberg, the statistical mechanics of Fermi, and Einstein's relativity theories. "A must for anyone doing serious study in the physical sciences," *J. of Franklin Inst.* 97 illustrations. 991pp. 2 volumes.

20003-5, 20004-3 Two volume set, paperbound $5.50

THE STRANGE STORY OF THE QUANTUM, AN ACCOUNT FOR THE GENERAL READER OF THE GROWTH OF IDEAS UNDERLYING OUR PRESENT ATOMIC KNOWLEDGE, *B. Hoffmann*

Presents lucidly and expertly, with barest amount of mathematics, the problems and theories which led to modern quantum physics. Dr. Hoffmann begins with the closing years of the 19th century, when certain trifling discrepancies were noticed, and with illuminating analogies and examples takes you through the brilliant concepts of Planck, Einstein, Pauli, de Broglie, Bohr, Schroedinger, Heisenberg, Dirac, Sommerfeld, Feynman, etc. This edition includes a new, long postscript carrying the story through 1958. "Of the books attempting an account of the history and contents of our modern atomic physics which have come to my attention, this is the best," H. Margenau, Yale University, in *American Journal of Physics.* 32 tables and line illustrations. Index. 275pp. 5⅜ x 8.

20518-5 Paperbound $2.00

GREAT IDEAS AND THEORIES OF MODERN COSMOLOGY, *Jagjit Singh*

The theories of Jeans, Eddington, Milne, Kant, Bondi, Gold, Newton, Einstein, Gamow, Hoyle, Dirac, Kuiper, Hubble, Weizsäcker and many others on such cosmological questions as the origin of the universe, space and time, planet formation, "continuous creation," the birth, life, and death of the stars, the origin of the galaxies, etc. By the author of the popular *Great Ideas of Modern Mathematics.* A gifted popularizer of science, he makes the most difficult abstractions crystal-clear even to the most non-mathematical reader. Index. xii + 276pp. 5⅜ x 8½. 20925-3 Paperbound $2.50

GREAT IDEAS OF MODERN MATHEMATICS: THEIR NATURE AND USE, *Jagjit Singh*

Reader with only high school math will understand main mathematical ideas of modern physics, astronomy, genetics, psychology, evolution, etc., better than many who use them as tools, but comprehend little of their basic structure. Author uses his wide knowledge of non-mathematical fields in brilliant exposition of differential equations, matrices, group theory, logic, statistics, problems of mathematical foundations, imaginary numbers, vectors, etc. Original publications, appendices. indexes. 65 illustr. 322pp. 5⅜ x 8. 20587-8 Paperbound $2.25

THE MATHEMATICS OF GREAT AMATEURS, *Julian L. Coolidge*

Great discoveries made by poets, theologians, philosophers, artists and other non-mathematicians: Omar Khayyam, Leonardo da Vinci, Albrecht Dürer, John Napier, Pascal, Diderot, Bolzano, etc. Surprising accounts of what can result from a non-professional preoccupation with the oldest of sciences. 56 figures. viii + 211pp. 5⅜ x 8½. 61009-8 Paperbound $2.00

College Algebra, *H. B. Fine*

Standard college text that gives a systematic and deductive structure to algebra; comprehensive, connected, with emphasis on theory. Discusses the commutative, associative, and distributive laws of number in unusual detail, and goes on with undetermined coefficients, quadratic equations, progressions, logarithms, permutations, probability, power series, and much more. Still most valuable elementary-intermediate text on the science and structure of algebra. Index. 1560 problems, all with answers. x + 631pp. 5⅜ x 8. 60211-7 Paperbound $2.75

Higher Mathematics for Students of Chemistry and Physics, *J. W. Mellor*

Not abstract, but practical, building its problems out of familiar laboratory material, this covers differential calculus, coordinate, analytical geometry, functions, integral calculus, infinite series, numerical equations, differential equations, Fourier's theorem, probability, theory of errors, calculus of variations, determinants. "If the reader is not familiar with this book, it will repay him to examine it," *Chem. & Engineering News.* 800 problems. 189 figures. Bibliography. xxi + 641pp. 5⅜ x 8. 60193-5 Paperbound $3.50

Trigonometry Refresher for Technical Men, *A. A. Klaf*

A modern question and answer text on plane and spherical trigonometry. Part I covers plane trigonometry: angles, quadrants, trigonometrical functions, graphical representation, interpolation, equations, logarithms, solution of triangles, slide rules, etc. Part II discusses applications to navigation, surveying, elasticity, architecture, and engineering. Small angles, periodic functions, vectors, polar coordinates, De Moivre's theorem, fully covered. Part III is devoted to spherical trigonometry and the solution of spherical triangles, with applications to terrestrial and astronomical problems. Special time-savers for numerical calculation. 913 questions answered for you! 1738 problems; answers to odd numbers. 494 figures. 14 pages of functions, formulae. Index. x + 629pp. 5⅜ x 8. 20371-9 Paperbound $3.00

Calculus Refresher for Technical Men, *A. A. Klaf*

Not an ordinary textbook but a unique refresher for engineers, technicians, and students. An examination of the most important aspects of differential and integral calculus by means of 756 key questions. Part I covers simple differential calculus: constants, variables, functions, increments, derivatives, logarithms, curvature, etc. Part II treats fundamental concepts of integration: inspection, substitution, transformation, reduction, areas and volumes, mean value, successive and partial integration, double and triple integration. Stresses practical aspects! A 50 page section gives applications to civil and nautical engineering, electricity, stress and strain, elasticity, industrial engineering, and similar fields. 756 questions answered. 556 problems; solutions to odd numbers. 36 pages of constants, formulae. Index. v + 431pp. 5⅜ x 8. 20370-0 Paperbound $2.25

Introduction to the Theory of Groups of Finite Order, *R. Carmichael*

Examines fundamental theorems and their application. Beginning with sets, systems, permutations, etc., it progresses in easy stages through important types of groups: Abelian, prime power, permutation, etc. Except 1 chapter where matrices are desirable, no higher math needed. 783 exercises, problems. Index. xvi + 447pp. 5⅜ x 8. 60300-8 Paperbound $3.00

FIVE VOLUME "THEORY OF FUNCTIONS" SET BY KONRAD KNOPP

This five-volume set, prepared by Konrad Knopp, provides a complete and readily followed account of theory of functions. Proofs are given concisely, yet without sacrifice of completeness or rigor. These volumes are used as texts by such universities as M.I.T., University of Chicago, N. Y. City College, and many others. "Excellent introduction . . . remarkably readable, concise, clear, rigorous," *Journal of the American Statistical Association.*

ELEMENTS OF THE THEORY OF FUNCTIONS,
Konrad Knopp
This book provides the student with background for further volumes in this set, or texts on a similar level. Partial contents: foundations, system of complex numbers and the Gaussian plane of numbers, Riemann sphere of numbers, mapping by linear functions, normal forms, the logarithm, the cyclometric functions and binomial series. "Not only for the young student, but also for the student who knows all about what is in it," *Mathematical Journal.* Bibliography. Index. 140pp. 5⅜ x 8. 60154-4 Paperbound $1.50

THEORY OF FUNCTIONS, PART I,
Konrad Knopp
With volume II, this book provides coverage of basic concepts and theorems. Partial contents: numbers and points, functions of a complex variable, integral of a continuous function, Cauchy's integral theorem, Cauchy's integral formulae, series with variable terms, expansion of analytic functions in power series, analytic continuation and complete definition of analytic functions, entire transcendental functions, Laurent expansion, types of singularities. Bibliography. Index. vii + 146pp. 5⅜ x 8. 60156-0 Paperbound $1.50

THEORY OF FUNCTIONS, PART II,
Konrad Knopp
Application and further development of general theory, special topics. Single valued functions. Entire, Weierstrass, Meromorphic functions. Riemann surfaces. Algebraic functions. Analytical configuration, Riemann surface. Bibliography. Index. x + 150pp. 5⅜ x 8. 60157-9 Paperbound $1.50

PROBLEM BOOK IN THE THEORY OF FUNCTIONS, VOLUME 1.
Konrad Knopp
Problems in elementary theory, for use with Knopp's *Theory of Functions,* or any other text, arranged according to increasing difficulty. Fundamental concepts, sequences of numbers and infinite series, complex variable, integral theorems, development in series, conformal mapping. 182 problems. Answers. viii + 126pp. 5⅜ x 8. 60158-7 Paperbound $1.50

PROBLEM BOOK IN THE THEORY OF FUNCTIONS, VOLUME 2,
Konrad Knopp
Advanced theory of functions, to be used either with Knopp's *Theory of Functions,* or any other comparable text. Singularities, entire & meromorphic functions, periodic, analytic, continuation, multiple-valued functions, Riemann surfaces, conformal mapping. Includes a section of additional elementary problems. "The difficult task of selecting from the immense material of the modern theory of functions the problems just within the reach of the beginner is here masterfully accomplished," *Am. Math. Soc.* Answers. 138pp. 5⅜ x 8.
60159-5 Paperbound $1.50

NUMERICAL SOLUTIONS OF DIFFERENTIAL EQUATIONS,
H. Levy & E. A. Baggott

Comprehensive collection of methods for solving ordinary differential equations of first and higher order. All must pass 2 requirements: easy to grasp and practical, more rapid than school methods. Partial contents: graphical integration of differential equations, graphical methods for detailed solution. Numerical solution. Simultaneous equations and equations of 2nd and higher orders. "Should be in the hands of all in research in applied mathematics, teaching," *Nature.* 21 figures. viii + 238pp. 5⅜ x 8. 60168-4 Paperbound $1.85

ELEMENTARY STATISTICS, WITH APPLICATIONS IN MEDICINE AND THE BIOLOGICAL SCIENCES, *F. E. Croxton*

A sound introduction to statistics for anyone in the physical sciences, assuming no prior acquaintance and requiring only a modest knowledge of math. All basic formulas carefully explained and illustrated; all necessary reference tables included. From basic terms and concepts, the study proceeds to frequency distribution, linear, non-linear, and multiple correlation, skewness, kurtosis, etc. A large section deals with reliability and significance of statistical methods. Containing concrete examples from medicine and biology, this book will prove unusually helpful to workers in those fields who increasingly must evaluate, check, and interpret statistics. Formerly titled "Elementary Statistics with Applications in Medicine." 101 charts. 57 tables. 14 appendices. Index. vi + 376pp. 5⅜ x 8. 60506-X Paperbound $2.25

INTRODUCTION TO SYMBOLIC LOGIC,
S. Langer

No special knowledge of math required — probably the clearest book ever written on symbolic logic, suitable for the layman, general scientist, and philosopher. You start with simple symbols and advance to a knowledge of the Boole-Schroeder and Russell-Whitehead systems. Forms, logical structure, classes, the calculus of propositions, logic of the syllogism, etc. are all covered. "One of the clearest and simplest introductions," *Mathematics Gazette.* Second enlarged, revised edition. 368pp. 5⅜ x 8. 60164-1 Paperbound $2.25

A SHORT ACCOUNT OF THE HISTORY OF MATHEMATICS,
W. W. R. Ball

Most readable non-technical history of mathematics treats lives, discoveries of every important figure from Egyptian, Phoenician, mathematicians to late 19th century. Discusses schools of Ionia, Pythagoras, Athens, Cyzicus, Alexandria, Byzantium, systems of numeration; primitive arithmetic; Middle Ages, Renaissance, including Arabs, Bacon, Regiomontanus, Tartaglia, Cardan, Stevinus, Galileo, Kepler; modern mathematics of Descartes, Pascal, Wallis, Huygens, Newton, Leibnitz, d'Alembert, Euler, Lambert, Laplace, Legendre, Gauss, Hermite, Weierstrass, scores more. Index. 25 figures. 546pp. 5⅜ x 8.
20630-0 Paperbound $2.75

INTRODUCTION TO NONLINEAR DIFFERENTIAL AND INTEGRAL EQUATIONS,
Harold T. Davis

Aspects of the problem of nonlinear equations, transformations that lead to equations solvable by classical means, results in special cases, and useful generalizations. Thorough, but easily followed by mathematically sophisticated reader who knows little about non-linear equations. 137 problems for student to solve. xv + 566pp. 5⅜ x 8½. 60971-5 Paperbound $2.75

An Introduction to the Geometry of N Dimensions, *D. H. Y. Sommerville*
An introduction presupposing no prior knowledge of the field, the only book in English devoted exclusively to higher dimensional geometry. Discusses fundamental ideas of incidence, parallelism, perpendicularity, angles between linear space; enumerative geometry; analytical geometry from projective and metric points of view; polytopes; elementary ideas in analysis situs; content of hyper-spacial figures. Bibliography. Index. 60 diagrams. 196pp. 5⅜ x 8.
60494-2 Paperbound $1.50

Elementary Concepts of Topology, *P. Alexandroff*
First English translation of the famous brief introduction to topology for the beginner or for the mathematician not undertaking extensive study. This unusually useful intuitive approach deals primarily with the concepts of complex, cycle, and homology, and is wholly consistent with current investigations. Ranges from basic concepts of set-theoretic topology to the concept of Betti groups. "Glowing example of harmony between intuition and thought," David Hilbert. Translated by A. E. Farley. Introduction by D. Hilbert. Index. 25 figures. 73pp. 5⅜ x 8. 60747-X Paperbound $1.25

Elements of Non-Euclidean Geometry, *D. M. Y. Sommerville*
Unique in proceeding step-by-step, in the manner of traditional geometry. Enables the student with only a good knowledge of high school algebra and geometry to grasp elementary hyperbolic, elliptic, analytic non-Euclidean geometries; space curvature and its philosophical implications; theory of radical axes; homothetic centres and systems of circles; parataxy and parallelism; absolute measure; Gauss' proof of the defect area theorem; geodesic representation; much more, all with exceptional clarity. 126 problems at chapter endings provide progressive practice and familiarity. 133 figures. Index. xvi + 274pp. 5⅜ x 8. 60460-8 Paperbound $2.00

Introduction to the Theory of Numbers, *L. E. Dickson*
Thorough, comprehensive approach with adequate coverage of classical literature, an introductory volume beginners can follow. Chapters on divisibility, congruences, quadratic residues & reciprocity. Diophantine equations, etc. Full treatment of binary quadratic forms without usual restriction to integral coefficients. Covers infinitude of primes, least residues. Fermat's theorem. Euler's phi function, Legendre's symbol, Gauss's lemma, automorphs, reduced forms, recent theorems of Thue & Siegel, many more. Much material not readily available elsewhere. 239 problems. Index. I figure. viii + 183pp. 5⅜ x 8.
60342-3 Paperbound $1.75

Mathematical Tables and Formulas, *compiled by Robert D. Carmichael and Edwin R. Smith*
Valuable collection for students, etc. Contains all tables necessary in college algebra and trigonometry, such as five-place common logarithms, logarithmic sines and tangents of small angles, logarithmic trigonometric functions, natural trigonometric functions, four-place antilogarithms, tables for changing from sexagesimal to circular and from circular to sexagesimal measure of angles, etc. Also many tables and formulas not ordinarily accessible, including powers, roots, and reciprocals, exponential and hyperbolic functions, ten-place logarithms of prime numbers, and formulas and theorems from analytical and elementary geometry and from calculus. Explanatory introduction. viii + 269pp. 5⅜ x 8½. 60111-0 Paperbound $1.50

A SOURCE BOOK IN MATHEMATICS,
D. E. Smith

Great discoveries in math, from Renaissance to end of 19th century, in English translation. Read announcements by Dedekind, Gauss, Delamain, Pascal, Fermat, Newton, Abel, Lobachevsky, Bolyai, Riemann, De Moivre, Legendre, Laplace, others of discoveries about imaginary numbers, number congruence, slide rule, equations, symbolism, cubic algebraic equations, non-Euclidean forms of geometry, calculus, function theory, quaternions, etc. Succinct selections from 125 different treatises, articles, most unavailable elsewhere in English. Each article preceded by biographical introduction. Vol. I: Fields of Number, Algebra. Index. 32 illus. 338pp. 5⅜ x 8. Vol. II: Fields of Geometry, Probability, Calculus, Functions, Quaternions. 83 illus. 432pp. 5⅜ x 8.

60552-3, 60553-1 Two volume set, paperbound $5.00

FOUNDATIONS OF PHYSICS,
R. B. Lindsay & H. Margenau

Excellent bridge between semi-popular works & technical treatises. A discussion of methods of physical description, construction of theory; valuable for physicist with elementary calculus who is interested in ideas that give meaning to data, tools of modern physics. Contents include symbolism; mathematical equations; space & time foundations of mechanics; probability; physics & continua; electron theory; special & general relativity; quantum mechanics; causality. "Thorough and yet not overdetailed. Unreservedly recommended," *Nature* (London). Unabridged, corrected edition. List of recommended readings. 35 illustrations. xi + 537pp. 5⅜ x 8. 60377-6 Paperbound $3.50

FUNDAMENTAL FORMULAS OF PHYSICS,
ed. by D. H. Menzel

High useful, full, inexpensive reference and study text, ranging from simple to highly sophisticated operations. Mathematics integrated into text—each chapter stands as short textbook of field represented. Vol. 1: Statistics, Physical Constants, Special Theory of Relativity, Hydrodynamics, Aerodynamics, Boundary Value Problems in Math, Physics, Viscosity, Electromagnetic Theory, etc. Vol. 2: Sound, Acoustics, Geometrical Optics, Electron Optics, High-Energy Phenomena, Magnetism, Biophysics, much more. Index. Total of 800pp. 5⅜ x 8.

60595-7, 60596-5 Two volume set, paperbound $4.75

THEORETICAL PHYSICS,
A. S. Kompaneyets

One of the very few thorough studies of the subject in this price range. Provides advanced students with a comprehensive theoretical background. Especially strong on recent experimentation and developments in quantum theory. Contents: Mechanics (Generalized Coordinates, Lagrange's Equation, Collision of Particles, etc.), Electrodynamics (Vector Analysis, Maxwell's equations, Transmission of Signals, Theory of Relativity, etc.), Quantum Mechanics (the Inadequacy of Classical Mechanics, the Wave Equation, Motion in a Central Field, Quantum Theory of Radiation, Quantum Theories of Dispersion and Scattering, etc.), and Statistical Physics (Equilibrium Distribution of Molecules in an Ideal Gas, Boltzmann Statistics, Bose and Fermi Distribution. Thermodynamic Quantities, etc.). Revised to 1961. Translated by George Yankovsky, authorized by Kompaneyets. 137 exercises. 56 figures. 529pp. 5⅜ x 8½.

60972-3 Paperbound $3.50

MATHEMATICAL PHYSICS, *D. H. Menzel*
Thorough one-volume treatment of the mathematical techniques vital for classical mechanics, electromagnetic theory, quantum theory, and relativity. Written by the Harvard Professor of Astrophysics for junior, senior, and graduate courses, it gives clear explanations of all those aspects of function theory, vectors, matrices, dyadics, tensors, partial differential equations, etc., necessary for the understanding of the various physical theories. Electron theory, relativity, and other topics seldom presented appear here in considerable detail. Scores of definition, conversion factors, dimensional constants, etc. "More detailed than normal for an advanced text . . . excellent set of sections on Dyadics, Matrices, and Tensors," *Journal of the Franklin Institute*. Index. 193 problems, with answers. x + 412pp. 5⅜ x 8. 60056-4 Paperbound $2.50

THE THEORY OF SOUND, *Lord Rayleigh*
Most vibrating systems likely to be encountered in practice can be tackled successfully by the methods set forth by the great Nobel laureate, Lord Rayleigh. Complete coverage of experimental, mathematical aspects of sound theory. Partial contents: Harmonic motions, vibrating systems in general, lateral vibrations of bars, curved plates or shells, applications of Laplace's functions to acoustical problems, fluid friction, plane vortex-sheet, vibrations of solid bodies, etc. This is the first inexpensive edition of this great reference and study work. Bibliography, Historical introduction by R. B. Lindsay. Total of 1040pp. 97 figures. 5⅜ x 8. 60292-3, 60293-1 Two volume set, paperbound $6.00

HYDRODYNAMICS, *Horace Lamb*
Internationally famous complete coverage of standard reference work on dynamics of liquids & gases. Fundamental theorems, equations, methods, solutions, background, for classical hydrodynamics. Chapters include Equations of Motion, Integration of Equations in Special Gases, Irrotational Motion, Motion of Liquid in 2 Dimensions, Motion of Solids through Liquid-Dynamical Theory, Vortex Motion, Tidal Waves, Surface Waves, Waves of Expansion, Viscosity, Rotating Masses of Liquids. Excellently planned, arranged; clear, lucid presentation. 6th enlarged, revised edition. Index. Over 900 footnotes, mostly bibliographical. 119 figures. xv + 738pp. 6⅛ x 9¼. 60256-7 Paperbound $4.00

DYNAMICAL THEORY OF GASES, *James Jeans*
Divided into mathematical and physical chapters for the convenience of those not expert in mathematics, this volume discusses the mathematical theory of gas in a steady state, thermodynamics, Boltzmann and Maxwell, kinetic theory, quantum theory, exponentials, etc. 4th enlarged edition, with new material on quantum theory, quantum dynamics, etc. Indexes. 28 figures. 444pp. 6⅛ x 9¼. 60136-6 Paperbound $2.75

THERMODYNAMICS, *Enrico Fermi*
Unabridged reproduction of 1937 edition. Elementary in treatment; remarkable for clarity, organization. Requires no knowledge of advanced math beyond calculus, only familiarity with fundamentals of thermometry, calorimetry. Partial Contents: Thermodynamic systems; First & Second laws of thermodynamics; Entropy; Thermodynamic potentials: phase rule, reversible electric cell; Gaseous reactions: van't Hoff reaction box, principle of LeChatelier; Thermodynamics of dilute solutions: osmotic & vapor pressures, boiling & freezing points; Entropy constant. Index. 25 problems. 24 illustrations. x + 160pp. 5⅜ x 8. 60361-X Paperbound $2.00

CELESTIAL OBJECTS FOR COMMON TELESCOPES,
Rev. T. W. Webb
Classic handbook for the use and pleasure of the amateur astronomer. Of inestimable aid in locating and identifying thousands of celestial objects. Vol I, The Solar System: discussions of the principle and operation of the telescope, procedures of observations and telescope-photography, spectroscopy, etc., precise location information of sun, moon, planets, meteors. Vol. II, The Stars: alphabetical listing of constellations, information on double stars, clusters, stars with unusual spectra, variables, and nebulae, etc. Nearly 4,000 objects noted. Edited and extensively revised by Margaret W. Mayall, director of the American Assn. of Variable Star Observers. New Index by Mrs. Mayall giving the location of all objects mentioned in the text for Epoch 2000. New Precession Table added. New appendices on the planetary satellites, constellation names and abbreviations, and solar system data. Total of 46 illustrations. Total of xxxix + 606pp. 5⅜ x 8. 20917-2, 20918-0 Two volume set, paperbound $5.00

PLANETARY THEORY,
E. W. Brown and C. A. Shook
Provides a clear presentation of basic methods for calculating planetary orbits for today's astronomer. Begins with a careful exposition of specialized mathematical topics essential for handling perturbation theory and then goes on to indicate how most of the previous methods reduce ultimately to two general calculation methods: obtaining expressions either for the coordinates of planetary positions or for the elements which determine the perturbed paths. An example of each is given and worked in detail. Corrected edition. Preface. Appendix. Index. xii + 302pp. 5⅜ x 8½. 61133-7 Paperbound $2.25

STAR NAMES AND THEIR MEANINGS,
Richard Hinckley Allen
An unusual book documenting the various attributions of names to the individual stars over the centuries. Here is a treasure-house of information on a topic not normally delved into even by professional astronomers; provides a fascinating background to the stars in folk-lore, literary references, ancient writings, star catalogs and maps over the centuries. Constellation-by-constellation analysis covers hundreds of stars and other asterisms, including the Pleiades, Hyades, Andromedan Nebula, etc. Introduction. Indices. List of authors and authorities. xx + 563pp. 5⅜ x 8½. 21079-0 Paperbound $3.00

A SHORT HISTORY OF ASTRONOMY, *A. Berry*
Popular standard work for over 50 years, this thorough and accurate volume covers the science from primitive times to the end of the 19th century. After the Greeks and the Middle Ages, individual chapters analyze Copernicus, Brahe, Galileo, Kepler, and Newton, and the mixed reception of their discoveries. Post-Newtonian achievements are then discussed in unusual detail: Halley, Bradley, Lagrange, Laplace, Herschel, Bessel, etc. 2 Indexes. 104 illustrations, 9 portraits. xxxi + 440pp. 5⅜ x 8. 20210-0 Paperbound $2.75

SOME THEORY OF SAMPLING, *W. E. Deming*
The purpose of this book is to make sampling techniques understandable to and useable by social scientists, industrial managers, and natural scientists who are finding statistics increasingly part of their work. Over 200 exercises, plus dozens of actual applications. 61 tables. 90 figs. xix + 602pp. 5⅜ x 8½. 61755-6 Paperbound $3.50

PRINCIPLES OF STRATIGRAPHY,
A. W. Grabau
Classic of 20th century geology, unmatched in scope and comprehensiveness. Nearly 600 pages cover the structure and origins of every kind of sedimentary, hydrogenic, oceanic, pyroclastic, atmoclastic, hydroclastic, marine hydroclastic, and bioclastic rock; metamorphism; erosion; etc. Includes also the constitution of the atmosphere; morphology of oceans, rivers, glaciers; volcanic activities; faults and earthquakes; and fundamental principles of paleontology (nearly 200 pages). New introduction by Prof. M. Kay, Columbia U. 1277 bibliographical entries. 264 diagrams. Tables, maps, etc. Two volume set. Total of xxxii + 1185pp. 5⅜ x 8. 60686-4, 60687-2 Two volume set, paperbound $6.25

SNOW CRYSTALS, *W. A. Bentley and W. J. Humphreys*
Over 200 pages of Bentley's famous microphotographs of snow flakes—the product of painstaking, methodical work at his Jericho, Vermont studio. The pictures, which also include plates of frost, glaze and dew on vegetation, spider webs, windowpanes; sleet; graupel or soft hail, were chosen both for their scientific interest and their aesthetic qualities. The wonder of nature's diversity is exhibited in the intricate, beautiful patterns of the snow flakes. Introductory text by W. J. Humphreys. Selected bibliography. 2,453 illustrations. 224pp. 8 x 10¼. 20287-9 Paperbound $3.25

THE BIRTH AND DEVELOPMENT OF THE GEOLOGICAL SCIENCES,
F. D. Adams
Most thorough history of the earth sciences ever written. Geological thought from earliest times to the end of the 19th century, covering over 300 early thinkers & systems: fossils & their explanation, vulcanists vs. neptunists, figured stones & paleontology, generation of stones, dozens of similar topics. 91 illustrations, including medieval, renaissance woodcuts, etc. Index. 632 footnotes, mostly bibliographical. 511pp. 5⅜ x 8. 20005-1 Paperbound $2.75

ORGANIC CHEMISTRY, *F. C. Whitmore*
The entire subject of organic chemistry for the practicing chemist and the advanced student. Storehouse of facts, theories, processes found elsewhere only in specialized journals. Covers aliphatic compounds (500 pages on the properties and synthetic preparation of hydrocarbons, halides, proteins, ketones, etc.), alicyclic compounds, aromatic compounds, heterocyclic compounds, organophosphorus and organometallic compounds. Methods of synthetic preparation analyzed critically throughout. Includes much of biochemical interest. "The scope of this volume is astonishing," *Industrial and Engineering Chemistry*. 12,000-reference index. 2387-item bibliography. Total of x + 1005pp. 5⅜ x 8. 60700-3, 60701-1 Two volume set, paperbound $4.50

THE PHASE RULE AND ITS APPLICATION,
Alexander Findlay
Covering chemical phenomena of 1, 2, 3, 4, and multiple component systems, this "standard work on the subject" (*Nature,* London), has been completely revised and brought up to date by A. N. Campbell and N. O. Smith. Brand new material has been added on such matters as binary, tertiary liquid equilibria, solid solutions in ternary systems, quinary systems of salts and water. Completely revised to triangular coordinates in ternary systems, clarified graphic representation, solid models, etc. 9th revised edition. Author, subject indexes. 236 figures. 505 footnotes, mostly bibliographic. xii + 494pp. 5⅜ x 8. 60091-2 Paperbound $2.75

A COURSE IN MATHEMATICAL ANALYSIS,
Edouard Goursat

Trans. by E. R. Hedrick, O. Dunkel, H. G. Bergmann. Classic study of fundamental material thoroughly treated. Extremely lucid exposition of wide range of subject matter for student with one year of calculus. Vol. 1: Derivatives and differentials, definite integrals, expansions in series, applications to geometry. 52 figures, 556pp. 60554-X Paperbound $3.00. Vol. 2, Part I: Functions of a complex variable, conformal representations, doubly periodic functions, natural boundaries, etc. 38 figures, 269pp. 60555-8 Paperbound $2.25. Vol. 2, Part II: Differential equations, Cauchy-Lipschitz method, nonlinear differential equations, simultaneous equations, etc. 308pp. 60556-6 Paperbound $2.50. Vol. 3, Part I: Variation of solutions, partial differential equations of the second order. 15 figures, 339pp. 61176-0 Paperbound $3.00. Vol. 3, Part II: Integral equations, calculus of variations. 13 figures, 389pp. 61177-9 Paperbound $3.00 60554-X, 60555-8, 60556-6 61176-0, 61177-9 Six volume set, paperbound $13.75

PLANETS, STARS AND GALAXIES,
A. E. Fanning

Descriptive astronomy for beginners: the solar system; neighboring galaxies; seasons; quasars; fly-by results from Mars, Venus, Moon; radio astronomy; etc. all simply explained. Revised up to 1966 by author and Prof. D. H. Menzel, former Director, Harvard College Observatory. 29 photos, 16 figures. 189pp. 5⅜ x 8½. 21680-2 Paperbound $1.50

GREAT IDEAS IN INFORMATION THEORY, LANGUAGE AND CYBERNETICS,
Jagjit Singh

Winner of Unesco's Kalinga Prize covers language, metalanguages, analog and digital computers, neural systems, work of McCulloch, Pitts, von Neumann, Turing, other important topics. No advanced mathematics needed, yet a full discussion without compromise or distortion. 118 figures. ix + 338pp. 5⅜ x 8½. 21694-2 Paperbound $2.25

GEOMETRIC EXERCISES IN PAPER FOLDING,
T. Sundara Row

Regular polygons, circles and other curves can be folded or pricked on paper, then used to demonstrate geometric propositions, work out proofs, set up well-known problems. 89 illustrations, photographs of actually folded sheets. xii + 148pp. 5⅜ x 8½. 21594-6 Paperbound $1.00

VISUAL ILLUSIONS, THEIR CAUSES, CHARACTERISTICS AND APPLICATIONS,
M. Luckiesh

The visual process, the structure of the eye, geometric, perspective illusions, influence of angles, illusions of depth and distance, color illusions, lighting effects, illusions in nature, special uses in painting, decoration, architecture, magic, camouflage. New introduction by W. H. Ittleson covers modern developments in this area. 100 illustrations. xxi + 252pp. 5⅜ x 8. 21530-X Paperbound $1.50

ATOMS AND MOLECULES SIMPLY EXPLAINED,
B. C. Saunders and R. E. D. Clark

Introduction to chemical phenomena and their applications: cohesion, particles, crystals, tailoring big molecules, chemist as architect, with applications in radioactivity, color photography, synthetics, biochemistry, polymers, and many other important areas. Non technical. 95 figures. x + 299pp. 5⅜ x 8½. 21282-3 Paperbound $1.50

THE PRINCIPLES OF ELECTROCHEMISTRY,
D. A. MacInnes
Basic equations for almost every subfield of electrochemistry from first principles, referring at all times to the soundest and most recent theories and results; unusually useful as text or as reference. Covers coulometers and Faraday's Law, electrolytic conductance, the Debye-Hueckel method for the theoretical calculation of activity coefficients, concentration cells, standard electrode potentials, thermodynamic ionization constants, pH, potentiometric titrations, irreversible phenomena. Planck's equation, and much more. 2 indices. Appendix. 585-item bibliography. 137 figures. 94 tables. ii + 478pp. 5⅝ x 8⅜.
60052-1 Paperbound $3.00

MATHEMATICS OF MODERN ENGINEERING,
E. G. Keller and R. E. Doherty
Written for the Advanced Course in Engineering of the General Electric Corporation, deals with the engineering use of determinants, tensors, the Heaviside operational calculus, dyadics, the calculus of variations, etc. Presents underlying principles fully, but emphasis is on the perennial engineering attack of set-up and solve. Indexes. Over 185 figures and tables. Hundreds of exercises, problems, and worked-out examples. References. Total of xxxiii + 623pp. 5⅜ x 8. 60734-8, 60735-6 Two volume set, paperbound $3.70

AERODYNAMIC THEORY: A GENERAL REVIEW OF PROGRESS,
William F. Durand, editor-in-chief
A monumental joint effort by the world's leading authorities prepared under a grant of the Guggenheim Fund for the Promotion of Aeronautics. Never equalled for breadth, depth, reliability. Contains discussions of special mathematical topics not usually taught in the engineering or technical courses. Also: an extended two-part treatise on Fluid Mechanics, discussions of aerodynamics of perfect fluids, analyses of experiments with wind tunnels, applied airfoil theory, the nonlifting system of the airplane, the air propeller, hydrodynamics of boats and floats, the aerodynamics of cooling, etc. Contributing experts include Munk, Giacomelli, Prandtl, Toussaint, Von Karman, Klemperer, among others. Unabridged republication. 6 volumes. Total of 1,012 figures, 12 plates, 2,186pp. Bibliographies. Notes. Indices. 5⅜ x 8½. 61709-2, 61710-6, 61711-4, 61712-2, 61713-0, 61715-9 Six volume set, paperbound $13.50

FUNDAMENTALS OF HYDRO- AND AEROMECHANICS,
L. Prandtl and O. G. Tietjens
The well-known standard work based upon Prandtl's lectures at Goettingen. Wherever possible hydrodynamics theory is referred to practical considerations in hydraulics, with the view of unifying theory and experience. Presentation is extremely clear and though primarily physical, mathematical proofs are rigorous and use vector analysis to a considerable extent. An Engineering Society Monograph, 1934. 186 figures. Index. xvi + 270pp. 5⅜ x 8.
60374-1 Paperbound $2.25

APPLIED HYDRO- AND AEROMECHANICS,
L. Prandtl and O. G. Tietjens
Presents for the most part methods which will be valuable to engineers. Covers flow in pipes, boundary layers, airfoil theory, entry conditions, turbulent flow in pipes, and the boundary layer, determining drag from measurements of pressure and velocity, etc. Unabridged, unaltered. An Engineering Society Monograph. 1934. Index. 226 figures, 28 photographic plates illustrating flow patterns. xvi + 311pp. 5⅜ x 8. 60375-X Paperbound $2.50

APPLIED OPTICS AND OPTICAL DESIGN,
A. E. Conrady
With publication of vol. 2, standard work for designers in optics is now complete for first time. Only work of its kind in English; only detailed work for practical designer and self-taught. Requires, for bulk of work, no math above trig. Step-by-step exposition, from fundamental concepts of geometrical, physical optics, to systematic study, design, of almost all types of optical systems. Vol. 1: all ordinary ray-tracing methods; primary aberrations; necessary higher aberration for design of telescopes, low-power microscopes, photographic equipment. Vol. 2: (Completed from author's notes by R. Kingslake, Dir. Optical Design, Eastman Kodak.) Special attention to high-power microscope, anastigmatic photographic objectives. "An indispensable work," *J., Optical Soc. of Amer.* Index. Bibliography. 193 diagrams. 852pp. 6⅛ x 9¼.
60611-2, 60612-0 Two volume set, paperbound $8.00

MECHANICS OF THE GYROSCOPE, THE DYNAMICS OF ROTATION,
R. F. Deimel, Professor of Mechanical Engineering at Stevens Institute of Technology
Elementary general treatment of dynamics of rotation, with special application of gyroscopic phenomena. No knowledge of vectors needed. Velocity of a moving curve, acceleration to a point, general equations of motion, gyroscopic horizon, free gyro, motion of discs, the damped gyro, 103 similar topics. Exercises. 75 figures. 208pp. 5⅜ x 8. 60066-1 Paperbound $1.75

STRENGTH OF MATERIALS,
J. P. Den Hartog
Full, clear treatment of elementary material (tension, torsion, bending, compound stresses, deflection of beams, etc.), plus much advanced material on engineering methods of great practical value: full treatment of the Mohr circle, lucid elementary discussions of the theory of the center of shear and the "Myosotis" method of calculating beam deflections, reinforced concrete, plastic deformations, photoelasticity, etc. In all sections, both general principles and concrete applications are given. Index. 186 figures (160 others in problem section). 350 problems, all with answers. List of formulas. viii + 323pp. 5⅜ x 8.
60755-0 Paperbound $2.50

HYDRAULIC TRANSIENTS,
G. R. Rich
The best text in hydraulics ever printed in English . . . by former Chief Design Engineer for T.V.A. Provides a transition from the basic differential equations of hydraulic transient theory to the arithmetic integration computation required by practicing engineers. Sections cover Water Hammer, Turbine Speed Regulation, Stability of Governing, Water-Hammer Pressures in Pump Discharge Lines, The Differential and Restricted Orifice Surge Tanks, The Normalized Surge Tank Charts of Calame and Gaden, Navigation Locks, Surges in Power Canals—Tidal Harmonics, etc. Revised and enlarged. Author's prefaces. Index. xiv + 409pp. 5⅜ x 8½. 60116-1 Paperbound $2.50

Prices subject to change without notice.

Available at your book dealer or write for free catalogue to Dept. Adsci, Dover Publications, Inc., 180 Varick St., N.Y., N.Y. 10014. Dover publishes more than 150 books each year on science, elementary and advanced mathematics, biology, music, art, literary history, social sciences and other areas.